Infrastructure and Safety in a Collaborative World

Evangelos Bekiaris • Marion Wiethoff •
Evangelia Gaitanidou
Editors

Infrastructure and Safety in a Collaborative World

Road Traffic Safety

Springer

Editors

Dr. Evangelos Bekiaris
Centre for Research and Technology
Hellas (CERTH)
Hellenic Insitute of Transport (HIT)
6th Km Charilaou – Thermi Rd.
570 01 Thermi, Thessaloniki
Greece
abek@certh.gr

Dr. Marion Wiethoff
Delft University of Technology
Fac. Technology, Policy & Management
Jaffalaan 5
2628 BX Delft
Netherlands

Evangelia Gaitanidou
Centre for Research and Technology
Hellas (CERTH)
Hellenic Insitute of Transport (HIT)
6th Km Charilaou – Thermi Rd.
570 01 Thermi, Thessaloniki
Greece
lgait@certh.gr

ISBN 978-3-642-43134-0 ISBN 978-3-642-18372-0 (eBook)
DOI 10.1007/978-3-642-18372-0
Springer Heidelberg Dordrecht London New York

Softcover re-print of the Hardcover 1st edition 2011

Cover design: eStudio Calamar, Girona/Berlin

Printed on acid-free paper

Springer is part of Springer Science+Business Media (www.springer.com)

Preface by the Editors

July 2011

Dear Readers,

After 3 years of collaborative research (and a lot of fun!) in the IN-SAFETY research project, we found that a lot of good knowledge has been created, that should not be left "at the shelves". Some embryonic concepts when we conceived the IN-SAFETY ideas (back in 2003!) were later mainstreamed. "Self-explanatory roads", "forgiving roads", "cooperative systems" were still vague ideas when the project started in 2005, but are in the focus of current research. Thus, we at IN-SAFETY, feel pioneers. In reporting our cumulative knowledge, we strived to add more know-how from distinguished researchers outside IN-SAFETY, thus offering added value to the reader.

The book is conceptually composed of 5 parts and 19 chapters. After a preface by one of the leading figures in ITS research in Europe and beyond, Prof. G. Giannopoulos, head of CERTH/HIT, an introduction on the importance of Transport Research for Europe by the father of eSafety initiative, Jean-Pierre Médevielle, and a short intervention by the Director of POLIS (the agglomeration of over 65 Municipalities with telematic applications in Europe), Mr. Haon, the first part of the book focuses upon the methodological approaches adopted and applied in IN-SAFETY and beyond. The five Chapters of this part introduce a holistic approach of how to structure implementation scenarios towards self-explaining and forgiving roads, which abide to actual problems (related to specific accident types) and are prioritized by stakeholders in a structured and transparent way. The presented methodologies may be utilized by the reader for any further research and are viewed by us as "best practices" in introducing and prioritizing new safety measures.

The second part of the book deals with new developments in safety related tools, such as micro and macro models, risk analysis tools, use of driving simulators and tool-assisted driver and operators training. Each of its four Chapters is attributed to one of the above areas. It is worth mentioning that micro/macro simulation has primarily been developed for traffic efficiency and capacity studies and then moved to environmental impact studies. Its application to traffic safety in IN-SAFETY is thus innovative and was followed-up by further research. Also, in the training chapter, training tools and curricula are presented for many stakeholder groups, such as driver trainees, professional drivers, road and other infrastructure operators, with emphasis on the use of multi-media and driving simulator tools for training optimization. Each of the proposed tool categories in these four chapters may thus be utilized in any relevant research.

The third and fourth parts present research results of specific forgiving (FOR) and self-explanatory (SER) road implementation scenarios (respectively), whether initiated and tested within IN-SAFETY, or not. There has been an effort to present a representative selection of measures. For FOR, the chapters cover lateral behavior, speed control, perception enhancement and several other individual interventions. For SER, the chapters focus upon standardization of icons and pictograms, context and text – into the so-called "Europeanisms" – as well as their ultimate personalization to the individual traveler language, needs and wants. There is a loose link with the implementation scenarios proposed in the first parts, the aim not being to present the own IN-SAFETY experiments but a selection of them, together with external ones, that altogether best represent the potential interventions in the FOR and SER domains.

The fifth and last part is presenting concepts and application examples on how to prioritize such implementation scenarios (given the limitations in funds and time of modern society), how to monetarily evaluate them, as well as suggestions on implementation guidelines and policies towards FOR and SER promotion. It should be emphasized that the future policy recommendations of Chapter 19 are the result of a wider stakeholders consultation within FERSI (Forum of European Road Safety Research Institutes, www.fersi.org), thus representing a consensus of over 21 Traffic Safety related Research Institutes Europewide.

For the rest parts, however, it should be also mentioned that the implementation scenarios priorities have taken into account only the views of the experts participating in the IN-SAFETY User Fora and rankings and opinions expressed there do not necessarily represent the views of the corresponding stakeholder communities. The same holds true for all the chapters, where all statements of each chapter represent the chapter authors' views and conclusions and for which no liability is assumed by the editors.

The editors wish hereby to thank all individual authors, as well as the IN-SAFETY Consortium in its entity and the corresponding EC services for their contributions to this book and the information included in it.

We hope that you, dear Reader, will enjoy reading it and will find something inside of your particular interest and value; a methodology, a tool, some data, a concept or an idea, that can facilitate your research needs, implementation or development plans and/or policy formulation interests.

Evangelos Bekiaris
Marion Wiethoff
Evangelia Gaitanidou

Preface by Prof. George Giannopoulos

The notion of safety in transportation goes hand in hand with efficiency, reliability, and other key concerns of our transport systems. In fact safety is the obvious top priority and primary concern in any transport system on land, sea, or air. So far, many years of research and practical experience from the implementation of specific policy actions, have given a wealth of information and data which can provide an invaluable source of reference material to all those involved in these issues in policy making and academia alike.

The IN-SAFETY research project (which my Institute had the honour of coordinating for all its three years of work), is certainly one such source. In fact it can rightfully be considered as one of the most original pieces of research in the field of transport safety and one that clearly aims at giving practical solutions and practical tools for improving road safety. The publication of this book, goes far beyond the normal obligations of the research Consortium in producing its contractual Deliverables, and shows clearly its commitment in pursuing their recommendations and making them available to the widest possible audience.

In its 19 chapters, the book provides the reader with a multitude of information, data, and suggestions for improving road safety. From the initial statistical analysis of safety data and accident statistics, to ways and means for improving the infrastructures, to ways to evaluate and prioritise road safety measures, and to the absolutely necessary training and education activities that must be the founding rock of any safety improvement effort.

Notions like the "self-explanatory roads", or the "forgiving roads", which were a cornerstone of the IN-SAFETY research work, are also presented. These, together with the views of experts (that participated in the various discussion fora organized by the project), and the policy recommendations that resulted from the consultation procedures with the 20 or so traffic safety related Institutes and Organisations that participated in the process, makes the material of the book quite unique, I would say.

In this way the reader is confronted with an array of ideas, opinions, and research results that will help him/her grasp the insides of this very difficult and complex problem.

As head of the National Transport research Organisation of Greece, a country with acute road and traffic safety problems, I am particularly happy to welcome this publication and I see it as a very useful tool for all those interested or involved in road safety.

G. A. Giannopoulos
Director, Hellenic Institute of Transport
Centre for Research and
Technology Hellas

Preface by Jean-Pierre Medevielle

As former Vice Chair of FERSI, current President of HUMANIST VCE, co-sponsored originally both by FERSI and ECTRI, both member of ERTRAC and eSafety Steering groups, it is a privilege to me to deliver some introductive remarks of the importance of Transport Research for Europe, including safety research as an integrated part.

Firstly, *the global competitiveness of Europe* is currently at stake; not only economic competitiveness of industries and operations, but also rule making and overall transport system as well as the transport research and/or education community competitiveness; obviously the new Grand Challenges are affecting this global competitiveness (climate change, energy, environment, sustainability, health and food).

Secondly, *transport is part of the problem and the solution*, so research and innovation have to be fostered and/or accelerated to solve or lighten the concrete problems in a vision and with a perspective towards the future.

Thirdly, *safety is an integrated part of transport in the road and aeronautics sectors*. As is obvious, both road and multimodal safety constitute a critical issue.

Fourthly, *on the scientific and technological side*, new technologies (ICT, nanotechnologies) are entering the transport system, the constituent components, services or products and their standardization. But there is also a need for integrated or system approach that draws on the agendas for soft research outcomes, such as HMI, economics, acceptance, innovation process and pathways, orgware research.[1] So, according to all models, "anticipation by research and the accompanying research for the transition" becomes a critical subject.

Fifthly, the Challenges lead to a requirement for an *increase of the European program dimensions* (both in their scope and funding) allocated to surface transport research. This should lead to greater contribution from and between EU programs, Joint Programming and national programming, and research innovation and education, as well as additions of new instruments in this perspective.

Sixthly, the needed response requires substantial investment for the *training and education* of professionals. If Europe *wants to keep or attract world R&D centres*,

[1]Research on organisational or institutional issues.

it is of overarching importance that a new generation of the *European transport scientific community* (scientists, engineers, and other staff) is created; replenishing and replacing but also growing industrial, commercial, public sector and academic research capacity.

Seventhly, it is of vital importance to *tackle the need for hard and soft research infrastructures*, providing the capabilities to address the challenges. This will include full-scale experimentation capabilities, addressing aspects, such as safety and climate change adaptation, as well as new simulation facilities and databases.

Eighthly, without *a strong integrated and dynamic European Research Area in the surface transport domain*, it could be difficult to tackle all these challenges.

IN-SAFETY project sets the scene for a new scientific holistic approach of promoting road safety through application of engineering sciences, ICT technologies, ergonomics and HMI sciences. As the definition and setting up of the next generation of transport infrastructure with its ICT and energy components is just ahead of us, strong research and innovation programs, such as IN-SAFETY, need to be created and funded in a proactive way, with the vehicle of tomorrow in mind and the driver.

But the *quest for excellence* in this domain is *not only the peer to peer evaluation of excellence*, but also the *relevance excellence,* determined by the outcomes achieved as well as the *governance and management excellence*, including scientific process or innovation processes.

Towards the future, we can see four critical issues of the Lyon Declaration, developing the Vision 2020 of the European Research Area for surface transport and confirmed for all scientific domains by the Lund Declaration and for ICT by the Visby Declaration:

- Mobility promotion as the fifth freedom
- New European Research Infrastructure creation
- Training and Education and European research partnership for the next generation of scientists and professionals
- Safety and Security Research intensification

Research priorities include the need for new methodologies, new databases, a new generation of field operational tests, and new scientific challenges for transport safety research: without a big development of Naturalistic Driving Studies and Distraction Research we cannot keep the European scientific competitiveness of transport safety research.

Jean-Pierre Medevielle
HUMANIST VCE Coordinator

Preface by Sylvain Haon

The European Road Safety Action Programme had set the objective of halving the number of deaths on European roads by 2010 in comparison to 2001. We now know that we have fallen short of achieving this target. Efforts to tackle this challenge should therefore be intensified. The main challenges of the years to come will be to bring more EU countries to the level of the Member States which have the safest roads and to continue to improve our overall ability to prevent accidents.

To do so, the focus of our concern should be on the individual, whether he/she is the driver, a passenger or another road user. Actions to prevent road accidents should continue to address all elements of the system, the vehicle, the infrastructure and the individual.

Awareness and educational campaigns should be reinforced and become systematic for all types of road users, as well as adapted to age and social conditions, in order to ensure maximum impact. The campaigns should cover the use of technology for all road users, especially car drivers, in view of the increasing range of technology available in the driving environment. This is necessary in order to ensure that the multiplication of messages given to the drivers, whether on board or by the infrastructure, does not induce dangerous distraction. They should be complemented by interventions in the infrastructure, of the type proposed in several chapters of this book, that lean towards the creation of forgiving road environments and self-explaining roads. This probably requires re-assessing the relationship between the vehicles and the infrastructure beyond traditional road intervention, and to consider holistically the three elements of road safety: infrastructure/driver/vehicles.

This book builds upon the research work conducted in the EC co-funded and sponsored (within the DG TREN workprogramme of the 6th Framework of the EC) IN-SAFETY (506716) project. The IN-SAFETY approach supports the safe use of information technology to strengthen road safety and, at the same time, contributes to more efficient mobility behaviour and pattern. It also acknowledges that work is required on the design and nature of messages to the driver. This should allow Europe to move progressively towards a pan-European electronic signing system, understandable by all drivers in all languages, including standardized pictograms and earcons for in-vehicle delivered messages and new/emerging ITS services.

The work of IN-SAFETY will facilitate the deployment of efficient road safety systems which are now still too rare. Intelligent Speed Adaption, for instance, has demonstrated its ability to reduce the number of accidents as well as, incidentally, to reduce CO_2 and local emissions from traffic. It has however made a very slow entrance on the market.

IN-SAFETY has demonstrated the possibility to make very significant progress. We can only call for further work on this topic, to accelerate research efforts, but also to support their more rapid deployment on the market. We call for European initiatives to pursue these efforts and stimulate the deployment of the solutions which have proved effective in reducing the number of accidents and fatalities on our roads. This should be supported by the necessary regulatory initiatives at the European level and by large-scale communication campaigns.

Sylvain Haon
Executive Director of POLIS
(European cities and regions networking for
innovative transport solutions)

Acknowledgements

This book draws research results from several Industrial, National and EC co-funded projects, which are dully recognized through references. However, a big amount of data stem from IN-SAFETY project (506716), which was an EC co-funded project within the Sixth Framework Program (1.6.2 Sustainable Surface Transport) of DG TREN of the EC. Therefore, the editors would like to recognize relevant support, thank all IN-SAFETY Partners and, above all, the relevant EC Officer, Mr. Sandro Francesconi, for their active support and close collaboration during the project.

Furthermore, IN-SAFETY was a project sponsored and promoted by FERSI, the Forum of European Road Safety Institutes; an Association of 21 Road Safety Research Institutes, from 21 European countries. Thus, FERSI Members' support is acknowledged too.

Contents

Abbreviations

ACC	Adaptive Cruise Control
ADAS	Advanced Driver Assistant System
ADR	Agreement on Dangerous Goods by Road
AHP	Analytical Hierarchical Process
AMI	Advanced Motorway Indicator
ARAS	Advanced Rider Assistant System
ATM	Active Traffic Management
AWS	Advanced Warning System
BC ratio	Benefic–Cost Ratio
CACC	Cooperative Advanced Cruise Control
CAS	Collision Avoidance System
CAT	Comprehension Test Animated Pictogram
CBA	Cost Benefit Analysis
CC	Cruise Control
CCD	Charge Coupled Device
CE ratio	Cost Effectiveness Ratio
CEA	Cost Effectiveness Analysis
CMOS	Complementary Metal Oxide Semiconductor
CRIM	Cluster of Repositories for IN-SAFETY Messages
CST	Content Structure Test
CWS	Collision Warning System
DALY	Disability – Adjusted Life-Year
DATEX	Data Exchange Service
DG	Dangerous Goods
DG	Directorate General (of the EC)
DLC	Distance to Line Crossing
DLL	Dynamic Link Library
DOT	Department of Transport (of the US)
DRAM	Darmstadt Risk Analysis Method
DRAT	Darmstadt Risk Analysis Tool
DSRC	Dedicated Short Range Communications
DSS	Decision Support System

DSuSy	Driving Support Systems
DVU	Driver – Vehicle Units (in traffic engineering simulation models)
EC	European Commission
EEA	European Economic Area
EEG	ElectroEncephaloGram
EFTA	European Free Trade Association
ELOT	Hellenic Organisation of Standards
ESC	Electronic Stability Control
ETSC	European Transport Safety Council
EU	European Union
FCW	Forward Collision Warning
FDW	Following Distance Warning
FEHRL	Forum of European National Highway Research Laboratories (http://www.fehrl.org/)
FERSI	Forum of European Road Safety Research Institutes (http://www.fersi.org/)
Fir	Far Infra - Red
FMEA	Failure Modes and Effects Analysis
FOT	Field Operational Test
FoV	Field of View
FP 6	Framework Program 6 (of the EC)
FOR	FOrgiving Road environment
FTA	Fault Tree Analysis
GDR	Group Decision Room
GNP	Gross National Product
GPRS	General Packet Radio Service
GPS	Global Positioning System
GUI	Graphical User Interface
HGV	Heavy Goods Vehicles
HMI	Human Machine Interaction
HMW	Headway Monitoring and Warning System
HOV	High Occupancy Vehicles
I2V	Infrastructure to Vehicle Communications
ICT	Information and Communication Technologies
ID	Identification Digit
INRAM	IN-SAFETY Risk Analysis Method
IN-SAFETY	Infrastructure and Safety, an EC co-funded project of the 6th Framework (ref. ROADS/506716/2003)
IR	Infra – Red
ISA	Intelligent Speed Adaptation
ISO	International Standardisation Organization
ITS	Intelligent Transport System
IVIS	In-Vehicle Information System (for cars)
LCD	Liquid Crystal Display

LDW(S	Lane Departure Warning (System)
LED	Light Emitting Diodes
LKS	Lane Keeping System
LRM	Lateral and Rear area Monitoring
MAMCA	Multi –Actor Multi-Criteria Analysis
MCA	Multi Criteria Analysis
MMT	Multi-Media Training Tool
MoA	Minute of Arc (for Road signs)
NAFTA	North American Free Trade Agreement
NDMPD	Numerical Described Multidimensional Probability Distributions
NIR	Near Infra – Red
NoE	Network of Excellence
NPV	Net Present Value
NVES	Night Vision Enhancement Systems
OBIS	On-Board Information System (for motorcycles)
OBU	On Board Unit
OD	Origin-Destination
OECD	Organisation for Economic Co-operation and Development
OEM	Original Equipment Manufacturer
ORN	Overall Risk Number
OS	Operating System
P + R	Park and Ride
PC	Personal Computer
PCC	Predictive Cruise Control
PIF	Performance Influencing Factors
PTW	Powered Two Wheeler
QCC	Quantitative Causal Chain
RACC	Responsive Advance Cruise Control
RDS	Radio Data System
RIPCORD – ISEREST	Road Infrastructure Safety Protection – Core-Research and Development for Road Safety in Europe; Increasing safety and reliability of secondary roads for a sustainable Surface Transport, an EC co-funded project of the 6th Framework (ref. 506184)
RNB	Behavioural Risk Number
RNL	Legal Risk Number
RNO	Organisational Risk Number
RNT	Technical Risk Number
RSA	Road Safety Audits
RSU	Roadside Unit
RT	Reaction Time
SER	Self-Explanatory Road environment
SGD	Smallest Graphical Detail (for Road signs)
SLIM	Success Likelihood Index Methodology
SMS	Short Message Service

SPLR	Standard Deviation of Lateral Position
STREP	Specific Targeted Research Project
TDD	Time Division Duplex
TERN	Trans-European Road Network
TET	Time Exposed Time-to-collision
TfL	Transport for London
THeading	Time Heading
TIT	Time Integrated Time-to-collision
TLC	Time to Line Crossing
TMC	Traffic Management Center
TMI	Traffic Management Information
TMIC	Traffic Management and Information Center
TMOT	Transportation Management Operations Technical Staff Development
TMS	Technology Management System
TTC	Time to Collision
TTI	Traffic and Traveller Information
US	United States
V2V	Vehicle to Vehicle communications
VMS	Variable Message Signs
VR	Virtual Reality
VRU	Vulnerable Road User
VSL	Variable Speed Limit
VVMS	Virtual Variable Message Sign
WHO	World Health Organization
WLAN	Wireless Local Area Network
WTO	World Trade Organization
XLS	XML for Location Services
XML	Extensive Markup Language

List of Tables

List of Figures

Part I
General Approach

Chapter 1
From Accidents to Measures

The General IN-SAFETY Approach

Karel A. Brookhuis, Marion Wiethoff, Evangelos Bekiaris, and Evangelia Gaitanidou

1.1 Traffic Safety in Europe and Its "Black-Spots"

In Europe alone, the annual total cost of road accidents amounts to more than 160 billion Euros, which is the equivalent of 2% of Europe's GNP. Moreover, not only financial costs are a daily bother, numerous deaths and injuries add to an intolerable social cost. Traffic participation is in fact a very unhealthy activity, and still aggravating if the vehicle is not halted. The rank of traffic injuries in the order of sources of diseases and injuries with respect to disability-adjusted life years (DALY) is rapidly rising, as demonstrated in Table 1.1, published by the World Health Organisation (2004).

Whilst the traffic system as a whole is least safe for vulnerable road users, car drivers still run the largest risk in terms of fatal or serious accidents leading to injuries. Motor vehicle occupants are the major part of the suffering, i.e. 57% of total European Union (EU) road deaths. For reliability reasons, fatal accident data are mostly used as an indication of how serious (un)safety is in the various EU countries. Table 1.2 gives an overview of such accident data concerning Heavy Goods Vehicles (HGV), available for 15 countries in the EU, separated by collision type. It is clear that the numbers indicate that there are large differences between countries, both in absolute and in relative numbers. The (relative) number of fatalities per billion vehicle kilometres, as far as available, ranges between

K.A. Brookhuis (✉)
Department of Transport Policy and Logistics, Delft University of Technology, Delft, The Netherlands
and
Department of Psychology, University of Groningen, Groningen, The Netherlands
e-mail: k.a.brookhuis@rug.nl

M. Wiethoff
Department of Transport Policy and Logistics, Delft University of Technology, Delft, The Netherlands

E. Bekiaris and E. Gaitanidou
Centre for Research and Technology Hellas, Hellenic Institute of Transport, Thessaloniki, Greece

E. Bekiaris et al. (eds.), *Infrastructure and Safety in a Collaborative World*,
DOI 10.1007/978-3-642-18372-0_1, © Springer-Verlag Berlin Heidelberg 2011

Table 1.1 Traffic safety in comparison to other causes of death, in 1990 and projected for 2020, world report on road traffic injury prevention (WHO 2004)

Change in rank order of DALYs for the ten leading causes of the global burden of disease			
1990		2020	
Rank	Disease or injury	Rank	Disease or injury
1	Lower respiratory infections	1	Ischaemic heart disease
2	Diarrhoeal diseases	2	Unipolar major depression
3	Perinatal conditions	***3***	***Road traffic injuries***
4	Unipolar major depression	4	Cerebrovascular disease
5	Ischaemic heart disease	5	Chronic obstructive pulmonary disease
6	Cerebrovascular disease	6	Lower respiratory infections
7	Tuberculosis	7	Tuberculosis
8	Measles	8	War
9	***Road traffic injuries***	9	Diarrhoeal diseases
10	Congenital abnormalities	10	HIV

DALY Disability-adjusted life year. A health-gap measure that combines information on the number of years lost from premature death with the loss of health from disability

7.6 and 26.7; the difference between countries cumulates to a factor of 3.5 for this type of motor vehicles.

Since the Second World-War, car-ownership and car-mileage has increased steadily in Europe. For example, in the Netherlands the number of motor-vehicles in this period has grown from less than 1 to about 8 million at the time of writing (and still rising), covering distances from less than 20 to almost 200 billion km these days. The number of accidents with fatal and/or severe injury outcome initially rose quickly as well, until the mid-1970s, when authorities, car manufacturers and research institutes started to combine forces in order to turn this dreadful increase successfully. A variety of accident reducing measures was developed and implemented, leading to a gradual fall in casualties.

Studies in Greece on the effectiveness of casualty reduction measures demonstrated that the largest reduction is to be expected from vehicle crash protection (15%). Measures against driving-while-intoxicated were second with 11% in this list, while road safety engineering measures were reported to result in a reduction of 6.5%. Due to the high cost of the latter type of measures, infrastructure improvements are not expected to significantly contribute to a major reduction of road fatalities. However, a suitable combination of new technologies with existing infrastructure, or with limited improvements of it, may lead to much more cost-effective solutions and may become the catalyst towards achieving the EU goal of halving the number of road accidents in 2010.

Strikingly persistent is the human involvement in accident causation, be it through impairment, errors, or inattention. Drivers are ever-fallible, make mistakes, encounter unexpected difficulties, make wrong judgements or decisions and miss relevant signals or objects while driving. Fortunately, only occasionally this collection of failures leads to accidents. For one reason, that is because of the ample margins in the traffic environment (nowadays). For instance, modern roads are normally wide, leaving lots of room for stray movements or swaying, and if moving

Table 1.2 Number of fatalities per country by collision type of heavy goods vehicles (HGV) (Lotz 2006)

Country code/year	Number of fatalities (car (HGV))						Total number of fatalities	Country	Fatalities per 1 billion vehicle km
	Single accidents (all)	Frontal collision	Lateral collision	Chain/rear collision	Collision with parked vehicle	Collision with animal			
BE 2001*	513 (8)	163 (0)	147 (5)	76 (15)	0 (0)	0 (0)	1,486	Belgium	16.2
DK 2001*	– (–)	– (–)	– (–)	– (–)	– (–)	– (–)	463	Denmark	9.7
EL 2001*	349 (19)	176 (1)	189 (3)	49 (2)	25 (0)	2 (0)	1,880	Greece	26.7
ES 2002*	1,437 (114)	631 (15)	715 (18)	241 (20)	26 (0)	7 (0)	5,347	Spain	N.R
FR 2002*	2,178 (82)	1,093 (11)	745 (8)	208 (15)	0 (0)	0 (0)	7,655	France	10.9
IE 2002*	88 (2)	81 (1)	16 (0)	5 (0)	0 (0)	0 (0)	378	Ireland	10.9
IT 1998*	440 (23)	755 (32)	973 (54)	279 (43)	23 (1)	0 (0)	6,314	Italy	N.R
LU 2002*	30 (1)	0 (0)	0 (0)	0 (0)	0 (1)	0 (0)	62	Luxemburg	N.R
NL 2002*	276 (5)	72 (2)	101 (2)	25 (1)	5 (1)	0 (0)	987	Netherlands	7.7
AT 2002*	254 (5)	145 (3)	21 (1)	36 (3)	2 (0)	1 (0)	956	Austria	11.7
PT 2002*	340 (11)	236 (3)	104 (1)	26 (2)	0 (0)	1 (0)	1,675	Portugal	N.R
FI 2002*	– (–)	– (–)	– (–)	– (–)	– (–)	– (–)	415	Finland	7.6
SE 2002*	147 (8)	128 (4)	68 (0)	13 (1)	0 (0)	6 (0)	560	Sweden	9.3
UK 2002*	– (–)	– (–)	– (–)	– (–)	– (–)	– (–)	3,581	England	7.6
DE 2003**	1,692 (–)	1,365 (–)	886 (–)	357 (–)	– (–)	1 (–)	6,613	Germany	9.7

*Data 2001

**Data 1998

across the line, in most cases there is a more or less forgiving (i.e. soft) border. Recently, intelligent driver support is added to the positive turn, mostly in the form of electronic driving aids that provide relevant information to the driver, or take over parts of the driving task in case the driver is in need.

The EU project IN-SAFETY[1] aimed to use intelligent, intuitive and cost-efficient combinations of new technologies and traditional infrastructure best practice applications, to enhance the forgiving and self-explanatory nature of roads, by a number of approaches. For instance, the potential and cost-effectiveness of combined use of new technologies and innovative Human Machine Interface (HMI) concepts, developing new simulation models, risk analysis tools, etc., were assessed. Additionally, part of the work was focussed in designing training tools for road traffic management and information centre and tunnel operators, harmonising signing and personalising information, as well as issuing priority implementation scenarios.

1.2 Supporting the Fallible Driver

A viable manner to identify concrete driver needs is to analyse accidents from the past and to figure out their causes. The put forward hypothesis is that each human failure corresponds to a non-satisfied need in the perception-decision-action loop, and that a specific accident could have been avoided, or at least the consequences minimised, if the need had been foreseen, or if the consequences had been attenuated by infrastructural or in-vehicle measures.

The rather small impact of road and infrastructure related measures on accident reduction until now may be well attributed to the high cost of such measures. Thus, although a study in Greece has identified hundreds of "black-spots" in the main national road network several years ago (TREDIT 2002), the authorities have intervened by local road reconstructions in only very few of them. Therefore, infrastructure improvements and enforcement campaigns are not expected to significantly contribute towards the projected 50% reduction of road fatalities, as is the target by EU for 2010. However, by the use of new technologies this goal might still be achieved, especially since the combination of new technologies with existing infrastructure, or with limited improvements of it, may lead to much more cost-effective solutions. Few data exist on the cost-efficiency level of installing new technologies on existing roads vs. traditional safety measures (e.g. separation barriers). Such data may only be produced by combining micro and macro models, which include actual driver behaviour parameters of Advanced Driver Assistance Systems (ADAS) equipped and non-equipped vehicles, and are able to predict safety impacts. The pre-requisites for such a successful implementation, however,

[1]6th Framework Programme, 1.6.2: Sustainable Surface Transport, nr. 506716.

can be found among the principles of self-explanatory roads and forgiving roads (Brookhuis et al. 2006).

1.3 Self-Explanatory Roads

Drivers have to cope with increasingly complex traffic environments, including different types of road layout and all kinds of signposting, of which an increasingly amount are already supported by telematics. The cases may impose a critical workload to the driver, as he/she may be:

- Striving to read the VMS (Variable Message Sign) message, while seeking the route in an unfamiliar environment (often in a foreign language and even with unfamiliar signs)
- Attempting to detect the required relevant piece of information among an abundance of information sources (like in-car navigation system, messages from TMIC or radio announcements, VMS signs, road signs, in-vehicle messages, etc.)

Thus, there is a considerable need for a self-explanatory road environment, preferably at a personalised level, which would offer intuitive guidance to the driver and information when this is needed, related to the driver's particular needs (route, disabilities, preferences, etc.) and, if possible, in the driver's own language. By way of example, urban tunnels are relevant, as they sometimes require swift decisions on direction selection by the driver, either under pressure, i.e., (sudden) information overflow, or during information "underflow", since tunnels are mostly quite disrupt of decoration and stimulation.

1.4 Forgiving Roads

Forgiving road environments are a basic issue in preventing or mitigating an important percentage of road accidents related to driving errors. More specifically, statistics show that about 25–30% of fatal accidents involve crashes with fixed roadside objects. Those accidents are mainly caused due to driving errors that lead to road departure. The existence of a forgiving road environment would have prevented accidents of this type (and generally accidents that involve driving errors) or, at least, reduced the seriousness of the consequences of an accident.

To develop a forgiving road environment certain characteristics must be included and measures should be taken. These measures involve applications related to either the infrastructure or telematics. It has been noted though that the combination of infrastructure and telematics measures can provide a more cost-efficient solution, as expensive infrastructure works can be substituted by telematics or other innovative systems.

1.5 Approach

To assess and evaluate the possible measures that can constitute an adequate self-explanatory road and/or an effective forgiving road environment, the following approach was devised within IN-SAFETY. First, a full set of alternative safety measures is generated, representing the self-explanatory or forgiven nature of a road. Next, the potential of these measures to contribute to road safety is estimated, applying advanced micro- and macro-safety modelling. Based upon these results, a set of most promising measures for implementation is considered, and the path to implementation for each of these promising measures may be developed. This involves an analysis of stakeholder opinions regarding different promising measures in terms of their preferences and, in case of (conflicting) differences in preferences among stakeholders (e.g. of different types or roles), looking for ways to bridge these differences.

The EU-project IN-SAFETY sets out to attain added value by developing intelligent, intuitive and cost-efficient combinations of new technologies and traditional infrastructure best practice applications, to enhance the forgiving and self-explanatory nature of roads. This chapter focuses upon the first step of the above process, the definition of forgiving and self-explanatory road environments, i.e. to define a set of quantitative and qualitative characteristics that constitute a forgiving and self-explanatory road environment. This involves all kinds of available and new Advanced Driver Assistance Systems (ADAS), In-Vehicle Information Systems (IVIS), new infrastructure elements, in particular standard VMS and newer, modern full colour versions, and their combinations, in enhancing road safety in highway, rural and (peri) urban areas, including tunnels. New, intuitive and innovative combinations of existing and new technologies are considered, so as to conceptualise the road environment of a forgiving and self-explanatory nature. The activity described here is finally designed to culminate in a first set of measures and priorities towards forgiving and self-explanatory roads.

1.6 Method

Devising the measures for self-explanatory or forgiving road environments, as they by definition aim at avoiding or mitigating negative consequences of driving errors, starts with listing possible driving errors to be supported, that in turn are related to accident statistics. The latter is important, since predictions about safety effects can only be based on supporting (avoiding, mitigating) safety-related errors.

The primary criterion is the contribution that each alternative has to attain the goal of avoiding or mitigating negative consequences of driving errors, which is de facto their effect on road safety. However, there are also a number of other (decisive) criteria. Some are typically important for the individual drivers, others more important for the society as a whole (public authorities) and yet others important for manufacturers. Amongst these are comfort, investment and user costs, technical readiness, etc. (eSafety Forum 2005). Since the measures with respect to self-explanatory and forgiving road environments are developed separately, it will be necessary to analyse whether they interact, for instance, whether measures that are highly prioritised as self-explanatory will have negative effects on the aims of forgiving road environments and/or vice versa.

1.7 General Approach of the Research

To identify and evaluate the possible measures that can constitute a working self-explanatory road and/or an effective forgiving road environment, an approach with four stages was chosen, which is analysed below:

1. First, analysis of accident statistics was performed, and a number of accidents prioritized, upon which measures for safety improvements are focused. German accident statistics were selected, for their quality and being representative for the EU as a whole (Lotz 2006). These were then translated in terms of five types of driver errors, e.g. speeding or wrong use of the lane. The project's Consortium added one more driver error: "driving too fast near an unexpected bend on rural roads".
2. Second, an extensive number of safety measures, to mitigate these driver errors, were generated for three different road types: urban roads, rural roads and motorways. The safety measures are associated specifically to the driver errors and the road types. For each error type a typical solution is defined, but in three technological varieties: infrastructural measures, in-vehicle measures, and combined measures (Wiethoff et al. 2007).
3. Third, these safety measures were specified on a number of characteristics in matrices and evaluated on their potential safety effects.
4. In the next stages, a set of 18 most promising safety measures was selected, representing the self-explanatory or forgiving character of a road. The set of 18 alternatives resulted from safety measures on six types of driver errors, in three different technological varieties. An initial prioritization of these 18 alternative scenarios was performed, taking into account their potential safety effects, as well as a wide range of other effects relevant to the stakeholders, such as e.g. full user cost, effects on travel time duration, socio-political acceptance, technical feasibility, etc. This initial prioritization was performed using the methodology of Multi-Actor Multicriteria-Analysis (MAMCA) and the Analytic Hierarchy Process (AHP). Three stakeholders were identified, namely users, society (public policy point of view) and manufacturers. For each of these a set of evaluation criteria (including weights) were identified.

5. Then, the potential of these measures to contribute to road safety was further estimated, applying advanced micro- and macro-safety modelling, or literature analysis of previous empirical data on experiments and pilot studies. The estimation was performed for different road, driver, and vehicle types.
6. Based upon the results of these studies a set of most promising measures for implementation was considered. These measures were then submitted to a final evaluation, using again the MAMCA methodology. Finally, a path to implementation for each of these promising measures was developed. This chapter contains a more detailed analysis for stage 3 and a more synthetic description of the analysis regarding stage 4. The first and the second stages are reported in Chaps. 2 and 3 of this book (Part I). The fifth stage is reported in Parts II, III and IV of this book and the sixth step is described in Part V of this book.

1.8 Accident Types and Driving Errors

As argued before, it was decided to take accident statistics as a starting point. In Fig. 1.1 an overview of fatal accidents by collision type is depicted, to provide an overview and a starting point on where to look for driving errors in different countries.

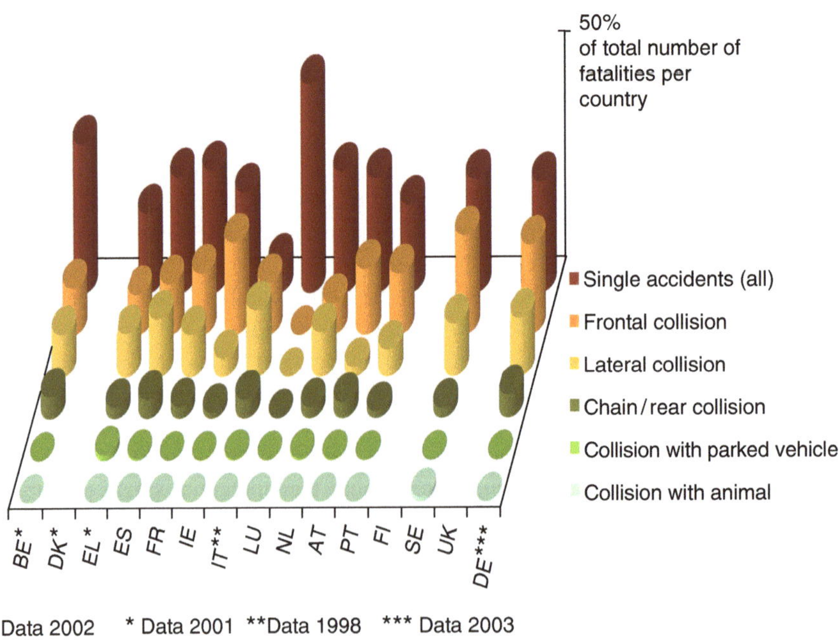

Fig. 1.1 Relative distribution of fatalities per country by type of collision (for country codes see Table 1.1 (Lotz et al. 2006))

The next step is to relate and prioritise accident statistics to driving errors. For this, data from three countries were selected, German and Dutch accident statistics and police reports from Sweden, for reasons of quality and being representative for the EU as a whole (Lotz 2006; Lotz and Wenzel 2006). Reasoning back from the accidents, the following ranking in order of frequency in driving errors could be made:

1.8.1 Driving Errors in Germany

1. Speeding
2. Wrong use of road (e.g. driving on the left lane in a curve, ghost riders)
3. Violation priority rules
4. Failure when overtaking
5. Failure when turning, entering
6. Insufficient safety distance

1.8.2 Driving Errors in Netherlands

1. Not giving way/priority (33%)
2. Lost vehicle control (14%)
3. Insufficient distance (9%)
4. Skidding (5%)
5. Failure while taking a curve (5%)

1.8.3 Police Reports in Sweden

One third of all police reported road traffic crashes with fatal or severe injuries were single vehicle crashes. These also accounted for the most serious injuries compared to other crashes. Out of 1,126 single crashes, 146 ended up in fatalities. These crashes are most common at the rural roads (75%). Crashes between motor vehicles accounted for more then 31% of all crashes with fatal or severely personal injuries. The accident typology among those 31% (1,089 crashes) was the following:

Type of accident	Number of crashes
Crossing/turning at different road	353
Rear end collision	283
Oncoming	241
Turning/two vehicles on the same road	114
Overtaking	24
Other	74

For practical reasons, the choice was made to use the German database as a starting point in the IN-SAFETY project. The reason for this was that these data were of sufficient quality to be able to relate the accident data to causes of accidents in a reliable manner. Furthermore, the Germany accident data may be regarded as sufficiently representative for the European Union as a whole, representing the median of European fatality figures (see Table 1.1).

Lotz et al. (2006) have made a categorisation of errors based on stages in the cause – effect chain (Table 1.3), using the German accident statistics as a starting point.

Table 1.3 shows, following the CARE database (SAFETYNET 2004), the Level 1 errors to be listed in decreasing fatalities in Germany. Level 2 errors can be distinguished by following the accident causes that are defined in the German accident databases. These are the observable driving errors, made by the human driver and leading to the accident. There are very many different examples of accident causes in Germany, the top six are listed in Table 1.3.

For the Level 3 errors, three information processing error types, i.e. information error, diagnostic error and performance error, stem from Rasmussen (1982) and Vollrath (2005). These errors refer to deficiencies in the human information processing. Level 4 errors can be distinguished by different causes for reduced psycho-physiological condition; pre-conditions that facilitate the occurrence of human errors. For the generation of alternatives, Level 2 errors were taken as a starting point.

Table 1.3 Categorisation of errors, level 1–3 in accordance to Hacker (2003), distinction within level 3 according to Rasmussen (1982) and Vollrath (2005)

Error level	Description	Errors
Level 1: "Accident type"	Result of the execution of an error	Single vehicle accident (with or without collision with an obstacle) Frontal collision Lateral collision Chain/rear collision Collision with parked vehicle Collision with animal
Level 2: "Driving error"	Action that has led to the accident	Driving too fast in an unexpected bend on rural roads (error 1) Speeding (error 2) Wrong use of the lane (error 3) Violation of priority rules (error 4) Failure when overtaking (error 5) Insufficient safety distance (error 6)
Level 3: "Human error"	Psychological process that is basis to the driving error	Information error (lack of perception: e.g. having not noticed the traffic sign while passing) Diagnostic error (incorrect evaluation of available information) Performance error (incorrect execution: e.g. having not found the brake pedal)
Level 4: "Psycho-physiological condition"	Condition that can influence the psychological process	Impairment Exhaustion, fatigue Disorders (neurological, cardio-vascular) Intoxication (alcohol, drugs)

Table 1.4 Conditions, constituting a scenario and for each condition the parameters chosen

Conditions	Parameters
The driver	Age, gender, driver type, driving experience
The vehicle	Type (light vehicle–heavy vehicle), status
The road infrastructure	Road type (urban, rural, motorway)
The traffic conditions	Density, speed
The environmental conditions	Environmental conditions: weather, road surface, lighting

1.9 Definition of a Scenario

A scenario is a key methodological element in defining the prevailing conditions outside (and inside) the vehicle, which strongly influence the outcomes of a safety measure. Scenario specification and assessment parameter definitions for outcome effects have to be included, when assessment of safety measures is developed. Here, a scenario is defined as a conglomerate of five types of elements (conditions), defining the conditions of the driver-vehicle system on the road in which an ADAS (Advanced Driver Assistance System) is implemented (Wiethoff 2003). The five conditions are shown in Table 1.4, first column.

The ADAS implementation may have very different effects (e.g. safety effects, effects on speed, network efficiency, and environmental load) for different scenarios. In the approach for assessment of safety measures to promote forgiving and self-explanatory road environments, scenarios are applied to define the specific circumstances and evaluate safety measures accordingly.

1.10 Driving Errors and Intelligent Measures

As will be argued extensively later in this book, the combination of infrastructure and telematics measures can provide added value in the form of a cost-efficient solution, avoiding expensive infrastructure works by providing the same function through ADAS or other, comparable, innovative systems. Examples of promising countermeasures for the selected errors may be conceived in different categories; in-vehicle measures, infrastructural measures, and co-operative vehicle-infrastructure measures (sometimes referred to as ambient intelligence). Countermeasures for each error will be described and each measure will be evaluated on effectiveness in subsequent chapters. Also, for each measure the focus in terms of driver's experience (skill), age, owner and type of vehicle, and the (road) environment (traffic density, road category, special sections, lighting condition, and weather) are necessary to be defined subsequently. Obvious intelligent countermeasures concern monitoring and controlling speed or speeding, manoeuvres relative to other traffic participants, responses to infrastructure, etc.; all dependent on person, circumstances and conditions. The IN-SAFETY approach prioritizes upon the support and safety benefits of integrated systems. At the same time, legal obstructions and manufacturer's interest are considered.

The opinion and expectations of the relevant stakeholders (users, authorities and manufacturers) must always be taken into account as well, even before prioritisation and implementation can actually start. Co-operative, integrated systems, as proposed in this book, are systems for the near future, but still have to be developed and prototyped for the main part. Their potential benefits have not been proven in practice yet. However, the co-operative systems that are ranked highly in the relevant chapters of this book, address the listed driver errors, and give direction for attention and resources focus. Further completion of the implementation scenarios of these measures may prove that these are cost-effective contributions, towards a solution of the car driving safety problem.

References

K.A. Brookhuis, D. De Waard, V.A.W.J. Marchau, M. Wierthoff, L. Walta, E. Bekiaris, Self-explaining and forgiving roads to improve traffic safety, in *Developments in Human Factors in Transportation, Design, and Evaluation*, ed. by D. de Waard, K.A. Brookhuis, A. Tofetti (Shaker Publishing, Maastricht, 2006), pp. 51–63

eSafety Forum, Draft final report and recommendations of the implementation road map working group. Meeting Report, July 2005

C. Lotz (ed.), *A1.1 Benchmarking of Forgiving Road Environments. In-Safety Report IN-Safety-BASt-WP1-R3-V5-Activity11* (Bast, Bergisch-Gladbach, 2006)

W. Hacker (2003) Action Regulation Theory: A practical tool for the design of modern work processes? European Journal of Work and Organizational Psychology 12(2):105–130

C. Lotz, G. Wenzel, IN-SAFETY: with co-operative systems towards forgiving road environments, in *ITS-Conference*, London, 2006

C. Lotz, K.A. Brookhuis, A. Bauer, M. Wiethoff, V.A.W.J. Marchau, D. de Waard, IN-SAFETY – towards 'Forgiving road environments': implementation scenarios for road design measures and ITS solutions, in *TRA, Europe 2006 Conference*, Goteborg, 2006

J. Rasmussen, Human errors. A taxonomy for describing human malfunction in industrial installations. J. Occup. Accid. **4**, 311–333 (1982)

SAFETYNET, Building the European road safety observatory. Workpackage 1 – Task 3, Deliverable No 1: Annual Statistical Report 2004 based on data from the CARE database, 2004

TREDIT, Constitution & specification of the road safety programme – Ministry of Environment, Land Planning and Public Works, Greece, 2002

M. Vollrath, Fehleranalysen bei Unfällen und Anforderungen an die Fahrerassistenz (Analyses of errors on accidents and requirements on driver assistance systems). Presentation at the Deutscher Verkehrsexpertentag, 2005

WHO (World Health Organisation): M. Peden, R. Scurfield, D. Sleet, D. Mohan, A.A. Hyder, E. Jarawan, C. Mathers (ed.), *World Report on Road Traffic Injury Prevention* (WHO (World Health Organisation), Geneva, 2004). ISBN 92 4 156260 9

M. Wiethoff, The ADVISORS Final Publishable Report. GRD1-10047 FP5 report, 2003, http://www.ADVISORS.iao.fhg.de

M. Wiethoff, K. Brookhuis, D. de Waard, V. Marchau, L. Walta, G. Wenzel, K. de Brucker, C. Macharis, New concepts for driver assistance systems to improve road safety, in *Proceedings of the BIVEC-GIBET Transport Research Day 2007*, ed. by P. Hilferink, P. Rietveld, T. van den Hanenberg (Nautilus Academic Books, Zelzate, 2007). ISBN 978-90-8756-014-X

Chapter 2
Towards Forgiving and Self-Explanatory Roads

Evangelos Bekiaris and Evangelia Gaitanidou

2.1 The Concept of Sustainable Safety

"Sustainable Safety" is a road safety concept, by which the entire traffic and transport system is adapted to human limitations. The aim is to prevent crashes and to limit their consequences. The infrastructure prevents road use involving large differences in direction, speed and mass, and directs the road user towards safe behaviour. Vehicles are constructed to simplify the driving task and offer protection in the event of a crash. Road users are educated and informed properly and their behaviour is tested regularly. The essence of the Sustainable Safety approach is: prevention is better than curement (IN-SAFETY DoW 2005). The Sustainable Safety vision of road safety is based on five principles. These five principles refer to the functionality of roads, the homogeneity of mass and/or speed and direction, physical and social forgivingness, recognition and predictability of roads and behaviour, and state awareness. The following points are the goals of the Sustainable Safety vision (Wegman and Aarts 2006; SWOV 2007):

- The prevention of (serious) crashes, and where this is not possible, the almost total elimination of the risk of severe injury.
- The notion that man is the measure of all things due to his/her physical vulnerability and cognitive capabilities and limitations (such as fallibility and offence behaviour).
- An integrated approach to the elements human-vehicle-road, which is tuned to the human measure.
- A proactive approach to bridging gaps in the traffic system.

More specifically, the principles of sustainable safety can be summarized in the following table (Table 2.1):

E. Bekiaris (✉) and E. Gaitanidou
Centre for Research and Technology Hellas/Hellenic Institute of Transport (CERTH/HIT), Thessaloniki, Greece
e-mail: abek@certh.gr

E. Bekiaris et al. (eds.), *Infrastructure and Safety in a Collaborative World*,
DOI 10.1007/978-3-642-18372-0_2, © Springer-Verlag Berlin Heidelberg 2011

Table 2.1 Description of the five sustainable safety principles (Wegman and Aarts 2006)

Sustainable safety principle	Description
Functionality of roads	Mono-functionality of roads as either through roads, distributor roads, or access roads in a hierarchically structured road network
Homogeneity of roads	Equality of speed, direction and mass at moderate and high speeds
Forgivingness of the environment and of road users	Injury limitation through a forgiving road environment and anticipation of road user behaviour
Predictability of road course and road user behaviour by a recognisable road design	Road environment and road user behaviour that support road user expectations through consistency and continuity of road design
State awareness by the road user	Ability to assess one's capacity to handle the driving task

As seen in the table above, two of the principles are referring to forgiving and self-explanatory road environments. Thus, striving to define the road environment of the future, these two characteristics should be secured. According to FEHRL (2001), the roads of the future will need to:

- Contribute to sustainability.
- Make wide use of innovation.
- Contribute to improvements in road safety, environment and road transport efficiency.
- Reduce to zero any contribution to accidents ("forgiving road infrastructure").
- Reduce traffic congestion.
- Reduce noise and vibration to the road environment.
- Reduce air and visual pollution.

To achieve the forgivingness and self-explainability of road environments, the EC has committed researchers and other related stakeholders, by means of research initiatives, so that such environments would be defined and further described, along with the pre-requisites for a road environment to be characterised as such, both in term of infrastructure based measures and the use of new technologies.

2.2 Forgiving Road Environments

Forgiving road environments constitute a basic tool in preventing or mitigating an important percentage of road accidents related to driving errors. As everybody makes mistakes, drivers will eventually keep doing erroneous manoeuvres or actions. Over 80% of accidents are related to driver's error. More specifically, statistics show that about 25–30% of fatal accidents involve crashes with fixed roadside objects. Those accidents are mainly caused due to driving errors that lead to lane/road departure. The existence of a forgiving road environment would have prevented accidents of this type (and generally accidents that involve driving errors) and/or reduced the seriousness of the consequences of such accidents.

Forgiving road environments may also take advantage of advanced telematic and in-vehicle systems, which will support the driver in case of an error. Those systems, in contrast to traditional and autonomous ADAS (Advanced Driver Assistance Systems), will not only support the driver by providing an adequate warning, but will supplement the road infrastructure. This, for example, can be achieved by simulating a rumble strips sound or using other haptic warnings, when the driver involuntarily crosses the road marking, overspeeds or initiates an erroneous overtaking.

2.2.1 *Definition*

> *In the context of this book, a forgiving road is defined as a road that is designed and built in such a way as to interfere with or block the development of driving errors and to avoid or mitigate negative consequences of driving errors, allowing the driver to regain control and either stop or return to the travel lane without injury or damage.*

Examples are roads that have structural layout elements that reduce the consequences of accidents or driving errors (e.g. when leaving the lane unintentionally) once they happen, or in-vehicle devices with the same function, like "Lane Departure Warning Assistant".

To develop a forgiving road environment certain characteristics must be included and measures should be taken, involving either the infrastructure itself or the use of telematic and other aids. Most notably, the combination of infrastructure and telematics measures can provide a more cost-efficient solution, as expensive infrastructure works may be substituted by telematics or other innovative systems.

2.2.2 *Forgiving Road Environments in Practice*

Devising the measures for forgiving road environments (FOR), as they by definition aim at avoiding or mitigating negative consequences of driving errors, starts with listing possible driving errors to be supported, that in turn are related to accident statistics. As various driving errors can be distinguished, usually some clustering or categorisation of errors is used. This procedure has been undertaken within the IN-SAFETY project (Wiethoff et al. 2006), where four levels of driving errors have been identified and relevant measures have been proposed for each error category:

1. Accident type errors: result of the execution of an error (e.g. collide with other vehicle).

Table 2.2 Errors and measures for FOR and SER measures

Measure error/scenario	In-vehicle	Infrastructure	Co-operative (based on vehicle-infrastructure and vehicle-to-vehicle communication and cooperation)
Speeding in an unexpected bend on rural roads	*Navigational aid*	*Variable message sign (VMS)*	*Electronic beacons, providing in-car info, merged into on-board navigation*
Over-speeding (in general)	*Speed alert system by speed sign recognition*	*VDS*	*Speed alert, based on digital maps, updated by road beacons*
Wrong use of road	*Lane departure warning system*	*Audio lane warning delineation*	*Adaptive LDWS*
Violation of priority rules	In-vehicle traffic sign recognition	Electronic traffic signs	*Traffic light status emitted to the car*
Overtaking failure	*Blind spot detector*	Rumble strips	Vehicle-to-vehicle communication
Insufficient safety distance	*A frontal warning system*	VMS with fog warning	*Adaptive frontal warning systems*

2. Driving errors: action that leads to an accident (e.g. inappropriate speed).
3. Human error: psychological process that forms the basis of a driving error (e.g. incorrect evaluation of speed and distance).
4. Psycho-physiological condition: condition that can influence the underlying psychological process (e.g. fatigue).

The safety potential of each measure has been estimated, followed by the construction of relevant scenarios (see Table 2.2) and their consecutive prioritisation, using the MCA/AHP methodology (more on these issues can be found in Chaps. 3 and 16).

2.3 Self-Explanatory Road Environments

The other basic principle of sustainable safety that is discussed in the present is this of self-explanatory roads (also referred to as self-explanatory roads). What this term implies is the interaction between the infrastructure (including the road, the road equipment and the whole roadside environment) and the road users. The key issue in this case is that the road succeeds (either by its layout, or by adequate signing) to communicate correctly to its users the necessary “messages”, so that they would be able to use it effectively, in the least distracting and risk-generating manner.

Examples are consistent pictograms and/or earcons, which are used in the traffic environment as well as employed by in-vehicle applications, to inform the driver or warn/alarm him/her upon the direction to follow, regarding danger ahead, etc. The multi-ethnic character of modern societies and the effects of globalization on the road network make it all the more important to substitute text at VMSes and on-board systems with internationally recognized symbols and sounds, many of which correspond to new functions (such as traffic congestions level, navigation, route guidance, lane deviation/departure, distance from frontal car, overspeeding, traffic management control signals, etc.) and thus are not included into the signs of the Vienna Convention.

But self-explanatory roads measures are not limited to standardization of the interaction elements because, no matter how standardized they become, they are still surely not suitable for everybody. Thus, a key element is that of information redundancy but also consistency and timeliness of provision and, ultimately, on info and warning adaptation and personalization, to match the individual participants own needs (Bekiaris et al. 2005).

2.3.1 Definition

In the context of this book, self-explanatory road is defined as one that is designed and constructed to evoke correct expectations from road users and elicit proper driving behaviour, thereby reducing the probability of driver errors and enhancing driving comfort.

A road accident is generally the end result of a multi-step process. The result of combinations and interactions between the three parts of the system (driver, road and vehicle) contribute to the traffic accidents. The aim is to understand the contribution of human factors and road characteristics to road accidents, in order to find the way to reduce accidents. For understanding the process of accidents the human factors and the road characteristics in the development of the accidents have to be examined. A clearer understanding of the role of these factors and characteristics will significantly contribute to the enhancement of road safety.

2.3.2 Self-Explanatory Road Environments in Practice

There are two main issues regarding self-explanatory roads (SER), on which IN-SAFETY (De Brucker et al. 2006) has focused: the first issue is related to the degree to which the total design of road environment, including road layout, contributes to creating a SER environment (through a process of prioritising road

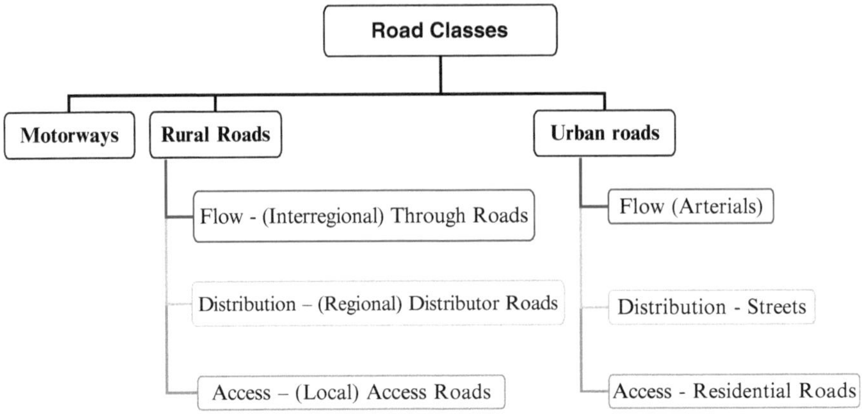

Fig. 2.1 Suggested road classes for self-explanatory roads (Matena et al. 2008)

accidents, followed by designing, choosing alternative measures to prevent these types of accidents and prioritising, using multicriteria analysis – MCA). The second issue is related to the readability and understandability of VMS messages (through an analysis of existing VMS, the design of alternative VMS, as well as the design of new VMS, followed by a user test).

The features that contribute to the creation of self-explanatory road (SER) environments were identified (and quantified) within the project and refer to (1) a sound road categorisation system, (2) assurance of sufficient time for the driver, (3) a safe field of vision offered to the driver and (4) respect for driver expectations. On the basis of these features, 14 recommendations for the development of variable message signs (VMS) have been formulated within the IN-SAFETY project. These refer to the size and design of pictograms, visual performance, text message and combined message recommendations, comprehensibility, route guidance, selection control, place of VMSes, distances between VMSes, combining several types of signals, changing messages in time and place, information overload and information absence. All these are further analyzed in Chaps. 13 and 14 of this book.

On the other hand, another EC funded research initiative, RIPCORD-ISEREST (506184), dealt with self-explanatory roads, merely from the infrastructure point of view. In it, among others, the concept and elements of self-explanatory roads were discussed, good practices identified and recommendations for self-explanatory road classes suggested (Matena et al. 2008) (Fig. 2.1).

2.4 Initial Concepts on Measures Promoting SER and FOR

The European transport system needs to be optimised to meet the demands of constant traffic enhancement and sustainable development. A modern transportation system must be sustainable from an economic and social as well as an environmental viewpoint. The principles of forgiving and self-explanatory road environments are among those which could contribute towards such an achievement.

In terms of forgiving road environments, the identification of error patterns that lead to accidents is the first step, in order to conclude to measures to be taken for rendering a road environment of forgiving nature. What is of outmost importance is to select the appropriate measure for each type of error, either in terms of infrastructure enhancement or application of telematics, or even their combinations, which are seen as the most promising solution, especially in terms of cost-efficiency.

As it has been seen, regarding self-explanatory road environments, several human factors depend on the traffic environment and there is no possibility to influence all of them. To lower the rate of accidents, the environment needs to be changed, most notably the road characteristics. Road characteristics that are suitable to human nature, and supply the driver with a clear, understandable picture about the given situation, have to be ensured. Such a road can be called a self-explanatory road.

Within IN-SAFETY, a set of measures have been proposed, as seen in the Table 2.2, covering both cases.

In Table 2.2, the alternatives that contribute to FOR only are represented in non-shaded cells and italics typeface. All the alternatives contributing to a SER environment are shaded cells. Those that contribute to a SER environment only are represented in black normal typeface. Those contributing to both SER and FOR (under specific circumstances) are represented in black, italics typeface, in shaded cells.

Reaching the deadline of 2010, set by the White Paper (COM 2001) road environments should, at the most possible degree, secure that people and goods can be transferred quickly, environmentally friendly and safely. This is a pre-requisite for the road transport to evolve towards the direction of sustainability, which is considered as the most promising feature for the future of transport.

References

E. Bekiaris, E. Gaitanidou, K. Kalogirou, IN-SAFETY project: towards road fatalities reduction through the enhancement of forgiving and self-explanatory roads, in *1st FERSI Scientific Road Safety Research Conference*, Bergish-Gladbach, Germany, 7–8 September 2005

COM, 370 final, White Paper: European transport policy for 2010: time to decide, European Commission, 12.9, 2001

F.C.M. Wegman, L.T. Aarts (red.), *Advancing Sustainable Safety; National Road Safety Outlook for 2005–2020* (SWOV, Leidschendam, The Netherlands, 2006)

SWOV Fact sheet, *Background of the five Sustainable Safety principles* (SWOV, Leidschendam, The Netherlands, October 2007)

M. Wiethoff, K. Macharis, C. Lotz et al., Implementation scenarios and concepts towards forgiving roads, Deliverable 1.1 IN-SAFETY project, 2006

K. De Brucker, M. Wiethoff et al., Implementation scenarios and concepts towards self-explaining roads, Deliverable 2.1 IN-SAFETY project, 2006

S. Matena, R. Louwerse, G. Schermers, P. Vaneerdewegh, P. Pokorny, E. Gaitanidou, R. Elvik (TOI), J. Cardoso, Road Design and Environment – Best Practice on Self-explaining and Forgiving Roads, Deliverable 3 RIPCORD-ISEREST project, 2008
IN-SAFETY project, Annex 1, Description of work 2005
FEHRL SERRP III (Strategic. European Road Research Programme III), 2001

Chapter 3
Structuring the Way

A New Approach on Multi-Criteria and Cost–Benefit Analysis to be Applied to Road Safety Measures

Klaas De Brucker, Cathy Macharis, and Knut Veisten

3.1 The Scope of Evaluation Tools

Several tools for the evaluation of intelligent-transport-systems (ITS)-based safety measures exist, such as cost–benefit analysis (CBA), cost-effectiveness analysis (CEA), financial analysis and multi-criteria analysis (MCA) or stakeholder analysis. The scope of these methods is, however, different when comparing one method with another and, hence, the conclusions may be different when using one method instead of another. In this chapter, the theoretical foundations, as well as the generic procedure to be followed when using these evaluation tools will be described. Specific applications of these evaluation tools to ITS-based safety measures will be presented in subsequent chapters of this book. In Chap. 16 the results of an MCA will be presented and in Chap. 17 the results of a CBA will be discussed, both applied to the evaluation of ITS-based safety measures.

3.2 The Use of Social Cost–Benefit Analysis for the Economic Analysis of Road Safety Measures

3.2.1 *Theoretical Foundations of Cost–Benefit Analysis*

Cost–benefit analysis (CBA) has its roots in traditional, neoclassical welfare economics. A societal perspective is taken, as opposed to e.g. financial analysis (or private investment analysis), where only the point of view of one person

K. De Brucker (✉)
Faculty of Economics and Management, Hogeschool-Universiteit Brussel (HUB), Brussels, Belgium
e-mail: klaas.debrucker@hubrussel.be

C. Macharis
Department MOSI-Transport and Logistics, Vrije Universiteit Brussel (VUB), Brussels, Belgium

K. Veisten
Institute of Transport Economics (TOI), Oslo, Norway

E. Bekiaris et al. (eds.), *Infrastructure and Safety in a Collaborative World*,
DOI 10.1007/978-3-642-18372-0_3,

or organisation is taken into account.[1] This means that, in CBA, all project related effects, whether benefits or costs, are taken into account irrespectively of the identity of the economic actors to whom these benefits or costs accrue. The relevant effects are obtained by comparing the effects that will occur by implementing the project with the effects that will occur in the absence of the project (i.e. the do-nothing scenario). The decision rule that implicitly underlies CBA, in other words the procedure used for aggregating the various effects a project has on society's members, is the criterion of potential Pareto improvement (Pareto 1927), also known as the Hicks–Kaldor compensation test (Hicks 1939; Kaldor 1939). This decision rule implies that a project will increase social welfare if the increases in utility levels (i.e. the benefits) for those who gain are higher than the decreases in utility levels (i.e. the costs) for those who lose. In other words, this test requires that winners should win more than losers lose. The compensation does not need to actually take place. If it were actually to take place, a pure Pareto improvement would result. When it is no longer possible to achieve a Pareto improvement, the situation is called "Pareto efficient" or simply "efficient".

When using CBA, a project's effects are given a monetary value, which is inferred from consumer behaviour in markets, as expressed by the consumers' willingness-to-pay. Economic values are recognised as expressions of individual/household preferences. The demand of consumers is, thus, assigned the leading role in deciding about the availability of goods and services. This is done without any judgment or corrections against those who demonstrate higher willingness-to-pay, for instance for time-savings and speed than for road safety. The principle of "consumer sovereignty" is, therefore, fundamental to CBA. The interaction between the diversity of preferences or tastes for marketable commodities, as expressed by consumers on the one hand and the production costs incurred by commodity producers on the other hand, results in a set of market prices. The thesis for a "perfect" (free) market states that price levels correspond to the point where marginal demand (or marginal willingness to pay) equals marginal supply. These prices are taken as the best indicators of the economic value for *private* goods. The competition in free markets also assures that a largest possible quantity is available for a lowest possible price (Mishan 1988; Varian 1992).

CBA is, therefore, not based on valuations given by politicians or decision-makers, but on individual/household valuations, as expressed by their willingness-to-pay revealed in markets or gauged in other ways (Mishan 1988; Hanley and Spash 1993). In case well functioning markets are absent for specific categories of benefits or costs, surrogate markets need, therefore, to be constructed and, by doing so, a monetary value for the relevant effects may be estimated.

[1] Some apply the term "social cost-benefit analysis (SCBA)", to stress the difference with respect to financial cost-benefit analysis. However, SCBA may also be understood as a stepwise analysis, starting by a financial appraisal, then adjusting prices and/or including non-market goods to produce an economic appraisal and finally also including a social appraisal in the meaning of an assessment of distributional effects (Thirlwall 2003).

3.2.2 The Applicability of Cost–Benefit Analysis to Road Safety Measures

Compared to infrastructure measures, use of ITS (Intelligent Transport Systems) measures attempts to influence individual behaviour and obtain positive safety, time-use and/or environmental effects without (necessarily) changing/extending road capacity or public transport capacity per se. Another issue is that most ITS-based technology is only recently introduced or at the brink of being introduced. All this may have implications for the economic analysis (Bekiaris and Nakanishi 2004; Gillen et al. 1999; Samstad and Markussen 2000).

Road safety can be regarded as a good with a mix of *private* and *public* aspects. Individuals can choose/buy travel modes or equipment that is considered "safe", like, e.g. advanced driver assistance systems (ADAS) or in-vehicle information systems (IVIS) (Panou and Bekiaris 2004). However, the infrastructure that enables the transport, including regulations, traffic controls, and map-based or infrastructure-based systems for information/navigation or positioning/speed, has clear public good aspects. The safety of the infrastructure cannot be portioned out to individual road-users. Although the choice of a car may influence the individual usability of some ITS-based devices, the safety of the infrastructure is (for most applications) a *non-excludable* good. The reason is that it cannot be denied or sold to the individual road users. People who do not pay the price for this good cannot be prevented from using it. Furthermore, the safety of the infrastructure may be regarded as less *congestible* (more *non-rival*) than infrastructure itself. The use of the infrastructure by one person (e.g. by occupying some space when driving the car) may, in some well-known situations, reduce the ability for other road-users to "consume" the same infrastructure (queues, congestion, *rivalry*). But "consumption" of the safety standards of the infrastructure and the safety regulations and the traffic control system does not reduce other road-users' "consumption" of the same goods. If provided at a given level, this public safety level of the infrastructure, including regulations and control, is, therefore, more or less equally available for all road users and has, therefore, a public good character.

3.2.3 Decision Criteria Used in Social Cost–Benefit Analysis

In CBA a variety of decision criteria exist to measure the efficiency of a project or policy measure. The most important ones are the *net present value* criterion and the *benefit–cost ratio*. The net present value (NPV) of a project is defined in formula (3.1).

$$NPV = Present\,value\,of\,all\,benefits - Present\,value\,of\,all\,costs \tag{3.1}$$

The benefit term should principally include all effects that are valued monetarily in an analysis. All benefits are usually added to obtain the total benefits. Negative benefits (or societal cost), such as for example increased travel time, if these were to be estimated, are subtracted from the benefits. The cost term usually refers to the implementation costs (i.e. the budgetary cost) of a measure, expressed in terms of the opportunity cost from a social point of view. To facilitate the comparison of projects of different scale/scope, a benefit–cost ratio (BC ratio) can be estimated, as presented in formula (3.2).

$$BC\,ratio = \frac{Present\,value\,of\,all\,benefits}{Present\,value\,of\,implementation\,costs} \tag{3.2}$$

As can easily be tested, there is a simple definitional relationship between the NPV and the BC ratio.[2] When the NPV is positive, the BC ratio exceeds the value of 1.

3.3 The Use of Cost-Effectiveness Analysis as Applied to Road Safety Measures

Cost-effectiveness analysis (CEA) can be described as an analysis by which a measure or alternative (c.q. safety measure) is selected that can achieve a policy objective (c.q. increasing road safety) at the lowest budgetary cost possible. Alternatively, CEA may examine how a fixed amount of resources (e.g. an acceptable or maximum cost) may be used to achieve a maximum level of effectiveness (e.g. a maximum reduction of accidents/risk) in realizing a specific policy objective (e.g. increasing road safety). The former approach corresponds to cost minimization, the latter to effect maximization (Vlakveld et al. 2005). In both approaches a cost-effectiveness ratio (CE ratio) needs to be calculated. Applied to road safety, this CE ratio may be defined as given in formula (3.3).

$$CE\,ratio = \frac{Number\,of\,fatalities\,prevented\,by\,a\,given\,measure}{Unit\,costs\,of\,implementation\,of\,measure} \tag{3.3}$$

The cost-effectiveness of a road safety measure can be defined as the number of fatalities, injuries or (injury) accidents prevented per unit cost of implementing the measure. Simply stated, the CE ratio could be e.g. prevented fatalities per one million euro spent.

Both CBA and CEA are methods of economic analysis that can be applied for the evaluation of public investment, where different projects (in road safety and

[2]Sometimes a criterion called "profitability index" (PI) is used, defined as the ratio of NPV to implementation cost. The relationship between the PI and the BC ratio is very simple, since PI = BC ratio − 1. Also the relation between PI and NPV is very simple: when the NPV is greater than 0, then also the PI exceeds 0.

other areas) are competing for scarce resources. There are two fundamental differences between both methods (Hakkert and Wesemann 2005). The first difference is that CEA takes a political objective as the point of departure and aims to select that measure or combination of measures which makes it possible to achieve this objective at the lowest budgetary cost possible. Thus, CEA is designed to find the most cost-effective (i.e. the cheapest) solution to realizing a given objective. The second difference is that CBA does not consider the political objective as absolute. Although performing a CBA initially is also guided by political objectives (such as e.g. the political decision to increase road safety), this method evaluates the fundamental desirability of achieving this objective. In CBA one will also search for the cheapest way to reach policy objectives, but this is done through weighing the social benefits and social costs of projects aimed at realizing this objective. Thus, a CBA shall indicate what measure (or combination of measures) provides the largest difference between benefits and costs (Gillen et al. 1999; Mishan 1988).

3.4 Input Data Required for Cost-Effectiveness and Cost–Benefit Analysis Applied to the Evaluation of Road Safety Measures

3.4.1 Introduction

In order to estimate the cost-effectiveness of a road safety measure (in a cost-effectiveness analysis (CEA)), the following information is generally needed:

- An estimate of the effectiveness of the safety measure, in terms of the number of accidents it can be expected to prevent per unit implemented of the measure (per vehicle with a given vehicle-based system and/or per kilometre road with a given infrastructure-based system).
- A definition of units of implementation for the measure (e.g. "in new cars in region/country *X*" or "on roads of class *Y* in region/country *X*").
- An estimate of the costs of implementing one unit of the measure (all private costs and public costs – if costs for some are profits/income for others they generally cancel out).
- A method for converting all costs of implementation to an annual basis (in order to make measures with different time spans comparable, e.g. using discounting and present costs).

In order to estimate the efficiency of a road safety measure through calculating the net present value (NPV) or benefit–cost ratio (BC ratio) in cost–benefit analysis (CBA), the same input regarding the effects and costs are needed as for CEA (listed above), plus one additional input, namely:

- Money values of fatalities and injuries (of different severity) avoided, of time-savings, of environmental effects, etc.

Ideally, a CBA should include "all benefits and costs, on all people, over all relevant areas and time periods" (Moore and Pozdena 2004). However, in many cases only some effects are within the scope of being quantified and valued (in our case these are the safety effects, i.e. the effects on expected injuries/fatalities), while other potential effects (e.g. on time use, environment, etc) are omitted from the calculations.

The accidents that are affected by a safety measure may be referred to as *target accidents*. In the case of general measures like speed limits, target accidents may include all accidents on a given road or in a given region/country. For measures related to in-vehicle systems or combined vehicle-infrastructure systems, however, only a share of all accidents is target accidents. The estimated percentage effect of the safety measure on target accidents defines the numerator of the cost-effectiveness ratio. To estimate the denominator, the first step is to define a suitable unit of implementation of the measure. In the case of infrastructure measures, the appropriate unit of implementation will often be one junction or 1 km of road. In the case of area-wide or more general measures, a suitable unit of implementation may be a typical area or a particular category of roads. In the case of vehicle safety measures, one vehicle will often be a suitable unit of implementation.

Once a suitable unit of implementation is defined, unit costs can be estimated. In order to make the CE ratios of different safety measures comparable, it is necessary to relate both the number of prevented accidents and the costs of implementing the measure to a certain time reference. This need arises because the relationship between costs and the duration of effects varies a lot between safety measures. In order to get comparable implementation costs for all safety measures, irrespective of the duration of their safety effects, investment costs can be converted to annual capital costs. Annuities can easily be obtained from the present value of costs of the investments plus operation/maintenance, using the inverse annuity factor.

3.4.2 Specific Data Requirements: The Ideal Case

There are several types of data that are necessary for economic analysis of ITS-based safety measures. As shown in Table 3.1 below, these may be divided into the following categories: "general accident data", "scenario-specific accident data", "safety effect estimates", "vehicle data", "cost data", "benefit value data" and "other data".

In addition to an application to selected countries, one could consider aggregated data/estimates for the whole EU (COWI[3] 2006). But, in the IN-SAFETY project the analysis was limited to the countries where accident data indicating accident cause, vehicle data and road data could be obtained.

[3]COWI is an international consultancy group based in Denmark. The abbreviation COWI stands for "Consultancy within Engineering, Environmental Science and Economics".

Table 3.1 Data needed for evaluation of ITS based safety measures

General accident data	
	No. of fatalities in country *X* (in year *Y*)
	No. of serious injuries in country *X* (in year *Y*)
	No. of slight injuries in country *X* (in year *Y*)
Scenario-specific accident data – target accidents (fatalities/injuries with a specific cause)	
	No. of fatalities due to accident cause *Z* in country *X* (in year *Y*)
	No. of serious injuries due to accident cause *Z* in country *X* (in year *Y*)
	No. of slight injuries due to accident cause *Z* in country *X* (in year *Y*)
Effect estimates (if full scale implementation from day 1, i.e. infrastructure/equipment on whole road length of relevant type, and all cars equipped)	
	% Reduction of fatalities if scenario S implemented in country *X* (in year *Y*)
	% Reduction of serious injuries if scenario S implemented in country *X* (in year *Y*)
	% Reduction of slight injuries if scenario S implemented in country *X* (in year *Y*)
Vehicle data (needed for in-vehicle/cooperative scenarios, assuming equipment installed in new cars)	
	No. of cars in country *X*
	No. of new cars per year in country *X* (if a penetration rate cannot be estimated based on car renewal, estimates for the market penetration rate for the first year in country *X* should be provided, as well as for the annual increase in this rate for the country studied)
	Average age of cars in country *X*
	Annual average mileage per car in country *X*
Cost data (estimates)	
	Unit investment cost for car equipment in country *X*
	Effective life/lifetime of car equipment
	Unit investment cost for infrastructure equipment in country *X*
	Effective life/lifetime of infrastructure equipment
	No. of kilometre/points with infrastructure equipment in country *X*
	Annual maintenance/operating costs for car equipment in country *X*
	Annual maintenance/operating costs for infrastructure equipment in country *X*
Benefit value data	
	Monetised value of an avoided fatality in country *X*
	Monetised value of an avoided serious injury in country *X*
	Monetised value of an avoided slight injury in country *X*
Other estimates (if full scale implementation from day 1, i.e. infrastructure/equipment on whole road length of relevant type, and all cars equipped)	
	Travel time changes
	Environmental changes (other than emission changes due to speed changes)

3.4.3 How to Cope with Lack of Data

A fairly general problem for the economic assessment of ITS-based measures (scenarios) is the lack of relevant data (Gillen et al. 1999). There are some few reports available with estimates of effects, particularly safety effects, from ITS-based systems. It is of course difficult to evaluate such measures that have been implemented only partly. A possible approach is to indicate target accidents through an "error-based approach", which is described below.

The (societal) benefit (welfare gain) from in-vehicle safety devices (IVIS) can be depicted as monetised valuations of the (safety, time-savings and other) effects.

In many European countries there exist unit prices for such effects, i.e. Euro values per fatality prevented or an hour of travel saved (Bickel et al. 2006; Nellthorp et al. 2001; Trawén et al. 2002; Hakkert and Wesemann 2005). National figures or proposed average European figures can be applied. In the CBA which is performed in Chap. 17 of this book, the values proposed by Bickel et al. (2006) are used. These represent some general European values that are only adapted to the specific country using purchasing power parity. The monetised values of the (safety) effects are needed only for the CBA and not for the CEA.

Also, the costs of the safety measures may be unknown or concealed. One possible approach here is to use what can be found in existing markets. Equipment costs may be estimated based on COWI (2006) and on information provided by the US Department of Transportation (DOT 2007). Vehicle equipment costs are available for installation in new cars only. It should be remarked that such equipment costs may fall sharply when the sales of the equipment pass certain levels. Such cost developments are, however, difficult to predict.

We are forced to work with approximate data regarding safety effects, other effects, as well as regarding the costs of the measures. The main approach for estimating the safety effects related to the measures proposed was the maximum impact *error-based approach,* using only existing general accident statistics. This approach is described more thoroughly in Chap. 17.

3.5 The Use of Multi-Criteria Analysis for the Evaluation of the Implementation of Road Safety Measures

3.5.1 Fundamental Differences Between Multi-Criteria Analysis and Social Cost–Benefit and Cost-Effectiveness Analysis

In contrast to social cost–benefit analysis (CBA), which is based on neo-Paretian welfare economics, multi-criteria analysis (MCA) has its roots in a different discipline, namely operations research (Charness and Cooper 1961). MCA does not necessarily rely on welfare economics concepts such as the consumer surplus, i.e. the Dupuit–Marshall surplus (Dupuit 1844; Marshall 1890). More recently, however, MCA has been applied in the context of economics-driven project evaluation. This appears useful especially when a neo-institutional approach to project evaluation is adopted and multiple stakeholders become relevant (Macharis 2004; De Brucker and Verbeke 2006, 2007). MCA can be considered as formal procedure (or a set of rules, i.e. an "institution") which allows comparing a number of actions (e.g. projects or policy measures, "alternatives" or "scenarios") in terms of specific criteria. These criteria represent the operationalization of the objectives and sub-objectives of decision makers and stakeholders participating in the decision-making process.

The MCA methodology is especially useful for the evaluation of ITS, since this method makes it possible to structure complex decision problems according to their constituent parts (objectives, sub-objectives as measured by criteria) and to make comparisons among project alternatives, even when effects cannot be monetised fully, nor even quantified. It is usually possible to link specific stakeholders with specific criteria in the MCA and, by doing so, stakeholder management may be performed and effective implementation strategies may be defined.

One fundamental difference with CBA is that the effects do not need to be assigned a monetary value. Benefits and even costs can be expressed in physical units or even in qualitative terms. Another fundamental difference with CBA is that MCA does not formally calculate the difference between benefits and costs (i.e. a net present value of a project) so as to make a statement about the fundamental desirability of a project.[4] Generally, MCA leads to a ranking or a selection of projects in terms of the decision makers' objectives. Yet another fundamental difference is that in MCA the values (and weights) of the effects are not derived from consumers' willingness-to-pay as expressed in markets, but are given by decision makers. In MCA, criteria result from policy makers' objectives and are weighed by policy makers. Experts may be involved when alternatives or scenarios need to be scored in terms of their contribution to the criteria.

In some respect MCA resembles cost-effectiveness analysis (CEA). The effectiveness score in the numerator of the cost-effectiveness (CE) ratio (formula (3.3)) is also expressed in physical units. Neither does CEA aim at making a statement about the fundamental desirability of a project (e.g. by calculating a net present value). The effectiveness score in numerator of the CE ratio is also derived from policy objectives and not from consumer willingness-to-pay. Also in MCA, the cost aspect is taken into account, since implementation cost usually is one of the criteria in MCA, either as a separate criterion or by dividing the effects or attributes by the implementation cost. Also the monetised benefits and costs may be integrated in a MCA in the form of one or more specific criteria. The MCA then becomes an eclectic evaluation tool (De Brucker 2000; De Brucker and Verbeke 2006, 2007). The difference between CEA and MCA is that, in the former, the effectiveness is a one-dimensional concept, whereas in MCA it is conceived as a multiple dimensional one. This means that in MCA a number of objectives which are additional to the main objective may be taken into account. For instance, regarding the evaluation of ITS based road safety measures, it is possible to take into account effects on time savings, environmental effects, investment risk, implementation barriers, etc. next to the main objective (c.q. reducing the number of fatalities).

Since the number of effects that can be taken into account in MCA is much larger than it is in CEA or CBA, it can be said that MCA transcends the realm of economic analysis. MCA (also called "multi-criteria decision *aid*"), therefore,

---

[4]It should be noted, however, that also in CBA the statement regarding the fundamental desirability of a project is a relative one, since CBA only makes it possible to compare the project with the doing-nothing alternative.

becomes a useful tool especially in the phase of decision, i.e. when a decision whether to implement a project or not needs to be made by political decision makers. CBA (and CEA) can then be used in the phase of analysis, i.e. the phase preceding the phase of political decision (De Brucker and Saitua-Nistal 2006).

3.5.2 Discussion of Subsequent Steps to be Followed in Multi-Criteria Analysis

In general terms, the process-related steps to be followed in MCA have a structure as shown in Fig. 3.1.

First, the nature of the problem is identified and analyzed. On the basis of this analysis, actions ("alternatives" or "scenarios") that may remedy the problem are formulated in the *second step*. In the *third step*, criteria are developed relevant to the evaluation of the actions to be studied. A criterion is a function that makes it possible to provide a score (quantitative or qualitative) for each action, measuring the contribution of that action to a relevant specific objective. By giving scores, a partial evaluation is performed (i.e. an evaluation in terms of one or more specific objectives as measured by criteria). The objectives identified in the MCA may correspond to the objectives of specific stakeholders identified in the decision-making process. Alternatively, it is possible to define objectives (and hence criteria) directly on the basis of stakeholder analysis. This is done e.g. in the multi-actor MCA (MAMCA) method, which will be presented in Sect. 3.5.4. The second and the third step as shown in Fig. 3.1 can also be reversed. When criteria are developed first and actions thereafter, value-focused thinking is adopted (Keeney 1996).

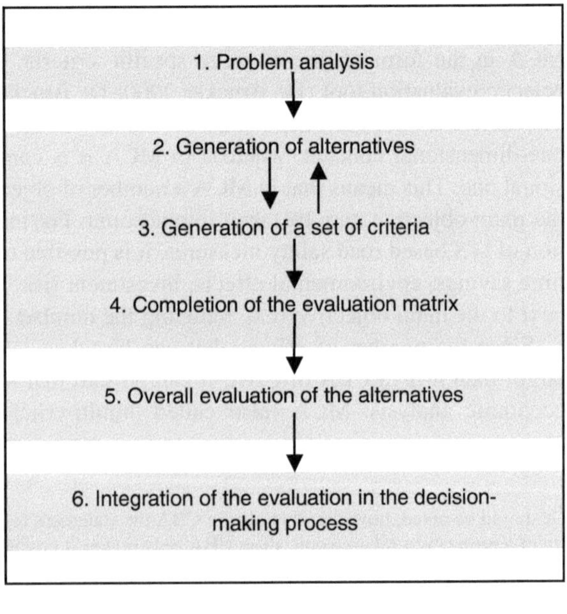

Fig. 3.1 Process-related steps in MCA
Source: Nijkamp et al. 1990:13, adapted by the authors

Table 3.2 Evaluation matrix

	g_1	g_2	...	g_j	...	g_m
a_1	e_{11}	e_{12}	...	e_{1j}	...	e_{1m}
a_2	e_{21}	e_{22}	...	e_{2j}	...	e_{2m}
...	...	...	...	...	...	...
a_i	e_{i1}	e_{i2}	...	e_{ij}	...	e_{im}
...	...	...	...	...	...	...
a_n	e_{n1}	e_{n2}	...	e_{nj}	...	e_{nm}

Source: Schärlig (1985:60), adapted by the authors

Values (to be measured by criteria) are made explicit from the outset. Only in the next step does one proactively attempt to identify actions that can contribute to these predefined values. The set of actions is thus "constructed" instead of being determined externally. This approach contrasts sharply with the method of alternative-focused thinking, which is often applied in practice. According to Keeney (1996), the latter approach reduces creativity and innovation, because the predetermined set of alternatives fundamentally constrains the evaluation process. The criteria are then typically selected based on thinking about the alternatives, not about the fundamental objectives (values) to be achieved. A possible way to reconcile these visions is to make the process iterative, as is suggested by the two opposite arrows in Fig. 3.1 (between steps 2 and 3).

The *fourth step* consists of constructing and completing the evaluation matrix. This is a matrix where all the actions (a_i) are evaluated in terms of all the criteria (g_j) as shown in Table 3.2 (whereby $i = 1,\ldots,n$ and $j = 1,\ldots,m$). Within the evaluation matrix, however, clusters of criteria can be distinguished. Criteria can be clustered in two ways. First, they can be clustered according to the type of effect or the way in which the effect was measured. Criteria may then be clustered into groups, such as a group that can be expressed in monetary units, another group related to non-monetary environmental or safety effects, still another group related to the non-monetary aspects of comfort, etc. A second way to cluster criteria is according to specific points of view, corresponding to specific stakeholder objectives, which is done in the MAMCA method (see Sect. 3.5.4).

In the *fifth step*, the information in the evaluation matrix needs to be aggregated. The information represented in the evaluation matrix seldom makes it possible to select one action in an unambiguous fashion. In most cases, the scores obtained by the actions on the various criteria (partial evaluations) are conflicting, which means that they do not unanimously point to a single "best" action, which would be superior in terms of all criteria. This situation is sometimes referred to as the "multi-criteria imbroglio" (Schärlig 1985). An aggregation method is, therefore, needed in most cases to synthesize the conflicting information. Each aggregation method relies on specific assumptions regarding the comparability of the partial evaluations and the relations between criteria. In most cases, criteria should be given explicit weights by policy makers. Here, analysts can introduce an interactive tool to help policy makers when reflecting on relative weights, but ultimately it is the decision makers themselves who must give the policy weights. Within each aggregation method, several MCA approaches can be used to aggregate the partial evaluations.

Within the limited scope of this contribution, it is not possible to give an overview of the various MCA methods that have been developed in the recent past. High quality overviews are provided in Belton and Stewart (2002) or in Figueira et al. (2007) (both in English), and De Brucker et al. (1998) (in Dutch). We shall, however, briefly discuss one specific MCA-method, namely the analytic hierarchy process (AHP) method of Saaty (1977, 1986, 1988, 1995) for a number of reasons. Firstly, this method has actually been applied already in various real life applications, including other EU funded research projects, such as e.g. the ADVISORS[5] project (De Brucker et al. 2002; Macharis et al. 2004, 2006). Secondly, it allows building ("constructing") a solution step by step, taking into account conflicting stakeholder objectives. Thirdly, the AHP method allows determining policy weights in a very logical way (through a set of pairwise comparisons). Fourthly, the AHP is the most widely used method for the evaluation of transport projects (Macharis and Ampe 2007). Fifthly, the AHP method makes it possible to obtain relative priorities for alternatives in terms of their contribution to specific criteria even when hard, i.e. quantified data, are hard to obtain for these criteria. This is done using expert judgment, as will be explained below. Sixthly, the AHP methodology will also be used in Chap. 16 of this book for the selection/prioritisation of actions ("alternatives" or "scenarios") in terms of possible future FOR and SER environments.

3.5.3 *The Method of the Analytic Hierarchy Process and its Ability to Process Qualitative Data*

The method of the analytic hierarchy process (AHP) is based on three principles: (1) construction of a hierarchy, (2) priority setting and (3) logical consistency.

A hierarchy (as shown in Fig. 3.2) is a complex system in which the constituent parts are hierarchically structured. The top of the hierarchy consists of a single element, which represents the overall objective or focus. The intermediate levels represent sub-objectives and their constituent parts (if possible, measured by operational criteria, i.e. $g_1 \ldots g_7$ in Fig. 3.2). The lowest level consists of the final actions considered (a_1, a_2 and a_3). The arrows represent causal relationships within the hierarchy. Hierarchies can be constructed top–down or bottom–up. Hierarchies can also be structured according to stakeholder groups or actors in the decision-making process as is the case in the MAMCA (which will be explained in Sect. 3.5.4 of this chapter and applied in Chap. 16 of this book).

The relative priorities given to each element in the hierarchy are determined by comparing all the elements at a lower level in pairs, in terms of contribution to the

[5]ADVISORS is the abbreviation for "Action for advanced Driver assistance and Vehicle control systems Implementation, Standardisation, Optimum use of the Road network and Safety", an EC co-funded project of the fifth Research Framework.

Fig. 3.2 Example of a hierarchy in the AHP
Source: designed by the authors, based on Saaty 1995

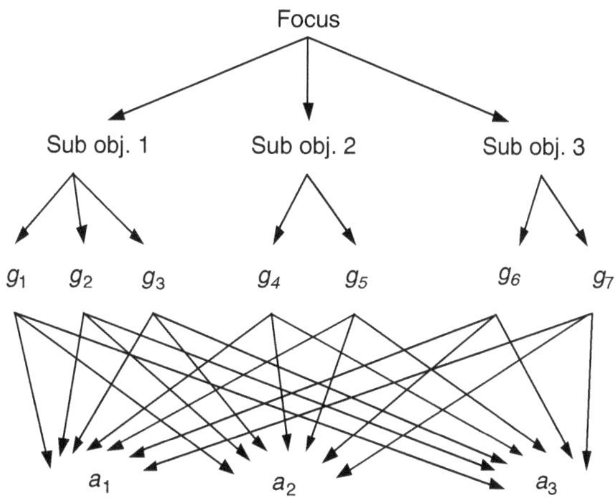

Table 3.3 Pairwise comparison matrix in the AHP

g_j	a_1	...	...	$a_{i'}$	...	a_n
a_1	1					
...		[1]				
a_i			[1]	$Pg_j(a_i,a_{i'})$		
...				[1]		
...					[1]	
a_n						1

Source: designed by the authors, based on Saaty (1995)

elements at a higher level with which a causal relationship exists, as illustrated in Table 3.3.

$Pg_j(a_i,a_{i'})$ represents the preference intensity for a specific pair of typical elements ([sub]-objectives, criteria or actions [a_i, $a_{i'}$]) in terms of the higher level element (objective or criteria [g_j]) with which a causal relationship exists. This preference intensity, $Pg_j(a_i,a_{i'})$, is measured on a scale from 1 to 9 as illustrated in Table 3.4. A similar approach is followed for the constituent components within each objective and sub-objective (criterion). For instance when the value of $Pg_j(a_i,a_{i'})$ is equal to 3, this means that the element mentioned at the extreme left of that row (i.e. the "row element" a_i) is considered to be of moderately higher importance than the element mentioned at the top of that column (i.e. the "column element" $a_{i'}$). The elements on the diagonal line of that matrix are all equal to 1, since any element is always considered as important as itself.

Within each subsystem of the hierarchy, the relative priorities of the elements are determined through the pairwise comparison mechanism described above (Tables 3.3 and 3.4). The relative priorities (weights) are given by the right eigenvector (W) corresponding to the highest eigenvalue (λ_{max}) as shown in formula (3.4). The pairwise comparison matrix is represented by the letter A. Its standard element is $Pg_j(a_i,a_{i'})$ (mentioned in Table 3.3).

Table 3.4 Pairwise comparison scale in the AHP

Intensity of importance $Pg_j(a_i,a_{i'})$	Definition	Explanation
1	Both elements have equal importance	Both elements contribute equally to the criterion considered
3	Moderately higher importance of row element (RE) as compared to column element (CE)	Experience and judgment reveal a slight preference of row element (RE) over column element (CE)
5	Higher importance of RE as compared to CE	Experience and judgment reveal a strong preference of RE over CE
7	Much higher importance of RE as compared to CE	RE is very strongly favoured over CE, and its dominance has been demonstrated in practice
9	Complete dominance in terms of importance of RE over CE	The evidence favouring RE over CE is of the highest possible order
2, 4, 6, 8 (intermediate values)		Intermediate position between two assessments
1/2, 1/3, 1/4, ... 1/9 (reciprocals)		When CE is compared with RE, it receives the reciprocal value of the RE/CE comparison
Rationals (ratios arising from the scale)		If consistency were to be forced by obtaining n numerical values to span the matrix
1.1–1.9 (for tied activities)		RE and CE are nearly indistinguishable; moderate is 1.3 and extreme is 1.9

Source: Saaty (1988:73), adapted by the authors

$$A.W = \lambda_{max}.W \tag{3.4}$$

Since in each pairwise comparison matrix, a number of pairwise comparisons are redundant, it is possible to neutralize possible estimation errors that may have occurred in the other pairwise comparisons of the same matrix on the one hand and to obtain *a measure of consistency* for the pairwise comparisons of the same matrix on the other hand. The latter is done using a mathematical technique based on the theory of eigenvectors and eigenvalues.[6]

The pairwise comparison scale presented in Table 3.4 for expressing the preference intensity, $Pg_j(a_i,a_{i'})$, presented in Table 3.3, makes it possible to derive a cardinal value function for each criterion. Three possible cases may be distinguished here. In case hard data are available and these hard data describe a cardinal

[6]In the event that the pairwise comparison matrix is completely consistent, then all eigenvalues are equal to 0, except 1, because all rows and columns of the matrix are linearly dependant (the rank of the matrix is equal to 1) in that case. The only eigenvalue different from 0 (λ_{max}) should then be equal to n (this is the number of rows and columns in the matrix), since the sum of all the eigenvalues in a square matrix is always equal to the "spur" of the matrix (this is the sum of the elements on its diagonal line). In case of a limited amount of inconsistency in the pairwise comparison matrix, λ_{max} will slightly differ from n. Hence, this difference ($\lambda_{max} - n$) can be used as the basis for a measure of inconsistency.

value function,[7] then no further operations are required and it is not necessary to use the pairwise comparison scale. In fact, the value function transforming the attribute scale (z_j) into a value scale (v_j) is a linear one in this case. In case hard data are available (even with underlying ratio scale) but these data do not describe a cardinal value function, then a cardinal value function may implicitly be constructed using the pairwise comparison scale.[8] In case the (hard) data can only be expressed on an ordinal scale (such as e.g. −−/−/0/+/++), then the pairwise comparison scale also needs to be used in order to obtain a cardinal value function.

In order to synthesize all local priorities, the various priority vectors are weighted by the global priorities of the parent criteria and synthesized. One starts this process at the top of the hierarchy. By doing so, the final or global relative priorities for the lowest level elements (i.e. the actions) are obtained. These final relative priorities indicate the degree to which the actions contribute to the focus. These global priorities form a synthesis of the local priorities, and thereby integrate the various inputs into the decision-making process. In that way, the various points of view are integrated into the final or global priorities, measuring the contribution of each action in terms of the overall objective or focus. In addition, one may as well perform a partial analysis (and synthesis) by doing the pairwise comparisons only from one specific point of view, i.e. taking into account only one sub-objective (or one stakeholder's point of view) (e.g. sub-objective 1 in Fig. 3.2).

The AHP is a powerful decision-making tool. This method makes it possible to decompose decision-making problems into their constituent parts. According to a carefully designed decision-making process, a decision is constructed step by step, by making pairwise comparisons. This step-by-step process eventually results in a synthesis in the form of overall or global relative priorities for the final actions. In spite of the very structured process, there is ample room for learning, creativity and interactions among the analyst, the decision maker and the stakeholders. In addition, it also allows for integrating qualitative data (e.g. obtained through expert judgment) and quantitative data (including monetized values), and the degree of conflict between various objectives or stakeholders can be analyzed through sensitivity analysis. The MCA-AHP makes it possible to take into account information that cannot easily be monetized or quantified.[9] Expert judgments and opinions

---

[7]A "cardinal value function" makes it possible to argue that a score of e.g. $2X$ is considered to be twice as good/bad as a score of X for each value of X (assuming that a natural zero point exists for this value scale). This may the case for a criterion or attribute such as "public expenditure", for which cost estimates may be available e.g. from a CBA.

[8]The relation between the attribute scale (z_j) and the cardinal value scale (v_j) is not necessarily a linear one. For instance, as regards income for a private person, it is generally accepted that the marginal utility of income is decreasing. A job that pays twice the salary of another job is not necessarily considered to be twice as good in terms of the criterion salary. In case the salary is low, the job with the double salary may be considered close to two times better than the other job. In case the salary is high, it may be considered e.g. only 1.5 times better.

[9]Indeed, Forman and Selly (2001) quoting Einstein synthesize this idea very well in the following statement: "Not everything that counts can be counted and not everything that can be counted, counts".

expressed by different experts may be synthesized using the pairwise comparison mechanism. By doing so, a final consensus on priorities and future research needs may be obtained, even when hard data are rather scarce.

3.5.4 The Approach of Multi-Actor Multi-Criteria Analysis for Assessing the Implementation Potential of Road Safety Measures

The multi-actor multi-criteria analysis (MAMCA) was developed by Macharis (2000, 2004) and Macharis et al. (2008) and was used in a number of applications, including other EU funded projects such as the ADVISORS project (De Brucker et al. 2002; Macharis et al. 2004, 2006). Basically, the method starts by mapping all relevant actors or stakeholders. This may be done at the same time or just after having defined the alternatives. Freeman (1984) defined a stakeholder as any individual or group who can affect an organization's performance or who is affected by the achievement of this organization's objectives. Then, in a next step, the objectives (as measured by criteria) that are considered relevant by each individual stakeholder need to be determined. These criteria are then clustered into groups that represent the points of view of the different stakeholders. Then, a separate MCA is performed according to each stakeholder's point of view. By doing so, it is possible to evaluate to which extent the various actions or scenarios contribute to the objectives of particular stakeholders in the decision-making process. In Fig. 3.3 this is illustrated based on the experience with a former EU funded research project (c.q., ADVISORS, EC: FP5 project GRD1 10047).

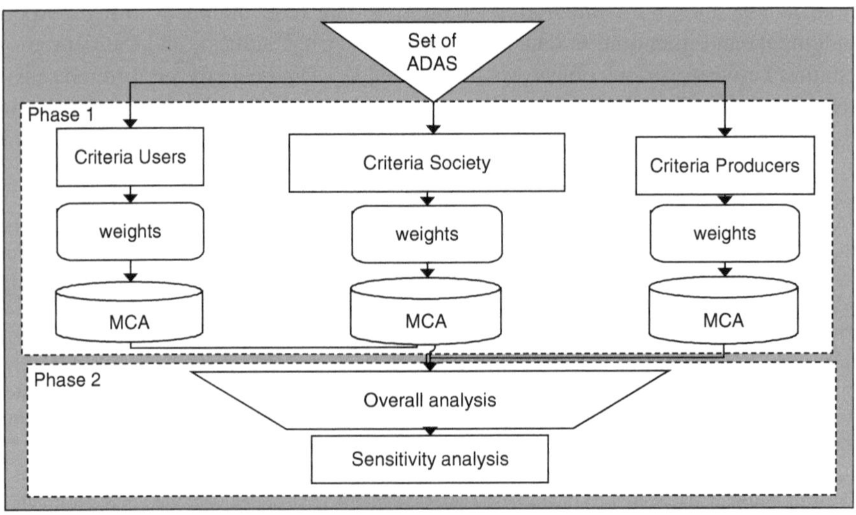

Fig. 3.3 Overview of the stakeholder approach using the MAMCA method

The point of view of society (i.e. the public policy point of view) may be integrated in the MAMCA as a separate point of view. This point of view is usually considered a very important one. The points of view of the two other stakeholders (c.q., the users' and manufacturers' points of view) are also important, especially to check the extent to which the priorities obtained from the societal point of view are in accord with those from users and manufacturers. In case they are, systems may easily be implemented as the result of market forces. In case they are not, government incentives may be necessary to stimulate demand or supply, or both. It is, therefore, very clear that the MAMCA may be used to perform stakeholder management and to derive implementation strategies. An actual application of the MAMCA method to the evaluation of road safety measures is given in Chap. 16 of this book.

3.6 A Toolkit to Use

Although the decision to perform a social cost–benefit analysis (CBA) may initially be a political one, CBA is primarily an evaluation tool for economic analysis, since it evaluates the economic efficiency (and hence the fundamental desirability) of achieving the political objective. Based on consumer willingness-to-pay, project effects are given a monetary value and ultimately the net present value of a project is calculated. An essential characteristic of CBA is that all effects, benefits and costs, need to be expressed in monetary terms, which is not always as easy.

Cost-effectiveness analysis (CEA), however, takes the political objective as given and aims to select that measure, or combination of measures, which makes it possible to achieve this objective at the lowest budgetary cost possible. By doing so, the most cost-effective solution to realizing a given objective can be selected. In CEA only the cost of a project needs to be expressed in monetary terms. Benefits are usually expressed in their physical units and, hence, willingness-to-pay for the benefits is not assessed in CEA.

Multi-criteria analysis (MCA), in some respect, resembles CEA. In MCA effects are also expressed in their physical units, neither does MCA aim at making a statement about the fundamental desirability of a project. The difference between CEA and MCA is that, in the former, the effectiveness is a one-dimensional concept, whereas in MCA it is conceived as a multiple dimensional one. This means that in MCA a number of objectives which are additional to the main objective (c.q. increasing road safety) may be taken into account, such as e.g. environmental effects, investment risk, implementation barriers, etc.

Since the number of effects that can be taken into account in MCA is much larger than in CEA or CBA, it can be said that MCA transcends the realm of economic analysis. MCA (also called "multi-criteria decision *aid*"), therefore, becomes a useful tool especially in the phase of decision, i.e. when a decision whether to implement a project or not needs to be made by political decision makers. MCA, especially multi-actor MCA (MAMCA), then makes it possible to successfully perform stakeholder management and to define effective implementation strategies as part of the phase of decision. CBA (and CEA) can then be used in the phase of analysis, i.e. the phase preceding the phase of political decision. Formal applications of CBA, CEA and MCA to the evaluation of innovative road safety measures will be presented in Chaps. 16 and 17 of this book.

References

E. Bekiaris, Y.J. Nakanishi (eds.), in *Economic Impacts of Intelligent Transportation Systems: Innovations and Case Studies*. Research in Transportation Economics, vol 8 (Elsevier, Amsterdam, 2004)

V. Belton, T. Stewart, *Multiple Criteria Decision Analysis. An Integrated Approach* (Kluwer Academic, Boston, 2002)

P. Bickel, R. Friedrich, A. Burgess, P. Fagiani, A. Hunt, G. De Jong, J. Laird, C. Lieb, G. Lindberg, P. Mackie, S. Navrud, T. Odgaard, A. Ricci, J. Shires, L. Tavasszy, Proposal for harmonised guidelines, Deliverable 5, Developing Harmonised European Approaches for Transport Costing and Project Assessment (HEATCO). Project funded by the European Commission under the Transport RTD Programme of the 6th Framework Programme, 2006

A. Charness, W.W. Cooper, *Management Models and Industrial Applications of Linear Programming* (Wiley, New York, 1961)

COWI, Cost-benefit assessment and prioritisation of vehicle safety technologies. Final Report, Economic assistance activities, Framework Contract TREN/A1/56-2004, Consultancy within Engineering, Environmental Science and Economics (COWI A/S), Kongens, Lyngby, 2006

K. De Brucker, *Ontwikkeling van een eclectisch evaluatie-instrument voor de sociaal-economische evaluatie van complexe investeringsprojecten, met een toepassing op het project Seine-Scheldeverbinding*. PhD thesis, Universiteit Antwerpen (RUCA), Antwerp, 2000

K. De Brucker, R. Saitua-Nistal, Naar een geïntegreerde methodiek voor de beoordeling van investeringsprojecten op vlak van mobiliteit en grootstedenbeleid. Een poging om de kloof tussen analysefase en beslissingsfase te dichten, in *Mobiliteit en Grootstedenbeleid, Referaten van het 27ste Vlaams Wetenschappelijk Economisch Congres*, ed. by C. Macharis, M. Despontin (VUBPress-Politeia, Brussels, 2006)

K. De Brucker, A. Verbeke, The eclectic multi-criteria analysis (EMCA): a tool for effective stakeholder management in project evaluation, in *Ports are More Than Piers. Liber Amicorum prof. dr. Willy Winkelmans*, ed. by T. Notteboom (DeLloyd, Antwerp, 2006)

K. De Brucker, A. Verbeke, The institutional theory approach to transport policy and evaluation. The collective benefits of a stakeholders' approach. Towards an eclectic multi-criteria analysis, in *Transport Project Evaluation: Extending the Social Cost-Benefit-Analysis*, ed. by E. Haezendonck (Edward Elgar, Aldershot, 2007)

K. De Brucker, A. Verbeke, W. Winkelmans, *Sociaal-economische evaluatie van overheidsinvesteringen in transportinfrastructuur. Kritische analyse van het bestaande instrumentarium. Ontwikkeling van een eclectisch evaluatie-instrument* (Garant, Leuven, 1998)

K. De Brucker, C. Macharis, A. Verbeke, B. Bekiaris, Integrated multicriteria analysis for advanced driver assistance systems. Final Deliverable of the research project "ADVISORS" (Commission of the European Union – Department Transport and Energy (DG TREN), FP5 project GRD11999 10047, 2002, Brussels), http://www.advisors.iao.fhg.de>Reports>Deliverables>D6.1. Accessed 31 July 2009

DOT 2007. U.S. Department of Transportation, Washington, DC. http://www.itscosts.its.dot.gov/. Accessed 9 May 2011

J. Dupuit, De la mesure de l'utilité des travaux publics. Ann. des Ponts et Chaussées (Ministère des Travaux Publics et des Transports, Paris, 1844)

J. Figueira, S. Greco, M. Ehrgott (eds.), *Multiple Criteria Decision Analysis State of the Art Surveys* (Springer, New York, 2007)
E.H. Forman, M.A. Selly, *Decision by Objectives. How to Convince Others That You Are Right* (World Scientific, Hackensack, 2001)
E.R. Freeman, *Strategic Management. A Stakeholder Approach* (Pitman/Ballinger, Boston, 1984)
D. Gillen, J. Li, J. Dahlgren, E. Chang, Assessing the benefits and costs of ITS projects: volume 1 methodology. Research Report, Institute of Transportation Studies, University of California, Berkeley, 1999
S. Hakkert, P. Wesemann (eds.), The use of efficiency assessment tools: solutions to barriers. SWOV Report R-2005-02, Institute for Road Safety Research (SWOV), Leidschendam, 2005
N. Hanley, C.L. Spash, *Cost-Benefit Analysis and the Environment* (Edward Elgar, Cheltenham, 1993)
J.R. Hicks, The foundations of welfare economics. Econ. J. **49**(196), 696–712 (1939)
N. Kaldor, Welfare comparisons of economics and interpersonal comparisons of utility. Econ. J. **49**(195), 549–552 (1939)
R. Keeney, *Value-Focused Thinking. A Path to Creative Decisionmaking* (Harvard University Press, Cambridge, 1996)
C. Macharis, The importance of stakeholder analysis in freight transport: the MAMCA methodology. Eur. Transp./Trasp. Eur. **25**(26), 114–120 (2004)
C. Macharis, *Strategische modellering voor intermodale terminals. Socio-economische evaluatie van de locatie van binnenvaart/weg terminals in Vlaanderen*. PhD thesis, Vrije Universiteit Brussel, Brussels, 2000
C. Macharis, A. Verbeke, K. De Brucker, The strategic evaluation of new technologies through multicriteria analysis: the ADVISORS case, in *Economic Impacts of Intelligent Transportation Systems. Innovations and Case Studies*, ed. by E. Bekiaris, Y.J. Nakanishi (Elsevier, Amsterdam, 2004)
C. Macharis, A. Stevens, K. De Brucker, A. Verbeke, A multicriteria approach to the strategic assessment of advanced driver assistance systems, in *Transportation Economics. Towards Better Performance Systems*, ed. by B. Jourquin, P. Rietveld, P. Westin (Routledge, Taylor & Francis Books, London, 2006)
C. Macharis, J. Ampe, The use of multi-criteria decision analysis (MCDA) for the evaluation of transport projects: a review. EURO 2007 Conference, Prague, 2007
C. Macharis, A. De Witte, J. Ampe, The multi-actor multi-criteria analysis methodology (MAMCA) for the evaluation of infrastructure projects: theory and practice. J. Adv. Transp. **43**(2), 183–202 (2008)
A. Marshall, *Principles of Economics* (Macmillan, London [1922], 1890)
E.J. Mishan, *Cost-Benefit Analysis: An Informal Introduction* (Unwin Hyman, London, 1988)
T. Moore, R. Pozdena, Framework for an economic evaluation of transportation investments, in *Economic Impacts of Intelligent Transportation Systems: Innovations and Case Studies*, ed. by E. Bekiaris, Y.J. Nakanishi. Research in Transportation Economics, vol. 8 (Elsevier, Amsterdam, 2004)
J. Nellthorp, T. Sansom, B. Bickel, C. Doll, G. Lindberg, Valuation conventions for UNITE. Unification of accounts and marginal costs for Transport Efficiency (UNITE). Working Funded by the 5th Framework RTD Programme, ITS, University of Leeds, Leeds, 2001
P. Nijkamp, P. Rietveld, H. Voogd, *Multicriteria Evaluation in Physical Planning* (North Holland, Amsterdam, 1990)
M. Panou, E. Bekiaris, ITS clustering and terminology: one concept with many meanings, in *Economic Impacts of Intelligent Transportation Systems: Innovations and Case Studies*, ed. by E. Bekiaris, Y.J. Nakanishi. Research in Transportation Economics, vol. 8 (Elsevier, Amsterdam, 2004)
V. Pareto, *Manuel d'Economie Politique, traduit sur l'édition italienne (1906) par A. Bonet, revue par l'auteur* (Giard, Paris, 1927)

T.L. Saaty, A scaling method for priorities in hierarchical structures. J. Math. Psychol. **15**, 234–281 (1977)
T.L. Saaty, Axiomatic foundation of the analytic hierarchy process. Manag. Sci. **32**(7), 841–855 (1986)
T.L. Saaty, *The Analytic Hierarchy Process* (McGraw-Hill, New York, 1988)
T.L. Saaty, *Decision Making for Leaders. The Analytic Hierarchy Process for Decisions in a Complex World* (RWS Publications, Pittsburgh, 1995)
H. Samstad, T.E. Markussen, Nytte-kostnadsanalyse som evalueringsverktøy for ITS-investeringer. TØI Report 501, Transportøkonomisk institutt, Oslo, 2000
A. Schärlig, *Décider sur plusieurs Critères* (Presses Polytechniques Romandes, Lausanne, 1985)
A.P. Thirlwall, *Growth and Development. With Special Reference to Developing Countries* (Palgrave Macmillan, London, 2003)
A. Trawén, P. Maraste, U. Persson, International comparison of costs of fatal casualty of road accidents in 1990 and 1999. Accid. Anal. Prev. **34**, 323–332 (2002)
H.R. Varian, *Microeconomic Analysis* (Norton & Company, New York, 1992)
W. Vlakveld, P. Wesemann, E. Devillers, R. Elvik, K. Veisten, Detailed cost-benefit analysis of potential impairment countermeasures. SWOV Report R-2005-10, Institute for Road Safety Research (SWOV), Leidschendam, 2005

Chapter 4
Putting the Legos in Place

A Selection of ITS for Enhancing Road Safety

Manfred Dangelmaier, Gunter Wenzel, Maria Gemou, Evangelos Bekiaris, Marion Wiethoff, Dick De Waard, Karel Brookhuis, Ewoud Spruijtenburg, and Vincent Marchau

4.1 Road Safety Functions

The traffic safety risk emanates from the cooperation of three main factors: driver-vehicle-traffic environment. Although several measures exist in order to support/improve any of these three contributing factors, they may have negative side-effects to the others. As an example, according to the risk homeostasis theory (Wilde 2001), the enhancement of safety level of a vehicle leads sometimes drivers to change their driving profile, undertaking more risky maneuvers, in order to keep their conceived level of risk constant. Thus, optimal are the measures which combine possible effects to all three contributors and these that build upon the strengths and interactions between each combined environment.

For the generation of alternatives for the safety hazards, both Autonomous (e.g., only Infrastructure or in-vehicle based) and Co-operative solutions (e.g., Vehicle-to-Vehicle communication, Infrastructure-to-Vehicle, or/and In-Vehicle ones) can be distinguished. The basic assumption of IN-SAFETY is that the combination of infrastructure and telematics measures can provide a more cost-efficient solution, avoiding performing expensive infrastructure works by providing the same function through a telematic or other innovative system. IN-SAFETY focused especially on the co-operative systems: in car-vehicle systems combined with infrastructural

M. Dangelmaier (✉) and G. Wenzel
University of Stuttgart, Institute for Human Factors and Technology Management (IAT), Stuttgart, Germany
e-mail: Manfred.dangelmaier@iat.uni-stuttgart.de

M. Gemou and E. Bekiaris
Hellenic Institute of Transport (HIT), Centre for Research and Technology Hellas (CERTH), Athens, Greece

M. Wiethoff, D. De Waard, K. Brookhuis, E. Spruijtenburg, and V. Marchau
Department of Transport Policy and Logistics, Delft University of Technology, Delft, The Netherlands

E. Bekiaris et al. (eds.), *Infrastructure and Safety in a Collaborative World*,
DOI 10.1007/978-3-642-18372-0_4, © Springer-Verlag Berlin Heidelberg 2011

systems and – as a matter of fact – it was one of the first conceivers of the cooperative systems concept.

Each existing or emerging technology, operating in any of the above levels/contributors, can be seen as a single "Lego" tool, to be used within holistic systems (Fig. 4.1).

In reality, in a modern car and a real traffic environment, these systems are combined, formulating integrated safety systems, in the following analogy (Fig. 4.2).

Each of these systems, whether autonomous, infrastructure-based or cooperative, needs to be thoroughly tested against its impact to each one of the main traffic safety contributors.

Within this chapter, promising technologies – legos will be reviewed against their potential traffic safety impact. Focus is on the following functionalities in line to the emerging accident-based priorities of Chap. 1:

- Lateral and rear area monitoring systems
- Lane Departure Warning/lane keeping Systems (LDWS)
- Collision Warning and Avoidance Systems (CWS/CAS), for the lateral and longitudinal area of the vehicle
- Longitudinal control systems, namely Cruise Control and Advanced Cruise Control (ACC) systems
- Intelligent Speed Adaptation (ISA) systems
- Stop&Go systems
- Vision enhancement and VRU (Vulnerable Road Users) detection (and protection) systems

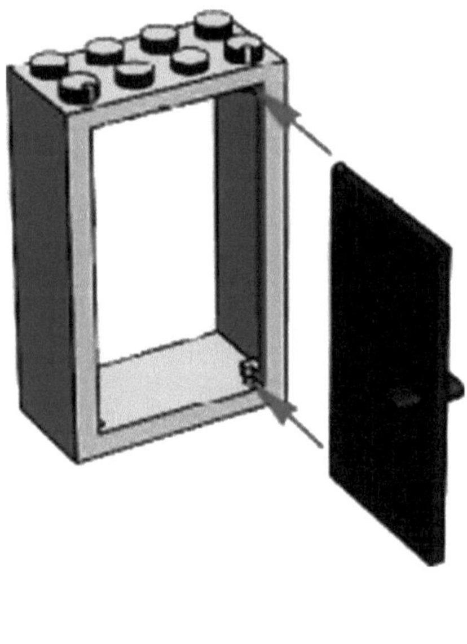

Fig. 4.1 Lego door-analogy to any ADAS/IVIS or other safety or traffic information provision system

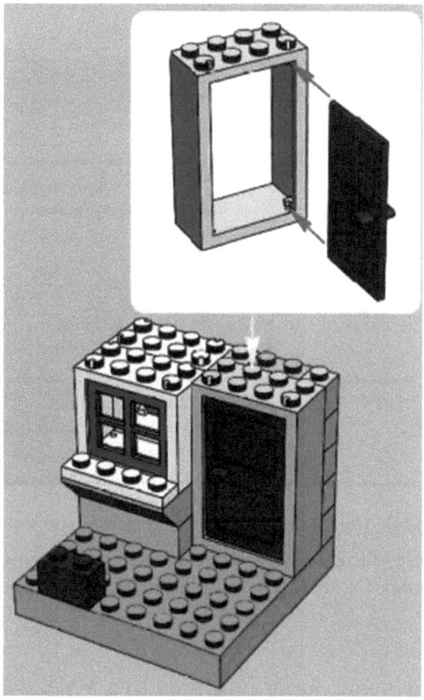

Fig. 4.2 Build with legos-analogy to integrated traffic safety systems

4.2 A Set of Evaluation Matrices

4.2.1 Matrix 1

As a method to evaluate the co-operative characteristics of the traffic safety enhancement measures and scenarios proposed, a matrix has been built within IN-SAFETY project to help characterising the solution for a scenario in terms of current solutions (usually infrastructure-based). The ratio behind the matrix is, for all those systems for which there is no empirical data available, that estimation is made of the expected safety effects, based on similarities in the functionality in the existing road infrastructure. In the matrix (Table 4.1), in-vehicle systems were put in 1D of the matrix (green color); while, on the other dimension, the infrastructural measures (orange color) were listed.

For instance, the functionality of a collision warning and avoidance system is close to the Infrastructural measure of an obstacle free zone, a median barrier, a safety barrier, dynamic hazard warning or dynamic speed VMS (Variable Message Sign). These infrastructural measures can be compared to the functionalities of ADAS, either existing, or still under development. Thus, a safety barrier can be compared to an obstacle and collision warning system (in-vehicle system), and a

Table 4.1 Extract from Matrix 1: characterization of road safety functions in terms of infrastructure measures and ADAS measures

Road Safety Functions	Type	Road Safety Measures	Communication	Road Safety Measures
Longitudinal support	Road	Protecting shoulder	In C	Speed Alert or ISA based on traffic sign recognition
	Road	Static traffic signs	C<->I	Speed Alert or ISA based on digital map data
	Road	Traffic lights	In C	ACC-Stop & Go
	Road	VMS	C<->C	ACC-Stop & Go + Foresight
	TM	Section speed management system (line or stretch control)		
Lateral support	Road	Lane width	In C	Blind Spot detection
	Road	Rumble Strips	In C	Lane change assistant
			In C	Lane departure warning
			In C	Lane keeping assistant

Only longitudinal support and lateral support are shown

Road Road design; In C autonomous in-vehicle system; C<-> C vehicle-to-vehicle communication; C<-> I vehicle-to-infrastructure communication

dynamic hazard warning to a VMS with local hazard warning (Vehicle-to-Vehicle communication; Vehicle-to-Infrastructure communication).

4.2.2 Matrix 2

In Matrix 2, (its extract can be found in Table 4.2), a total of 55 different specific scenarios with proposed safety solutions are listed, generated by open discussions with experts. Each column is associated with one solution per scenario. Each row either defines an aspect of the scenario (e.g., type of road or type of vehicle) or the result of the evaluation of the solution for that scenario (e.g., safety level). This matrix gives an overview of the expected safety effects for each solution over a number of aspects.

Then, the scenarios from Matrix 2 are defined in terms of the involved infrastructure based systems (e.g., obstacle free zone, rumble strips, reflecting road markings, protective shoulder) and the ADAS, as described in Matrix 1, and, subsequently, for each of the different scenarios safety measures are selected, that are considered most effective in increasing safety for those scenarios.

The main focus is here on the co-operative systems, defined by the combinations of infrastructural and in-vehicle systems by way of communication. However, autonomous system measures have been kept in the list to provide a comparison between all measures during the further process. To reduce the amount of measures for further evaluation the measures have been consolidated. For the current chapter,

Table 4.2 Extract of Matrix 2: three scenarios and their characterisation in terms of scenario elements

Measure description	When leaving Road Constructions the driver is reminded to reactivate his/her LDW-System.	The LDW-System is deactivated at the beginning of long construction sites to avoid false alarms.	Head Up display showing road curvature when road is not properly visible.
Problem	Accidents due to lane departure.	Accidents due to lane departure.	Run-off road or head-on accidents due to inappropriate speed in unexpected bends.
Errors	wrong use of lane	wrong use of lane	bends
Other errors	distraction	distraction	visibility
Urban			
Rural	x	x	x
Highway	x	x	x
Technology	communication with the local infrastructure	communication with the local infrastructure	GPS info, head up display
Driver	all	all	all
Vehicle	all	all	all
Traffic conditions	all	all	all
Environmental	road construction sites	road construction sites	adverse weather, darkness

LDW Lane departure warning; *VMS* variable message sign; *GPS* global positioning system

it was decided to focus only on those systems listed under "Lateral support" and "Longitudinal support".

4.3 Safety Effects of Various Systems

Below follows an indicative list of promising systems ("lego blocks") to be used in FOR (Forgiving Road Environment) applications. The presented systems and functionalities is rather indicative than exhaustive and intends to help the reader understand the key components of the proposed implementation scenarios, but also to get own ideas on further traffic safety enhancement scenarios; just as anyone can build with lego kits more structures than the ones suggested by the instructions.

4.3.1 Lateral Support

Lateral support systems are mainly discerned into vehicle lateral and rear area monitoring systems, Lane Departure Warning/Lane Keeping Systems (LDWS) and

Lateral Collision Avoidance Systems (LCAS), for the lateral area, including lane change support systems.

Lateral Support is mostly important for averting the following driver errors: Wrong use of the lane (vehicle area monitoring systems, LDWS) and head-on or head-tail accidents due to driver errors when overtaking (LCAS). Table 4.5 suggests further correlations between driver errors and individual systems.

4.3.1.1 Vehicle Lateral and Rear Area Monitoring Systems

Usually, the vehicle lateral and rear area monitoring systems work as a complement and extension of the rear view mirror, enabling the driver to monitor the surrounding traffic in the lateral and rear area of the vehicle. The position, the speed (or relevant speed) and the size of other vehicles are presented to the driver via the HMI (usually through a visual display). Ultrasonic sensors or microwaves sensors, mounted on the lateral and rear areas of the vehicle, are the major technologies deployed in this case.

In the Lateral and Rear Area Monitoring (LRM) applications, the driver is informed about vehicles to the sides of, and behind, his/her own vehicle. Vehicles, in particularly dangerous positions, are highlighted by alternative colours. The information is meant to enhance the driver's understanding of the traffic situation, thereby reducing the likelihood of him/her making a dangerous manoeuvre, particularly in cases of limited visibility or heavy mental workload.

Such applications exist also for trucks. It is typically mounted on the truck dashboard, showing the objects in the surrounding area from a bird's eye view (Figs. 4.3 and 4.4).

4.3.1.2 Lane Departure Warning System (LDWS)

Lane Departure Warning (LDW) systems (in some cases mentioned as Lane Keeping and Warning Systems or Lane Drift Warning Systems) stand for systems

Fig. 4.3 HMI solution (rear view mirror leds) addressing the LRM application for passenger cars (CRF Fiat Stilo demonstrator of LATERAL SAFE project)
Source: Danielsson et al. (2007)

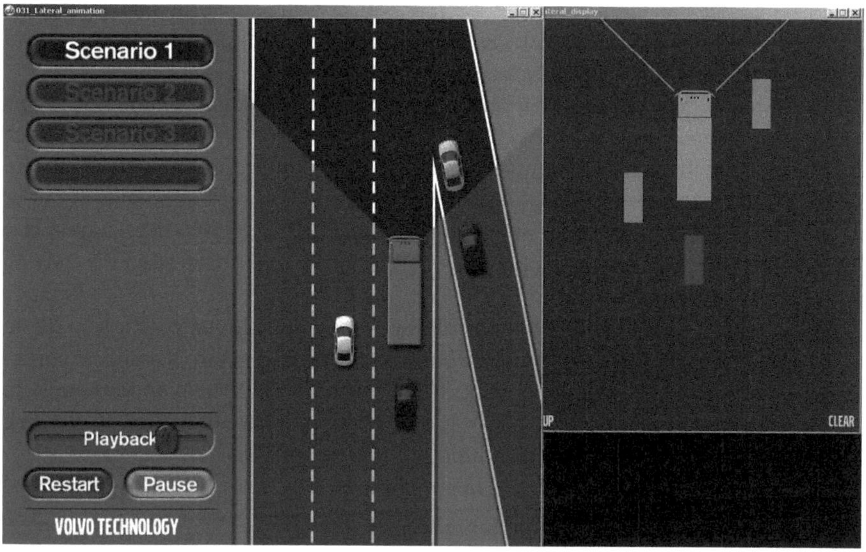

Fig. 4.4 LRM scenario description and information presented to the driver in a truck (Volvo FH12 fixed base truck simulator; demonstrator of LATERAL SAFE project)
Source: Danielsson et al. (2007)

embedding advanced technology that can help prevent crashes resulting from an unintentional drift (when the turn signal is not in use) of the vehicle out of its travel lane. They are in-vehicle electronic systems that monitor the position of a vehicle within a roadway lane and warn a driver if the vehicle deviates or is about to deviate outside the lane. Some authors distinguish Lane Departure Warning Systems and Lane Keeping Systems according to the availability of intervention mode; see for more Chap. 10.

Currently available LDWS are forward looking, vision-based systems, which use algorithms to interpret video images to estimate vehicle state (lateral position, lateral velocity, heading, etc.) and roadway alignment (lane width, road curvature, etc.). These LDW systems use a typically forward-facing camera that is mounted to the windshield, behind the rear view mirror, in the cab of the vehicle.[1] This also implies high requirements for lane markings recognition, as the system has to be able to deal with different types of lane markings as well as with gaps in markings.

LDWS warn the driver of a lane departure when the vehicle is travelling above a certain speed threshold and the vehicle's indicator is not used. In addition, LDWS notify the driver when lane markings are inadequate for detection, or if the system malfunctions.[2] Alternatively to the video systems, a number of infrared sensors may

[1]ITS Decision, http://www.calccit.org/itsdecision/serv_and_tech/AVCSS-section-one/lane-departure.html.

[2]Federal Motor Carrier Safety Administration, http://www.fmcsa.dot.gov/facts-research/research-technology/report/lane-departure-warning-systems.htm.

be installed on the bumper of the vehicle. The sensors monitor the white lane markings and whenever the vehicle crosses the lane, while the indicator is not switched on, an alarm signal is given.

The systems include an electronic control unit and a warning indicator. Some LDW systems may issue directional warnings to alert the driver regarding to which side of the lane the vehicle is unintentionally drifting, indicating towards where the driver should steer. A directional warning may be audible, such as sounds in left or right in-cab speakers, or haptic (or a combination of both). LDWS may graphically indicate on a user interface display how well the vehicle is centered in the lane on a time-averaged basis.

Although traditional LDW systems do not take any automatic action to avoid a lane departure or to control the vehicle, some emerging systems actually incorporate the use of a steering shaft actuator. If the driver does not react appropriately to a lane departure warning, the system automatically applies torque to the steering wheel and steers the vehicle back within the lane boundaries.[1]

LDWS are already implemented in commercial cars. The system requires good contrast between the road and delineation. LDWS are reported to cause a 20% decrease in line-crossings and an increase of the driver's use of the direction indicator of about 20% (University of Twente 2007). Katteler (2003) found that fewer startle reactions among drivers equipped with LDWS occur; fewer startle reactions among other road users in the view of the drivers with LDWS experience, and shortened reaction time. These positive safety effects together probably result in a decrease of accidents (estimated 1.7%; Hoogendoorn et al. 2007). The above data are collected from different types of studies, also pilot studies on the road (further reporting on safety analyses of LDWS, including also truck LDWS is available in Chap. 10). Their certainty level is considered as relatively high. The needed penetration rate to reach the safety level is considered to be moderate. Potential negative effects of LDWS might be a decrease in driving skills due to overreliance on the system. Another negative effect may be that the presence of a warning system in the vehicle may also cause drivers' inattention to other (concurrent) stimuli (Katteler 2003).

As aforementioned, one of the most common user interfaces of LDWS is the audible directional warnings. In parallel, there are two significant extensions of LDWS, namely the adaptive LDWS, adapted to specific traffic conditions and utilising info coming from the infrastructure (could be seen as cooperative LDWS), as well as the future extension of LDWS that would be integrated with Adaptive Cruise Control (ACC). A few details on these main subclusters of LDWS are provided below.

LDWS Operating as Virtual Rumble Strips

"Virtual rumble strips" refer to the type of HMI embedded in the system, while the system function is the same, as described above.

Standard rumble strips give auditory feedback to non-alert drivers, leading to reduced crossing of the line (Räsänen 2005). On motorways they are used as countermeasure to fatigue induced accidents and on rural roads as centre lane marking to discourage overtaking. Practically, in the context of the LDWS and in order to achieve the same results, the sound of the rumble strips is simulated and is usually provided directionally to the driver (to indicate the side of drifting), in order to offer him/her the same experience as standard rumble strips.

The safety level of a virtual variant of the rumble strip is moderate, just as for the conventional variant of the rumble strip. Because it is only a newer technology with a higher chance on technical problems, its certainty level is also moderate. The needed penetration rate is rather high because many vehicles have to make use of the system to make it really efficient.

Within the framework of AWAKE project (IST-2000-26089), an experiment had been performed (Bekiaris et al. 2002) to analyse the effect of "virtual" rumble strips (with the sound of rumble strips crossing) upon the amplitude and the duration of the incursions into the emergency lane by fully awake and drowsy drivers during simulated driving. The analysis has been made on data collected in simulator with 12 subjects during daytime driving and 12 subjects during night time driving. During daytime driving, 2 out of the 12 subjects were diagnosed as drowsy, while this ratio was changed to 7 drowsy subjects (out of the 12) during night time driving.

The study of the intrusions into the emergency lane has shown that the average duration of these incursions was quite similar whatever the vigilance level of the driver was. A quite similar conclusion was made about the average amplitude of these incursions, although there was a tendency to observe the largest amplitudes in the most drowsy drivers. However, due to the small size of the experimental sample, these results have to be confirmed in further studies. Nevertheless, the basic hypothesis made at the beginning can partly be retained. That is: the absence of large differences between the two groups of subjects (drowsy or not drowsy) is in favour of an "awakening" effect of the rumble strip in the case of drowsy drivers who, therefore, returned as quickly as the non-drowsy ones in the normal driving lane.

Then, the size (amplitude and duration) of the incursions into the emergency lane with the presence of the rumble strips sound (sound "on") and without the presence of this sound (sound "off") was compared. The statistical analysis made on the different incursions into the emergency lane, with or without the rumble strip sound, has lead to the following results:

- A significant reduction of the number of incursions into the emergency lane in the curves when the sound was "on" compared to the "off" situation (50% reduction). The subjects, probably, were trying to avoid the uncomfortable sound generated by driving on the protuberances. On the other hand, there was no difference between the "on" and "off" situations in the straight portions of the road, the lowering in vigilance level being independent of the actual situation.

- A significant decrease (in the order of 30%) of the duration and amplitude of incursions into the emergency lane when the sound was "on" compared to the situation when the sound was "off". Thus, driving on the protuberances reactivated the driver without leading to expected errors such as over correction of the trajectory of the car.

The above test results indicate that the simulation of rumble strip effect by only its sound may be enough to warn effectively drowsy drivers (should not be neglected that drowsiness is the main root cause for run-off accidents, such as lane departure accidents).

Nevertheless, the effective use of virtual rumble strip warnings for generic lane deviation/departure was still not tried and, thus, was introduced into an implementation scenario and tested within IN-SAFETY project.

Adaptive LDWS

Adaptive LDWS work similarly as the LDWS described above, but are particularly adapted to specific traffic situations (i.e., tunnels, road constructions). In these cases, the systems are GPS-based. They use GPS position data, combined with a high resolution map database, to determine where the vehicle is within the lane. Information may be in addition received from roadside beacons at road works and other road sections without (reliable) lane markings. A processor in the vehicle calculates the forecasted vehicle position, using additional information from speed and steering wheel movements.

A more advanced extension of such systems encompasses novel alarm decision models, which take into account road geometry and past driver behaviour, in order to adapt to the driving style of individual drivers.

This is the reason why the certainty for these systems is lower in comparison to normal LDWS. The whole situation is just more complex than on normal roads. There are no extensive experiments with these systems so far; some comparisons can only be made with normal LDWS. The advantages of such systems are that they can be useful in all weather conditions (bright sun, rain, snow, fog, etc.), whereas they enable a series of other functionalities (such as Route Navigation, etc.). On the other hand, very highly detailed map databases that must be continuously updated for high availability are required, whereas updates would require DSRC or similar infrastructure. Finally, GPS dropouts from bridges and other objects may affect availability and current GPS accuracy does not match the lane position requirements (without DSRC input or Galileo accuracy in the future). Potential negative effects of adaptive LDWS might be the misconception by drivers of their changing (adaptive) functionality as well as their reliability level, as they require a big number of data to operate properly.

4.3.1.3 Lateral Collision Avoidance Systems (LCAS)

The major lateral collision avoidance systems are namely the Blind Spot Detection Systems, the Lateral Collision Warning Systems, and the Lane Change Assistance Systems. Lately, overtaking systems, although not practically implemented yet, at least on commercial level, are considered as part of lateral collision avoidance systems.

Blind Spot Detection Systems

The aim of the blind spot systems (that are also considered as forgiving systems) is the detection of obstacles (short obstacles) in the lateral and rear area of the vehicle, which are not detected from the rear view mirror (that is actually the "blind spot" area). The vehicles are equipped with passive infrared sensors, whereas more recent systems use ultrasonic sensors in the outer part of the vehicle or radar in each vehicle side or behind the bumpers or CMOS cameras, mounted behind the side mirrors (Fig. 4.5).

In this way, the traffic approaching from behind is detected, in order to assist the driver in cases that s/he intends to overtake (understood by the system through the use of turn indicators and/or steering wheel movements). As an example, a relevant system in the market works with six short range radar sensors on a short distance. They are used for detection of the immediate environment behind and lateral (left and right hand side) of the vehicle. Whenever the system notices a vehicle within

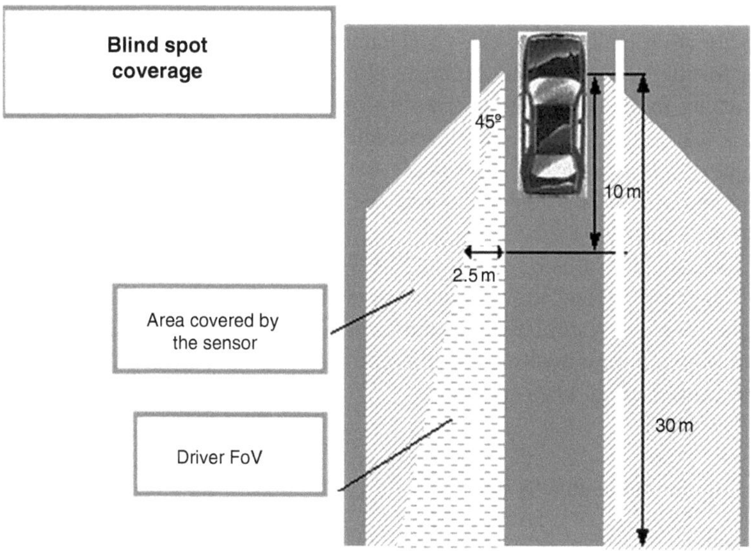

Fig. 4.5 Blind Spot coverage

the blind spot area, a red warning symbol (led) appears in the side mirror. If the driver ignores the warning and switches on the indicators, the red warning symbol starts to blink and an auditory warning is activated. Furthermore, in some systems, vehicles are also equipped with GPS and a digital map to identify whether the vehicle is driving on a road with more than one lane per direction or not and take into account the interference of curves.

Through the use of blind spot detectors, a relevant number of run-over accidents of specifically vulnerable road users could be diminished. The relative accident reduction on rural and urban roads is estimated to be in the magnitude of 3% (Louwerse 2005), which results in a moderate safety level. This is mainly significant for heavy vehicles with a huge blind spot area. On motorways the system will also lead to a decrease of "typical" overtaking accidents. Drivers' behaviour is not expected to significantly reduce the system benefits or may even further enhance them (Wiethoff 2003a, b). The certainty level is high because there have been several experiments with blind spot detectors. The needed penetration rate is low because one vehicle does not need other vehicles to be equipped to reach the enhanced safety level. Risks of the system include possible breakdown without failure indication, as well as possible potential driver overreliance on the system. Furthermore, the detector should also detect other traffic users, in particular vulnerable road users (i.e., motorcycles or bicycles at the blind spot area).

Lateral Collision Warning Systems

Lateral Collision Warning Systems aim to warn drivers for obstacles moving (or detected) in the lateral area of the vehicle. Such systems are especially useful in lane merging situations in the highway or in cases of high traffic density when driver attention is disrupted, etc. CCD cameras are mounted behind the rear view mirror for the lane detection. Alternatively, infrared sensors are used. Also, CCD cameras are mounted in side mirrors for the detection of vehicles moving in the adjacent lanes (up to 20 m behind the vehicle) and microwave radars are mounted in the right or left rear part of the vehicle for the monitoring of the rear area of the vehicle and provide the relevant speed and distance of the vehicles (up to 100 m behind the vehicle).

In the Lateral Collision Warning (LCW) application, the driver is informed about dangerous lateral movements towards obstacles in left and right side area of his/her-vehicle. A directional sound/light warning is given when his/her vehicle is approaching a vehicle/obstacle to the side with risk of collision. There is typically no intervention (Fig. 4.6).

Lane Change Assistance Systems

Lane Change Assistance systems detect the vehicles and obstacles in the adjacent lanes and warn the driver respectively; could be seen as a subcategory of the lateral

Fig. 4.6 HMI solution (pillar-shaped sign) addressing the LCW application for car (Volvo Cars demonstrator of LATERAL SAFE project) *Source*: Danielsson et al. (2007)

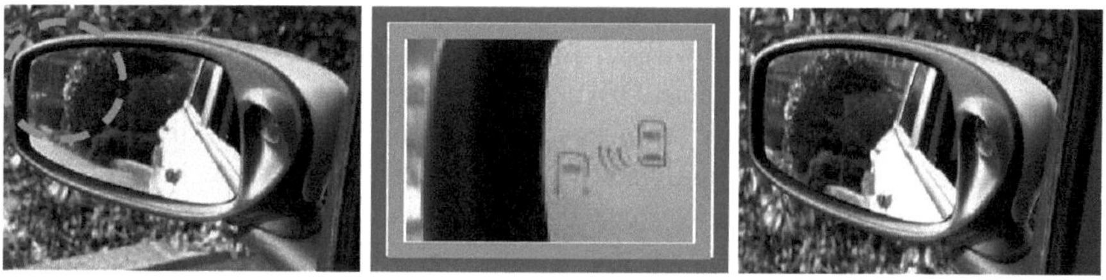

Fig. 4.7 HMI solution (side-mirror led) for the lateral area of the vehicle addressing LCA (Fiat Stilo demonstrator of LATERAL SAFE project)
Source: Danielsson et al. (2007)

collision warning systems. Next generation systems will be able to advice the driver for the actions that have to be done and warn them about obstacles/vehicles in the adjacent lanes which could be considered as a potential risk, whereas fully advanced systems may automatically control the speed and the direction of the vehicle in order to avoid any collision. Lateral cameras (in side mirrors) are combined with radars in the rear part of the vehicle. The system field of view is typically up to 20 m in the right side of the vehicle and up to 70 m in the left side of the vehicle. The system range and accuracy is lower in turns. Audio or visual warnings are used in most cases. Directional information about threats is given with different levels of urgency (from visual to combined visual and auditory), depending on how high the risk for collision is. There is typically no intervention (Fig. 4.7).

Overtaking Systems

Existing prototypes aim at avoiding accidents on rural roads with one driving lane per direction and no median barrier, in which a vehicle overtakes in spite of meeting traffic.

Cars should be equipped with sensors that detect meeting traffic, and with vehicle-to-vehicle communication that enables vehicles to detect approaching vehicles also if these are not (yet) within sight distance. This system requires information on speed and location of all vehicles, and a digital map with information on infrastructure (number of lanes, road curvature) and speed limits. In case of identified dangerous overtaking maneuvers a warning is given to the driver.

Table 4.3 Benefits estimations by diverse studies for LCAS

	Collision reduction		Fatalities reduction	
Speed range (km/h)	Not defined	100–120	50–90	0–50
Najm (lane change)	47.7%			
de Visser (lane change)		2%	1%	2%
Najm (lane departure)	64.9%			
de Visser (lane departure)		35%	33%	12%

The system must be designed in such a way as to avoid the interpretation of "no warning" as "nothing in front". There may be approaching vehicles that were not detected (e.g., motor vehicles without functioning vehicle-to-vehicle equipment), vulnerable road users, and objects. Drivers should therefore remain vigilant. Implementation of this system, however, needs a 100% penetration rate. This system has a potential to decrease the number of overtaking accidents, especially on rural roads. This will depend on how much the system will take over the driving task. If the drivers are only warned by the system (most likely), the safety level and certainty level are probably low. The penetration rate of the measure must be very high; in fact every vehicle must use the system to make communication possible all the time between all vehicles (Wiethoff 2003a, b; Dragutinovic et al. 2005). Like other ADAS (i.e., LDWS), this system will also have the risks of drivers' mistrust and over-reliance.

Some benefits' estimations by diverse studies for systems providing lateral support are provided in Table 4.3. It is questionable if the collision reduction rates of Najm for lane changing and lane departure match with the fatality reduction rates de Visser found for the same types of collision avoidance (Jagtman et al. 2001).

Finally, Table 4.4 summarises the information identified in relation to the effectiveness of LCA systems.

4.3.1.4 Lateral Support Systems as Implementation Scenarios Legos

Table 4.5 constitutes a first approach towards which lateral support technologies may be employed in order to support various sustainable safety implementation scenarios.

4.3.2 Longitudinal Support

Longitudinal support systems are mainly discerned into longitudinal control systems, namely Cruise Control and Advanced Cruise Control (ACC) systems, Collision Warning and Avoidance Systems (CWS/CAS) for the frontal area of the vehicle, Intelligent Speed Adaptation (ISA) systems, Stop&Go systems, vision enhancement and VRU (Vulnerable Road Users) detection (and protection) systems.

Table 4.4 Effectiveness of LCA systems

System	Effectiveness	Source
LDW and LCA	Avoidance 25% (15–35%)[a] Fatalities 25% (15–35%)[a] Severe injuries 25% (15–35%)[a] Slight injuries Mitigation 15% (10–20%)[a] Fatal to severe 15% (10–20%)[a] Severe to slight	COWI (2006) effectiveness based on literature which included Abele et al. (2005)
LCA	43 ± 22% of Right lane changes for proximity sensing system only (USA) 32 ± 22% of Right lane changes for proximity sensing system only (USA)	Talmadge et al. (2000) estimated effectiveness using drivers errors as a surrogate for collisions during road tests
LCA	9% Fatalities[b] 9% Severe injuries[b] 9% of Slight injuries[b]	Bosch (2005a, b, c) in COWI (2006) estimated casualty savings in Germany.
LCA	60% Avoidance and 10% mitigation of side collisions	Abele et al. (2005) estimated effectiveness of the systems based on the reduced reaction time.
LCA	15–40% reduction in side collisions	Malone et al. (2006) reported a review of literature and comparison with expert opinions from a workshop that was carried out under the ADASE2 project. A simulator study (Wang et al. 2003 cited in Malone et al. 2006) reduction in accidents of 15% whereas other literature reviews and mathematical modelling quoted 40% reduction in side collisions

[a]Ranges used for sensitivity analysis
[b]COWI present only number of casualities saved for LCA. However number and proportion of casualties saved by LDW are presented and therefore the proportion of casualties prevented by LCA is estimated
Source: Visvikis et al. (2008)

Longitudinal Support is mostly important for averting the following driver errors: “Driving too fast in an unexpected bend on rural roads”, “Speeding” (CC and ISA systems), “Errors due to inattentiveness of the driver” (CAS, Stop&Go systems) and “Insufficient safety distance” (ACC). Table 4.9 shows a few specific solutions.

4.3.2.1 Cruise Control and Advanced Cruise Control (ACC) Systems

The aim of vehicle cruise control systems is the maintenance of a constant vehicle speed. A decade ago, the car industry introduced the first Advanced Cruise Control systems (ACC) to the market as an extension of the ‘conventional cruise control’. ACC is marketed as a comfort and convenience system rather than a safety system.

Table 4.5 Scenarios for which several different lateral support functions are considered effective

Solutions for the scenarios	Vehicle lateral and real monitoring systems	Lane Departure Warning Systems (LDWS)		Collision Avoidance Systems (CAS)			
		LDWS operating as virtual rumble strips	Adaptive LDWS	Blind Spot Detection Systems	Lateral Collision Warning Systems	Lane Change Assistance Systems	Over taking systems
General							
Preventing unintended lane change	√	√	√				
Preventing overtaking manoeuvres when vehicles approaching from behind or opposite direction	√ (Only for vehicles approaching from behind)			√ (Only for vehicles approaching from behind)	√ (Only for vehicles approaching from behind)	√ (Only for vehicles approaching from behind)	√
Preventing accidents when turning right or left due to other road users in the blind spot area				√			
Accidents due to lane departure							
When leaving road constructions the driver is reminded to reactivate his/her LDW-system			√				
The LDW-system is deactivated at the beginning of long construction sites to avoid false alarms			√				
Adaptive LDWS; sensitivity of lane departure warning assistant is adapted in special conditions, such as road works, tunnels			√				

In-vehicle warning of oncoming vehicles in curves		√				√
Head-on or head-tail accidents due to driver failure when overtaking						
Driver receives a warning signal from a vehicle that is catching up fast from behind			√	√	√	√
"Telematic barrier line": the driver receives a rumble strip vibration/sound when planned overtaking is not safe	√ (If not planned)	√ (If not planned)	√	√	√	√
Warning not to overtake when being overtaken and switching on the indicator (blind spot warning/lane change assistant)			√		√	√

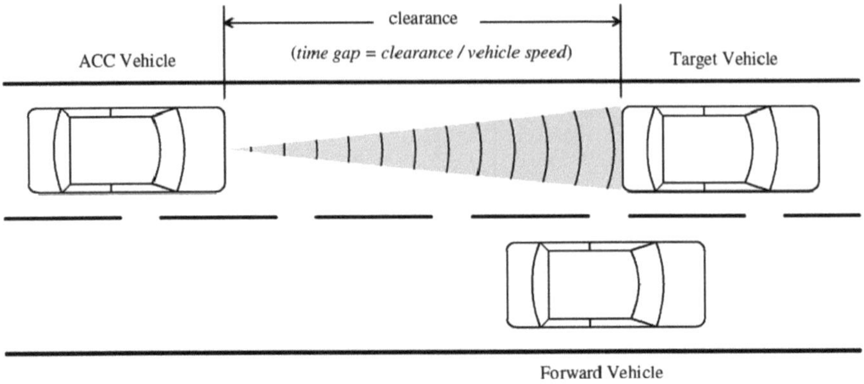

Fig. 4.8 ACC vehicle relationships
Source: 5th Meeting of the U.S. Software System Safety Working Group (2005)

However ACC systems may have positive effects on road safety, as well as on traffic efficiency and the environment (Fig. 4.8).

Advanced Cruise Control (ACC), also known as adaptive or intelligent cruise control, is an extension of conventional cruise control systems. ACC not only maintains the driver-set vehicle speed, but also adjusts the vehicle's speed to that of a preceding vehicle, and helps to maintain a pre-selected headway time to the vehicle ahead. ACC uses a frontal radar/laser sensor to detect vehicles in front and subsequently adjusts the vehicle's speed and headway by controlling fuel flow or by slightly braking. Active braking carried out by ACC can usually reach up to 30% of the vehicle's maximum deceleration. When a stronger deceleration is needed, the driver is warned by an auditory signal. Once the preceding, slower vehicle has moved out of the lane, the vehicle's speed will return to the driver-set cruise speed (SWOV fact sheet 2008).

Ten years ago, the first ACC systems that were introduced to the market were a rather expensive option for top-of-the line vehicle models. Today, ACC can be found on a rather wide range of vehicle models of various car manufacturers (ADAS Management Consulting & Bishop Consulting 2004; Bishop 2005; Alkim, et al. 2007). However, the equipment rate within the entire vehicle fleet is still very low. Most of the ACC systems which now are available, function for speeds above 30 km/h, have a detection range of 120–150 m, and allow for a manually set headway time between 1 and 3 s.

ACC systems may have a favourable effect on road safety when used on motorways with non-congested traffic. In these conditions ACC has a moderating effect on the driving speed, and decreases the percentage of very short headway times.

In the past decade several studies of ACC effects on driving behaviour were reported, but different studies showed different results. Some studies showed that ACC could have positive impact on road safety, for instance by a reduction of the mean driving speed (Hogema and Jansen 1996; Hoedemaeker 1999), a reduction of the maximum speed (Bjørkli et al. 2003) a reduction of speed differences, i.e., increased speed homogeneity (Törnros et al. 2002; Hoedemaeker 1999), and a reduction of very short headway times (Alkim et al. 2007). However, negative ACC safety effects were also found, like for example increased lane position variability (Hoedemaeker and Brookhuis 1998), delayed braking (Hogema et al. 1994), and colliding with a stationary queue more frequently (Nilsson 1996). Differences in operational characteristics of various ACC may result in different effects on driving behaviour. When driving with ACC types that take over more of the driving task and offer more support to drivers in more critical situations (e.g., capability of a complete stop in every situation), drivers seem to adapt their behaviour by increasing their speed (Dragutinovic et al. 2005). Besides, the traffic conditions, i.e., traffic density and road type, play a role in the observed effects. When ACC is used in low-density traffic conditions, the mean driving speed could be expected to decrease and speeds to become more homogeneous. On the other hand, when driving with ACC in high-density traffic conditions, the mean driving speed could be expected to increase, and there are some indications that speeds will be less homogeneous. However, these indications are not as clear as in the case of low-density traffic condition.

Regarding the type of the road, from the road safety point of view, the use of ACC should be avoided on rural roads (with curves and intersections), as well as on the urban roads, due to difficulty in detecting small silhouettes and vehicles out of the line of sight (Hoetink 2003).

A recent ACC field trial performed in the Netherlands showed that ACC could decrease the number of traffic crashes on motorways by about 13% and those on provincial main roads by 3.4%, assuming all vehicles are equipped with ACC (Alkim et al. 2007).

Several simulation studies investigated the potential impact of ACC on traffic flow. The simulation studies used different ACC algorithms, for instance to get different headway times, applied them in different environments and used different behavioural models. Furthermore, different penetration rates of the ACC technology were used. All these differences strongly influenced the outcomes on traffic capacity and speed, and therefore make comparison between these studies and their results very difficult.

With a 40% ACC equipment rate and a 1 s headway time, Broqua et al. (1991) estimated throughput gains at 13%. van Arem et al. (1996) and Minderhoud and Bovy (1998) found a decrease in average speed as a result of a collapse of speed in the fast lane when ACC with headway times of 1.4 s and above were used. Minderhoud and Bovy (1999) performed simulations with headway times as low as 0.8 s and concluded that current ACC using a 1 s headway time could achieve capacity gains of 4%.

ACC has the effect of decreasing the standard deviation of speed up to as much as 50%, which results in more homogeneous driving speeds. This is the main reason that ACC is generally expected to lead to a decrease in fuel consumption and hence to a decrease of harmful emissions.

Bose and Ioannou (2001) used field experiments and simulation models to quantify the environmental effects of ACC. Their results showed that an ACC equipment rate of 10% smoothed traffic flow, resulting in less fuel consumption and lower pollution levels in comparison to manual driving. The recent Dutch 'Rij Assistent' field study (Alkim et al. 2007) found a 3% reduction of fuel consumption. In general, drivers consider ACC to be a useful and comfortable system. Some characteristics of the system itself, like having the freedom to choose different headway times, can significantly affect the acceptance of the system (Hoedemaeker 1999). Drivers find ACC reliable and easy to use and to drive with, although objective data about the process of learning to drive with ACC is very limited. It seems that 2 or 3 weeks of intensive driving are needed to master the operation of ACC and the assessment of the takeover situations (Weinberg et al. 2001). Expressed in distance driven, it seems to take approximately 400 km of driving with ACC to know, understand and anticipate ACC reactions (Brouwer and Hoedemaeker 2006). Unfortunately, learning to drive with ACC is not part of the official driver training as yet and it seems that, like for the conventional cruise control, most of the drivers do not read the manuals and therefore the most common familiarization method is the salesman's explanation (Portouli et al. 2006).

The use of ACC by drivers is related to the type of the road they are driving on and the traffic conditions. ACC is most extensively used on motorways, somewhat less on provincial roads and almost never in urban areas. And on motorways, drivers primarily use ACC in free flow conditions (speeds higher than 90 km/h), less so in dense traffic conditions (speeds between 70 and 90 km/h) and hardly at all in congested conditions (speeds lower than 70 km/h) (Alkim et al. 2007).

4.3.2.2 Stop&Go Systems and Other ACC Extensions

Progressing technological developments may eventually result in a new generation of ACC systems that may overcome some of the limitations of today's ACC, regarding functionality and driver behaviour. Relevant developments in this field deal with the upgrading of the autonomous ACC system, and with combining different Advanced Driver Assistance Systems (ADAS) functionalities into a more integrated driver assistance system. Some examples of initiatives, aiming at features such as better speed support, and better anticipation to dangerous situations beyond the human senses, are reported in Morsink et al. (2006) and are described below.

"Stop and go" systems are considered to be the next generation ACC, and they are close to market introduction. Unlike common ACC, this system has the possibility to slow down the vehicle to a complete standstill. To do this "Stop and go" ACC, among other things, has to be capable of detecting other road users or stationary objects at a much closer range than the common ACC. ACC which operates from standstill to the maximum speed is also called 'Full-range ACC'.

Another extended type of ACC that is close to market introduction is called Predictive Cruise Control (PCC). This system issues location specific warnings, such as speed while approaching a dangerous curve (e.g., making use of the onboard navigation system).

The combination of ACC and Intelligent Speed Adaptation (ISA) (see following section), in which the ACC takes the current speed limit as its default value, is promising. Where ISA reduces the average speed, ACC could reduce tailgating and further reduce speed variations. ETSC (2005) reports a market initiative to launch such a combined system. In fact PCC can also be considered a form of ISA that is functionally integrated in ACC. Another, more advanced, combination of ACC and ISA is known as Responsive ACC (RACC) (Bishop 2005). The system receives a speed advice from a traffic control centre, taking into account local speed limits and traffic flow in the network. This information is used to adjust the vehicle's speed, allowing fine changes in speed (beyond the control of the driver), independent of time and location. At the moment RACC only exists at a conceptual level.

A combination of ACC and Lane Departure Warning (LDW) was tested in the Alkim et al. (2007) study. Although LDW was found much less effective than ACC, some of the test drivers reported an interesting positive integration effect. With ACC activated, a slight increase in the variation of lateral position in the driving lane was found. The test drivers, however, claimed that the warning issued by LDW effectively compensated for this, and increased their alertness.

Communication between vehicles and between vehicle and roadside is considered the technology that will make a whole new generation of ADAS possible. Several European research projects (e.g., INVENT in Germany, ADASE 2, CAR-TALK, PReVENT and CVIS) have already worked on these so-called cooperative systems, and research activities are increasingly expected worldwide. Cooperative ACC (CACC) makes use of communication between a series of successive ACC-equipped vehicles in the same lane. The vehicles exchange their position, speed and deceleration (De Bruin et al. 2004). This may increase safety as the ACC system can optimise its speed support and drivers can get early warnings of braking or of slow vehicles ahead. The potential road safety benefit may be accompanied by a better performance on traffic throughput and emissions on main roads (Malone and van Arem 2004). For instance, CACC with 0.5 s headway time would almost enable double the traffic flow at 100% penetration rate (Vanderwerf et al. 2002). On the longer term, assuming a further automation of driving tasks, this may be achieved without compromising safety.

4.3.2.3 Intelligent Speed Adaptation (ISA) Systems

Intelligent Speed Adaptation is a generic term for a class of ITS in which the driver is warned and/or vehicle speed is automatically limited when the driver is, intentionally or inadvertently, travelling over the posted speed limit for a given location. In general, ISA systems establish the position of a vehicle, compare the speed of the vehicle with the posted speed limit at a given location, and then give in-vehicle feedback about that speed limit to the driver or even restrict the vehicle's speed according to the speed limit in force.

There is a wide range of ISA systems that differ in the level of support and the kind of feedback they give to the driver, as showed in Table 4.6.

Table 4.6 Overview of different variants of ISA systems

Level of support	Type of feedback	Feedback
Informing	Mostly visual	The speed limit is displayed and the driver is reminded of changes in the speed limit
Warning (open)	Visual/auditory	The system warns the driver when s/he exceeds the posted speed limit at a given location. The driver him/herself decides whether to use or ignore this information and to adjust his/her speed
Intervening (half-open)	Haptic throttle (moderate/ low force feedback)	The driver gets a force feedback through the accelerator if s/he tries to exceed the speed limit. If applying sufficient force, it is possible to drive faster than the limit
Automatic control i.e. speed limiter (closed)	Haptic throttle (strong force feedback) and Dead throttle	The maximum speed of the vehicle is automatically limited to the speed limit in force. Driver's request for speeds beyond the speed limit is simply ignored

Source: SWOV Fact sheet (2007)

The system is usually not mandatory, which means that the user has the option to activate of deactivate it. ISA systems use three types of speed limits:

1. Fixed speed limits – The driver is informed of the posted speed limits.
2. Variable speed limits – The driver is additionally informed about (lower) speed limits at special locations, like road construction sites, pedestrian crossings, sharp curves, etc. Therefore, the speed limits are dependent on the location.
3. Dynamic speed limits – The dynamic ISA system uses speed limits that take account of the actual road and traffic conditions (weather, traffic density). Therefore, besides being determined by location, the dynamic speed limits are also dependent on time.

Information regarding the current position of the vehicle may be provided deploying global positioning system (GPS) technology, whereas the speed limit that applies to that location can be obtained either by means of electronic signals to the vehicle from beacons or transmitters attached to speed signs or other roadside infrastructure (i.e., lampposts), by means of on-board systems (i.e., cameras) that read themselves the speed attached to speed signs or through web services; in this case, communication with Traffic Management Centres is required, that allow the dynamic update of the info stored in the in-vehicle digital maps (alternatively, if no web services are deployed, the info in the digital maps database is static).

Whichever is the way the road network and posted speed limit information is provided, it is stored in a digital map database within the vehicle. A GPS receiver-fitted to the vehicle locates vehicle position. An on-board computer continuously analyses the location of the vehicle and compares the posted speed limit with the current speed of the vehicle. Warnings are issued when the vehicle is exceeding the speed limit or some other nominated speed threshold for a given location (Fig. 4.9).

According to the above, there are three main categories of ISA (speed alert) systems, in terms of technology deployed.

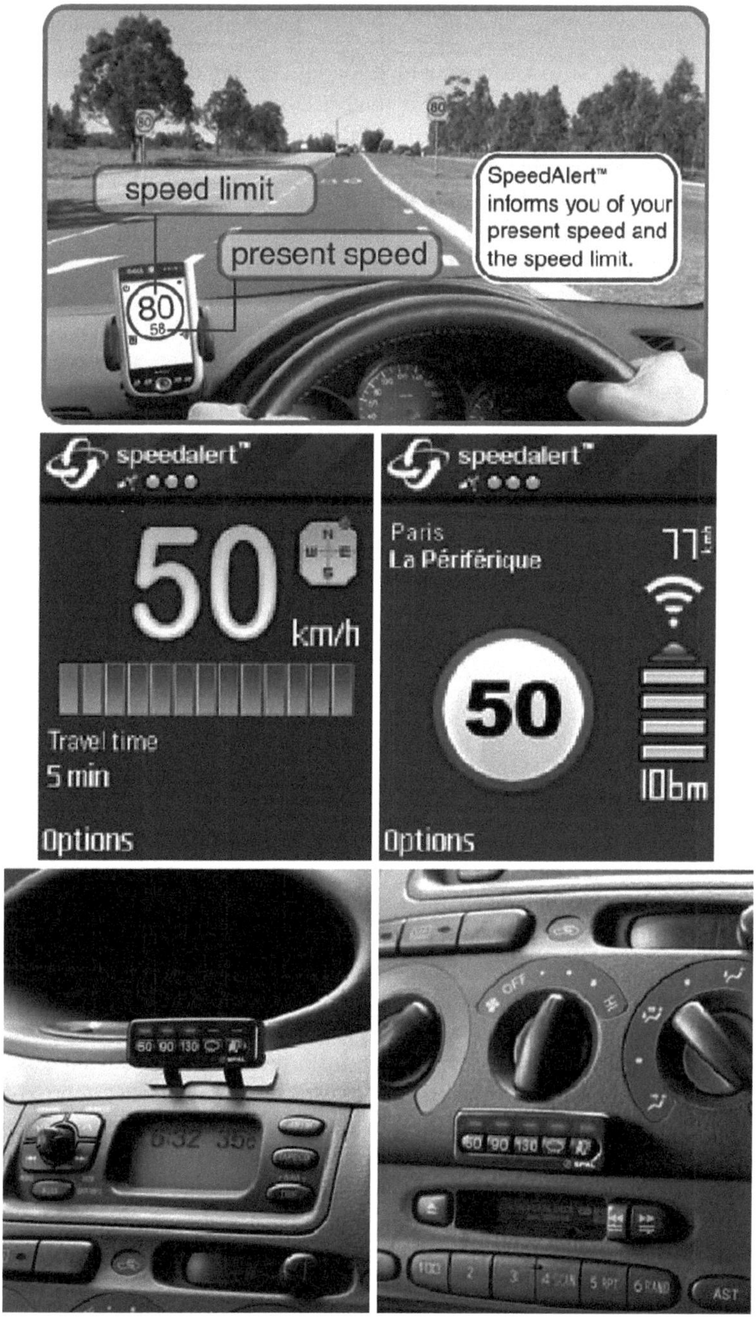

Fig. 4.9 ISA systems alternative HMIs'
Sources: http://www.speedalert.com.au/; http://www.wayfinder.com/?id=3764; http://www.spal.it

Speed Alert Based on GPS and Digital Maps (Static and Dynamic)

In order for the vehicle to know on which road it is being driven, where on which road it is being driven, where on that road it is currently located and in which direction along that road it is travelling, the solution, adopted most widely in ISA trials around the world, utilises global positioning system (GPS) technology combined with map-matching and dead reckoning techniques. The accuracy of an uncorrected GPS receiver ranges from 5 to 15 m. A differential Global Positioning System (dGPS) is also often used and improves the accuracy of the position determination to within a metre. The dGPS receiver is a special FM ratio receiver and usually requires a subscription with a service provider. A GPS-based navigation system can supplement information acquired from the GPS with dead reckoning (from compass and yaw sensor) and map-matching. The new European correction system, GNSS, will provide additional accuracy, as will the new satellite system, Galileo, when it becomes operational (Jamson et al. 2006).

GPS allied to digital speed limit maps allows ISA technology to continuously update the vehicle speed limit to the road speed limit. In specific, systems based on dynamic digital maps decrease the risk of outdated information about speed limits.

Speed Alert Based on I2V (i.e., Road Beacons)

In this solution, electronic signals are transmitted to the vehicle from roadside beacons attached to speed signs or other roadside infrastructure, such as lampposts. These beacons transmit information regarding the posted speed limit to the vehicle and an on-board computer triggers the warning and/or limiting system if the vehicle exceeds this limit (Jamson et al. 2006). The Swedish (Umeå) trial adopted this approach, using in-vehicle equipment consisting of a RF receiver, a microprocessor, electronic compass for direction, a pulse-counter for distance measurement and an in-vehicle display. The advantage of the beacon system is that it is immediately operational as the vehicle passes it, with little delay. The disadvantages relate to maintenance and initial set-up costs. The Swedish system cost approximately around 112€ for the initial installation of the beacon, not counting the on-going maintenance. Furthermore, there is the possibility that road network information may fail to download into passing vehicles, which means that these vehicles would be travelling at an incorrect speed until they pass the next beacon (Carsten and Tate 2005). In the Umeå trial there were also problems experienced with the electronic compass due to magnetism.

The certainty level of electronic beacons (or other transmitters) is supposed to be high and the needed penetration rate to reach the safety level is low because if only a few drivers adapt their speeds, the others have to follow (Fig. 4.10).

Fig. 4.10 Speed alert based on I2V
Source: Fowkes (2007)

Speed Alert Based on Road Sign Reading

This alternative (relatively recent solution) deals with the problem of recognising signs in a complex dynamic traffic environment, which is especially useful in situations of bad visibility.

German manufacturers BMW, Mercedes-Benz and Audi are racing to import the first car into Australia that can read speed limit signposts and display a visible warning to drivers if they exceed the speed limit.[3]

BMW's system is driven by a camera fitted to the inside of the windscreen near the rear-view mirror that reads roadside signs, including variable speed limits, and compares this with data contained in the car's navigation system. In the case of speed signs indicating temporary speed limits – such as roadworks – priority is given to the camera reading.

The speed is displayed on the instrument cluster, or in the head-up display on the windscreen, reducing the risk of the driver breaking the speed limit by mistake. The system is designed to inform the driver, not to take control of the car.

According to BMW, the system can read both painted metal signs and dynamic displays, and will be especially useful to drivers in built-up areas where speed limits frequently change.

Generally speaking about ISA systems, and according to SWOV Fact sheet (2007), ISA equipped vehicles show an average speed reduction of approximately

---

[3]http://www.drive.com.au/Editorial/ArticleDetail.aspx?ArticleID=63948, last update: 26 June 2009.

2–7 km/h, as well as a reduction in speed variance and speed violations (see Table 4.9). The size of these reductions depends on the type of ISA, with more controlling ISA types being more effective. Only one study found an increase in average speed; (Peltola 2000) who investigated the effects of ISA on icy roads. The ISA system gave a speed advice that was lower than the general speed limit in force. It appeared, however, that the mean speed of ISA drivers was higher than that of drivers without ISA. A possible explanation could be that the ISA speed advice exceeded the speed that drivers would have chosen themselves (Table 4.7).

In a SWOV driving simulator experiment (van Nes et al. 2007), the effects of warning ISA on speed behaviour were investigated in combination with credible speed limits. As in previous studies, the results of this experiment showed that ISA has a significant speed reducing effect. A new observation in this study was that the effect was especially significant in situations where speed limits were of low credibility. In addition there were fewer speed violations and smaller differences in speed when driving with ISA.

Testing by Veilig Verkeer Nederland and Senter Novem of one particular warning ISA system named SpeedAlert, also shows that it is effective in reduction of driving speed. In 80% of the situations where the speed limit was exceeded, participants adjusted their driving speed after being warned by the SpeedAlert system (http://www.veiligverkeernederland.nl).

On the other hand, it is not simple to determine the effect of ISA on traffic crashes. The proportion of vehicles equipped with ISA in the field trials was relatively small, while, in order to measure the effect on traffic crashes, a substantial number of ISA vehicles are required. Therefore, studies making use of a driving simulator and traffic simulation studies are used for effects estimates based on current best knowledge.

Table 4.7 Overview of the ISA effects on mean speed and standard deviation of speed in various studies (↓ decrease, ↑ increase, ? not investigated)

Study	Methodology	Country	Effect on mean speed	Effect on standard deviation of speed	Speed violations
Comte (2000)	Driving simulator	UK	↓	↓	?
Peltola (2000)	Driving simulator	FIN	↑	↓	?
Hogema and Rook (2004)	Driving simulator	NL	↓	↓	↓
Van Nes et al. (2007)	Driving simulator	NL	↓	↓	↓
Brookhuis and De Waard (1999)	Instrumented vehicle	NL	↓	↓	↓
Paeaetalo et al. (2001)	Instrumented vehicle	FIN	↓	?	↓
AVV (2001a, b)	Field trial	NL	↓	↓	?
Lahrmann et al. (2001)	Field trial	DK	↓	?	?
Biding and Lind (2002)	Field trial	S	↓	↓	↓
Regan et al. (2006)	Field trial	AUS	↓	↓	↓
Vlassenroot et al. (2007)	Field trial	B	↓	↓	↓

Source: SWOV Fact sheet (2007)

Based on the found reductions of mean speed, speed distribution and the percentage of speeding, ISA systems are assumed to achieve substantial reductions in the incidence and severity of road crashes (Várhelyi 1996; de Kievit and Hanneman 2002; Louwerse and Hoogendoorn 2004; Carsten and Tate 2005). There is a large variation in effects, depending on the system type, the type of speed limit and the ISA penetration rate in the vehicle fleet. When comparing the effectiveness of various ISA systems, advisory or informative systems have a much smaller effect than mandatory systems. In addition, the effect of ISA based on fixed or variable speed limits is smaller than ISA based on dynamic speed limits. Table 4.8 presents the results of the Carsten and Tate study. Their estimates assume a 100% ISA penetration level and no behavioural adaptation to ISA and therefore they represent a 'best case scenario'. The results study by Carsten and Tate (2005) gives an expected estimate of a 36% reduction of injury crashes and a 59% reduction of fatal crashes from mandatory forms of ISA and dynamic speed limits.

Similarly, assuming all vehicles being equipped with an ISA system that would not allow exceeding the (fixed) speed limit, Oei (2001) estimated the reduction in annual fatalities and injuries in the Netherlands. Based on detected speed violations on different road types and using Nilsson's formula on the relation between driving speed and the number of traffic victims (Nilsson 1981), Oei estimated the reduction to be 25% which is well in line with the estimated 29% for mandatory ISA with fixed speed limits in Table 4.8.

It is not yet clear whether these large effects would also be realised in reality. Although limited, there is evidence that drivers could develop certain risky driving behaviour like adapting closer following distances, accepting smaller gaps when merging or braking relatively late when driving with ISA (Comte 2000). Furthermore, the long-term ISA effects on driving behaviour are not as yet unknown, as is the behavioural response of other drivers towards ISA drivers.

The expectations of the effects of ISA on traffic efficiency and environment are based on the reduction and the homogenization of driving speeds.

The results of micro-simulation modeling of the ISA effect on network efficiency showed that in high traffic density conditions, ISA would not have a

Table 4.8 Best estimates of crash savings by ISA type and crash severity

System type	Type of speed limit	Best estimate of injury accident reduction (%)	Best estimate of fatal and serious accident reduction (%)	Best estimate of fatal accident reduction (%)
Advisory	Fixed	10	14	18
	Variable	10	14	19
	Dynamic	13	18	24
Driver select (voluntary)	Fixed	10	15	19
	Variable	11	16	20
	Dynamic	18	26	32
Mandatory	Fixed	20	29	37
	Variable	22	31	39
	Dynamic	36	48	59

Source: Carsten and Tate (2005)

significant effect on network total travel time, because driving speeds are already largely limited by congestion in high traffic density conditions. However, in lower traffic density conditions, the travel time would increase due to lower average speeds, especially with increasing ISA penetration rates (Liu et al. 1999).

The micro-simulation study by Liu et al. (1999) showed that the emissions of CO, NO_x and HC varied by only ±2% for all ISA penetration rates. The total fuel consumption gradually decreased with increasing penetration levels of ISA equipped vehicles and a total of 8% reduction in fuel consumption was achieved.

The data about the real effect of ISA on the environment is very limited. The Dutch ISA trial (AVV 2001a, b) resulted in data that was insufficient to come to an indicative conclusion about the ISA effect on emissions. The results of the Swedish trial in the city of Lund, showed that there were reductions in the emission volumes mainly for dual carriage ways and 50 km/h speed limits. The average reduction for CO volumes was 11%, for NO_x 7% and for HC 8%. On the other road types there were no significant changes and on arterial streets with a 70 km/h speed limit emissions increased (Várhelyi et al. 2004).

In 2002, over 24,000 European drivers were questioned about how useful they find a system which prevents exceeding the speed limit (Cauzard 2004). More than 50% of the drivers questioned found that such a system would be very of fairly useful and an even higher percentage of drivers were in favour of fitting such devices to a car.

Acceptance is critical for the potential success and effectiveness of ISA. Several factors seem to be significant for the users' acceptance of ISA: the type of ISA system, the type of the road environment and the driver's character.

Regarding the type of ISA system, the more intruding and controlling a system is, the less it will be accepted by the drivers. At the same time, however, the more intruding and controlling, the larger the effects on speed and on traffic safety in general. Evidently, there is the trade-off between the effectiveness and the acceptance by drivers of the ISA systems. The characteristics of the specific feedback given by the ISA system are also important for the acceptance. In general, continuous visual and auditory feedback is preferred over the haptic feedback.

It seems that drivers, whose speed behaviour would benefit most from ISA, accept it least. Hence, there is a danger of a self-selection bias when ISA is introduced on a voluntary basis. Drivers who "need" ISA most would be least willing to use it.

The acceptance of ISA differs for different road types, their related speed limits and the driving speeds. The acceptance is the highest for urban roads with 30 and 50 km/h speed limits (AVV 2001b; Wiethoff 2003a, b).

In general, test drivers initially did not have a very positive attitude towards ISA systems and they favoured normal, unsupported driving. However, drivers' attitudes turned out to be more positive after testing the system. Especially the "usefulness" and "satisfaction" offered by the system were more appreciated by the test drivers after driving with ISA than before having gained experience with the system. Eventually, a combination of ISA features, like fewer tickets for speeding, more comfortable and economic driving, and optimization of travel times, may increase the product image and improve the attractiveness for individual drivers.

The conclusions of the EU PROSPER project which assessed road speed management methods (http://www.prosper-eu.nl), focused on the identification of obstacles to ISA implementation. The most important general barriers to the ISA implementation were found to be the technical functioning of the system, the applicability to the whole road network and the benefits to the users. However, for some countries the cost price is also a very important barrier, as well as the public and political acceptance.

Because of the complexities and uncertainties surrounding the implementation of ISA, one of the suggested approaches is flexible or adaptive policy making (Marchau and Walker 2003). This adaptive approach suggests taking some actions immediately and creating a framework for future actions that allows for adaptations over time as knowledge about ISA accumulates and critical events for ISA implementation take place (Marchau and Walker 2003).

4.3.2.4 (Forward) Collision Warning and Avoidance Systems (CWS/CAS)

Collision Warning Systems are in-vehicle electronic systems that monitor the roadway in front of the host vehicle and warn the driver when a potential collision risk exists. For example, currently available radar-based CWS use algorithms to interpret transmitted and received radar signals to determine distance, azimuth, and relative speed between the host vehicle with the CWS and the vehicle or object ahead of it in the lane. When the host vehicle is traveling along the roadway, the CWS can warn the driver when a vehicle or object is in its lane within a predefined closing time threshold.

In most cases, CWS do not take any automatic action to avoid a collision or to control the vehicle; therefore, drivers remain responsible for the safe operation of their vehicles using both steering and braking, if safe to do so, to avoid a crash. As the time interval to the vehicle ahead decreases, CWS issue a progressively more urgent warning. The system's beam width/field of view forms an isosceles triangle with its apex at the front center of the vehicle. As an object gets closer to the front of the vehicle, a different range or time interval is reached, and the system issues a different type of alarm. The system manufacturers set these warning thresholds. In most cases, CWS also warn the driver if the system malfunctions and may be also integrated with ACC systems.

In some cases, CWS automatically apply the brakes if the driver doesn't act in time to avoid a crash. Though the technology first appeared on luxury cars, crash prevention systems have started to trickle down to more reasonably priced vehicles.

The HMI's used include visual indicators projected on the windshield, audible warnings through the stereo system's speakers, whereas the brakes are pre-charged and electronic brake is activated to assist to help the driver stop more quickly. Some systems tug the seatbelt and lightly engage the brakes, and if they determine that a crash is imminent, they cinch the seatbelts and apply brakeforce to mitigate impact velocity and the force of the collision (Fig. 4.11).

In Najm et al. (2006) 66 subjects participated in a FOT for a period of 4 weeks for the purpose to evaluate a combination of forward crash warning and adaptive cruise control. The study indicates that the system might prevent 3–17% of all rear-end-crashes, expressed as a "conservative estimate".

In parallel, according to the studies executed in the context of the TAC SafeCar project by the Monash University Accident Research Centre (MUARC) and Ford Australia, where, among other systems (ISA, Seatbelt Reminder and Reverse Collision Warning system), the effects on driving of long-term exposure to a Following Distance Warning system were examined (Regan et al. 2006). Fifteen specially equipped vehicles, called "SafeCars" were sub-leased to nine public and private companies in and around Melbourne. Twenty-three drivers each drove one of the vehicles for a distance of at least 16,500 km. The Following Distance Warning used frontal radar technology to compute elapsed time between the SafeCar and the vehicle directly in front. Graded visual and auditory warnings were issued when the driver was travelling 2 s or less from the car in front.

As observed in this study, when the Following Distance Warning system was active, drivers left a greater time gap between the SafeCar and the car in front and spent less time travelling at very small gaps of less than 1 s. The system increased the minimum gap between the SafeCar and vehicle in front on each trip. The system appeared to be particularly effective at increasing time headways in higher speed zones (80 and 100 km/h), with the increases in mean time headway and reductions in the percentage of time spent at headways below 0.8 s greater in the 80 km/h and, in particular, 100 km/h zones. This finding most likcly results from the fact that drivers maintained shorter time headways in the higher speed zones and the Following Distance Warning system therefore had more opportunity to influence following distance in these zones compared to the lower speed zones.

Interestingly, the speed reduction effects of the ISA system were more pronounced when it operated in conjunction with the Following Distance Warning

Fig. 4.11 Collision warning head-up display
Source: Coelingh et al. (2006)

system. However, in terms of following behavior, the two systems combined were no more effective than the Following Distance Warning system alone in increasing following distance. Another consistent finding was that both the ISA and Following Distance Warning systems were effective only while turned on; when they turned off, drivers reverted to their usual driving behaviours, indicating the importance of drivers having exposure to these systems. However, this finding is inconsistent with previous research by Shinar and Schechtman (2002), which found the drivers drove at longer time headways when using a Following Distance Warning system and that these increased headways were maintained up to 6 months after exposure to the system (even though the drivers in that study were exposed to the system for far less time than drivers in the SafeCar study).

It was also sometimes found that drivers drove at significantly greater time headways when the Following Distance Warning system was active compared to the After period, but not the Before period (it is unclear why this would have occurred, but it may result from the drivers becoming over-reliant on the Following Distance Warning system warning over the course of the During period and, thus, driving at shorter headways in the After period because they were expecting to receive warnings to alert them when they were driving too close to the vehicle ahead).

Finally, the ISA and Following Distance Warning systems appeared to be equally effective at night and during the day and for younger (aged less than 45 years) and older drivers (aged 45 years and over). Another positive finding from the study was that there was little evidence of any "negative behavioural adaptation" to the systems. That is, the drivers did not compensate for the added safety benefits derived from the systems by engaging in increased risk taking. Also, a significant reduction in fuel consumption, Carbon Dioxide, Nitrogen Oxide and Hydrocarbonates emissions was found, in 80 km/h zones and only when the ISA and Following Distance Warning systems were jointly active. The positive changes in driving behaviour observed in the study translate into large crash reduction benefits. For the Following Distance Warning system, the percentage of driving distance spent in rear-end collision mode (that is, when the vehicle would collide with the lead vehicle if it braked suddenly) is expected to reduce by up to 34% when the system is active and when the lead vehicle is braking at a moderate rate. However, the Following Distance Warning system was rated as more effective at the end of the study, although after experience with the system, less of the participants were inclined to keep it. The system did not increase the perceived level of workload participants experienced while driving, whereas there was also a positive effect on road safety awareness: after using the system, drivers were more likely to adhere to the recommended following distance. Potential barriers were found to be the cost of the system and the auditory warning system considered annoying.

4.3.2.5 Traffic Light Status Emitted to the Vehicle

Red light running causes about 260,000 crashes and 750 fatalities each year in USA alone (Retting et al. 2003, 2007).

Systems providing an early warning system for vehicles approaching traffic lights, using at least one transmitter, or transceiver, on the traffic light and a receiver, or transceiver, in each vehicle are already available in the market.[4] The transmitters transmit a signal prior to the traffic light changing color, during a predetermined cycle, generally a single sequence of red, green, and amber light changes. The vehicle receiver activates at least one alarm or indicator thereby notifying the driver of the action required. The transmitter and receiver can contain algorithms that control information transmitted and the response of the display panel. The vehicle transceiver can be connected to the on board computer system to receive and read data from the vehicle. When conditions, such as ice, prevent the vehicle from operating correctly, the algorithm transmits to the light receiver the need to delay the color change.

At present, it is not certain what the effects of this measure will be. If red light running is caused by inattentiveness, the measure could work out really well. The safety level is high because the system is in principle comparable to a Collision Avoidance System. The certainty level is low because there have not been enough experiments with a system like this. The needed penetration rate to reach the safety level is moderate because if only one driver runs the red light this will immediately result in a dangerous situation. On the other hand, if one driver stops for the red light, the vehicles behind it have to stop anyway.

In case of multiple lights at an intersection it may not always be clear to whom a specific traffic light applies.

4.3.2.6 Vision Enhancement Systems

Such systems enhance visibility in cases of intense fog, rain, snow or darkness (therefore, are often called "Night Vision Enhancement Systems-NVES") and provide additional information to the driver about the road as well as not yet visible critical obstacles ahead of the vehicle.

They are capable of detecting objects beyond the reach of normal headlamps and help with other unfavorable conditions, like blinding headlamp glare from oncoming vehicles. They are not intended to be stared at by the driver, but rather serve as a supplemental aid for periodic glancing and peripheral view.

Vision enhancement systems typically use infrared (IR) technology to aid drivers in bad visibility situations. They work by detecting either near infrared (NIR) or far infrared (FIR) light waves (this type of technology may also be referred to as thermal imaging and is ideal for detecting living things like people and animals), invisible to the human eye, and displaying a brightened image of the dark road ahead. Both types use special cameras that are able to collect small amounts of infrared light and amplify it so it can be seen with human eyes.

---

[4]http://www.freepatentsonline.com/y2002/0070880.html.

Warnings are provided through in-vehicle displays (virtual or not) in combination with audio and haptic warnings. Alternative HMI configurations are shown in Figs. 4.12 and 4.13.

The results of the simulator trials in Linköping University of various configurations of simulated Vision Enhancement Systems confirm the findings from previous experiments that these systems produce an indisputable improvement in drivers' anticipatory control, and hence have considerable safety potential. The earlier detection of objects made possible by such a system buys the driver valuable time to assess the situation and prepare a measured response. There also seems to be an advantage to having a wider field of view and to placing the Vision Enhancement System in the normal line of sight of the driver, although there were no strong negative effects of a lateral displacement. The consensus in the driving research community, as summarised by Rumar (2002), is that in night driving visual

Fig. 4.12 Display technologies (PR LED and PR MHUD)
Source: Krems et al. (2006)

Fig. 4.13 *Left*: BMW displays the monitor on the centre console to show Night Vision image. *Right*: Mercedes-Benz uses a high-resolution virtual instrument cluster for displaying the Night View Assist image
Source: Technical Information For The Collision Industry (2006)

guidance is less impaired than target detection. An NVES should therefore help with the detection of objects rather than with the recognition of objects. Indeed, in driving – either during day or night – the first priority is to know something is on the road ahead, while the second is to know what it is. This clearly has consequences for how the information should best be presented (Hollnagel and Källhammer 2003).

4.3.2.7 VRU (Vulnerable Road Users) Detection (and Protection) Systems

According to recent studies (SCANIA 2009), around 30% of all deaths related to accidents with heavy vehicles involved are pedestrians and two wheelers, also denoted as vulnerable road users. The sheer number of vulnerable road users who died in Euro-15 during 2007 was a staggering 15,000, with at least 150,000 more being injured. Measures to reduce these figures for the vulnerable road user are recognised as necessary when considered in parallel with the increasing encouragement to make use of public transport and cycling as alternatives to the use of the car (COM(2007) 560 final SEC(2007) 1245).

With increasing efforts from vehicle manufacturers and society to reduce the total number of deaths on the roads, the problem of vulnerable road users has been addressed from many points of views, with safer road designs, pedestrian friendlier car hoods, bicycle lane boxes and advanced driver assistance systems being some examples.

Currently most of the VRU protection systems in vehicles utilise forward looking cameras, which are jointly used with other applications, such as Lane Departure Warning. By processing the video from these cameras, computer vision methods permit to detect and classify objects, perhaps a VRU, before impact is made. Image sensor or video camera systems are superior to radar and lidar scanners in their capability for detecting VRUs. In this category of video cameras, near-wave infrared systems appear to be preferred over the more costly thermal imagers, though both spectral ranges are deployed in current products.

Detecting humans automatically in the video is a challenging problem, owing to the motion of the subjects, the camera and the background and to variations in pose, appearance, clothing, illumination and background clutter. The literature concerning pedestrian detection algorithms is vast and contains many approaches.

Pedestrian tracking and trajectory estimations need to be computed in 3D, in order to obtain a measure of the expected collision risk. An important concept in the context of tracking is that of dynamic filtering. It is a tool which recursively estimates the current state of the system given a new measurement, in the presence of noise.

Video based systems also complement radar which is used in automotive applications, such as Adaptive Cruise Control. Radar accurately determines an object's position and velocity, while video determines the object's type.

Recently, new sensor fusion approaches are attempted which combine video detection with real time positioning systems operating at different radio frequencies and various band widths to determine the relative localisation from the vehicle to the VRU.

This kind of complementary sensor combination was in the focus of WATCH-OVER[5] with successful tests performed using IEEE 802.15.4a communication protocol for ranging and azimuth measurement of VRUs. Of interest to the industry is also IEEE 802.11p, a protocol which is adding wireless access in the vehicular environment. It can be used for ad-hoc communication between vehicle and VRU-based terminals for the exchange of self-localisation and identity data. However, it should be noted that 802.11p is mainly targeted to vehicle-to-vehicle communication.

In cases of darkness or visibility constraints, due to meteorological conditions, combined video and telecommunication based systems are expected to perform best. This holds true also for the detection of VRU behind objects which obstruct the direct view. Acoustic and optical warning provided in night-vision systems could be provided more reliably and the functions of the vision system also extended to day-time operations, adding to the perceived customer value.

Thus, significant technological advances are expected before a full scale market deployment of such systems. Prototype systems seem to enhance the detection rate up to 80 m but still up to limited vehicle speed. The best commercial performance of such a system seems to be to detect a pedestrian in a range of 5–25 m, and up to 4 m lateral, in real time, for a vehicle speed of up to 40 km/h (operating at 7–15 Hz).

Figure 4.14 presents the HMI of the on-board system of the WATCH-OVER system (Pieve et al. 2008). Once the VRU is detected, a threat assessment is performed and the driver is warned in case of risk. For the pre-crash scenario the warning is acoustic and supported by the actuation of the automatic braking system in case of high risk. For the preventive safety scenario, since the time to collision is less critical, the approach has been to enhance warning strategies: the combination of visual and acoustic warning in the ego-vehicle, as well as the warning to the VRUs.

Fig. 4.14 HMI of the on-board VRU detection system of WATCH-OVER
Source: Pieve et al. (2008)

[5]FP6-2004-IST-4 – 027014; http://www.watchover-eu.org/.

An EC study[6] shows that pedestrian protection can be significantly improved by use of a combination of active and passive safety measures. Passive measures help to reduce injury levels on impact by provision of softer vehicle surfaces. Active measures alleviate the conditions under which impact may take place, e.g., by reduction of impact speed. According to this study such a combination of measures could afford an 80% higher level of protection than the previously existing provisions.

4.3.2.8 Longitudinal Support Systems as Implementation Scenarios Legos

Table 4.9 constitutes a first approach towards which longitudinal support technologies may be employed in order to support various sustainable safety implementation scenarios.

4.4 Preliminary Selections

In this chapter a preliminary benchmarking of a number of potential systems that may contribute to the design of a more forgiving road (FOR) and a more self-explanatory road (SER) environment has been performed. The preliminary evaluation was done both in terms of the expected safety effects, as well as in terms of a more general scope of effects. The results of these evaluations may serve as a useful input for further in-depth studies, pilot tests and policy recommendations.

Within the in-safety project, a number of existing (and future) applications from the ones presented in this chapter were selected, based on an analysis of accident data, and evaluated for their expected effectiveness on safety. Their selection occurred within the context of the specifications of scenarios that aimed to create more FOR and SER environments. It turned out that specific combinations of infrastructural and in-vehicle systems show the highest potential towards cost effective solutions in terms of safety effects. The most relevant scenarios with respect to the likelihood of contributing to a substantial decrease in severe accidents were included in the analyses, estimating actual accident reductions under various conditions. In this way, a testing and validation program could be devised for future pilot studies to be conducted. We can conclude that for many of the systems studied in this chapter, the safety effects are at least moderate for various types of scenarios. For longitudinal support, the safety effects are expected to be higher as compared to lateral support. On the other hand, it has become obvious that quite a number of different dangerous situations can be envisaged too, that will benefit from lateral support. Therefore, the opportunities of lateral support functions for improving road safety should not be underestimated.

[6] http://ec.europa.eu/enterprise/automotive/pagesbackground/pedestrianprotection/index.htm.

Table 4.9 Scenarios for which several different longitudinal support functions are considered effective

Solutions for the scenarios	Cruise Control (CC)	Advanced Cruise Control systems (ACC)	Stop&Go systems and other ACC extensions	Intelligent Speed Adaptation (ISA) systems. Speed alert, based on			(Forward) Collision Warning and Avoidance Systems (CWS/CAS)	Traffic light status emitted to the vehicle	Vision enhancement systems	VRU (Vulnerable Road Users) detection (and protection) systems
				GPS and digital maps	Road sign reading	I2V				
General										
Speed control in order to prevent speeding or unsafe speed	√	√	√	√	√	√		√	√	√
Distance control in order to prevent too narrow distance with the vehicle in front		√	√				√		√	
Accidents due to violation of priority rules										
Displaying traffic signs in the car also when they are invisible (tree, fog. . .)					√					
In-vehicle warning from infrastructure in case of driving too fast towards red traffic light								√		
Head-on or head-tail accidents										
The safety distance of ACC is automatically increased in critical situations (fog, tunnels, traffic jam ahead, etc.) and the braking force is adapted		√	√						√ (Through input to ACC on visibility level)	√ (Through input to ACC on visibility level)

References

5th Meeting of the U.S. Software System Safety Working Group April 12th–14th 2005 @ Anaheim, California USA, Adaptive Cruise Control System Overview

J. Abele, C. Kerlen, S. Krueger, H. Baum, T. Geissler, S. Grawenhoff, J. Schneider, W.H. Schulz, *Exploratory study on the potential socio-economic impact of the introduction of intelligent safety systems in road vehicles*. Tetlow, Germany: VDI/VDE Innovation (2005)

ADAS Management Consulting & Bishop Consulting, The Assisted Driver' Pilot program: advice regarding project planning and partnership development. Final report, ADAS Management Consulting, Bregenz, Austria/Bishop Consulting, Granite, MD, 2004

T. Alkim, G. Bootsma, P. Looman, *De Rij-Assistent; systemen die het autorijden ondersteunen* (Studio Wegen naar de Toekomst (WnT), Directoraat-Generaal Rijkswaterstaat, Delft, The Netherlands, 2007)

Audible stop light advance warning system, United States Patent Application 20020070880, Kind Code: A1, http://www.freepatentsonline.com/y2002/0070880.html. Accessed 20 Sept 2009

AVV, *Evaluatie Intelligent SnelheidsAanpassing (ISA): het effect op het rijgedrag in Tilburg* (AVV, Nieuwegein, The Netherlands, 2001a)

AVV, *Eindrapportage praktijkproef Intelligent Snelheidsaanpassing* (AVV, Rotterdam, The Netherlands, 2001b)

E. Bekiaris, S. Nikolaou, M. Panou et al., User needs analysis per category of drivers group. Final Deliverable (Del.1.1) of the AWAKE research project, Brussels, Commission of the European Union-DG Information Society and Media (DG INFSO) (contract no.28062), 2002

Biding T, Lind G (2002) Intelligent Speed Adaptation (ISA); Results of large-scale trials in Borlänge, Lidköping, Lund and Umeå during 1999–2002. Publication 2002:89 E. Swedish National Road Administration SNRA, Borlänge, Sweden

R. Bishop, *Intelligent Vehicle Technology and Trends* (Artech House, Norwood, MA, 2005)

C. Bjørkli, G. Jenssen, T. Moen et al., Adaptive Cruise Control (ACC) and driver performance: effects on objective and subjective measures, in *Solutions for Today... and Tomorrow; Proceedings of the 10th World Congress and Exhibition on Intelligent Transportation Systems and Services ITS*, Madrid, Spain, 16–20 Nov 2003

Bosch, Advanced Driver Assistance Systems – and essential contribution to achieve the European Road Safety Target. Presentation (2005a)

Bosch, Estimated potential for avoiding and mitigating traffic accidents with driver assistance systems and for reducing the economic damage they cause in the Federal Republic of Germany. March 2005 (2005b)

Bosch, Predictive Safety Systems - From Convenience towards Collision Avoidance and Collision Mitigation. Presentation, ADAS, Nivelies, July 2005 (2005c)

A. Bose, P. Ioannou, Evaluating of the environmental effects of intelligent cruise control vehicles. Transport. Res. Rec. **1774**, 90–97 (2001)

K. Brookhuis, D. de Waard, *Limiting speed, towards an intelligent speed adapter (ISA)*, Transportation Research F, **2**(2), 81–90 (1999)

F. Broqua, G. Lerner, V. Mauro et al., Co-operative driving: basic concepts and a first assessment of "Intelligent Cruise Control" strategies, in *Advanced Telematics in Road Transport; Proceedings of the DRIVE Conference*, Brussels, Belgium, 4–6 February 1991, vol. II, pp. 908–929

R.F.T. Brouwer, D.M. Hoedemaeker, Driver support and information systems; Experiments on learning, appropriation and effects on adaptiveness. Final Deliverable (Del.1.2.3) of the AIDE IP research project, Brussels, Commission of the European Union-DG Information Society and Media (DG INFSO) (contract no. 507674), 2006

O.M.J. Carsten, F.N. Tate, Intelligent speed adaptation: accident savings and cost-benefit analysis. Accid. Anal. Prev. **37**(3), 407–416 (2005)

J.-P. Cauzard (ed.), European drivers and road risk. Part1, report on principal results. Project on Social Attitudes to Road Traffic Risk in Europe, SARTRE 3, INRETS, Paris, France, 2004

E. Coelingh, H. Lind, W. Birk, Collision Warning with AutoBrake, in *FISITA World Congress, F2006V130*, Yokohama, Japan, 2006

COWI, *Cost-benefit assessment and prioritisation of vehicle safety technologies (TREN-ECON2-002)*. Brussels, Belgium: European Commission Directorate General Energy and Transport (2006)

Commission staff working document – Accompanying document to the Proposal for a Regulation of the European Parliament and of the Council on the protection of pedestrians and other vulnerable road users – Impact Assessment {COM(2007)560 final SEC(2007)1245} /* SEC/2007/1244 final */

S.L. Comte, New systems: new behaviour? Transp. Res. F **3**(2), 95–111 (2000)

L. Danielsson, H. Lind, E. Bekiaris et al., HMI principles for Lateral Safe applications, in *Universal Access in Human Computer Interaction, Ambient Interaction*. Lecture Notes in Computer Science, vol. 6, part II (Springer, Heidelberg, 2007), pp. 330–338, ISSN 0302-9743

Daytime Running Light. http://ec.europa.eu/enterprise/automotive/pagesbackground/pedestrian protection/index.htm. Accessed 20 Nov 2009

D. de Bruin, J. Kroon, R. van Klaveren et al., Design and test of a cooperative adaptive cruise control system, in *Proceedings of the IEEE International Conference on Systems, Man and Cybernetics*, Parma, Italy, 14–17 June 2004, pp. 392–396

E. de Kievit, F. Hanneman, Setting up an ISA trial from a traffic safety perspective methodology research in the Netherlands, in *Proceedings of the ICTCT Workshop*, Nagoya, Japan, 2002, http://www.ictct.oprg/workshops/02-Nagoya/Kiewit.pdf

ITS Decision, hosted by the Institute of Transportation Studies at the University of California at Berkeley and Caltrans, http://www.calccit.org/itsdecision/serv_and_tech/AVCSS-section-one/lane-departure.html. Accessed 6 Aug 2007

N. Dragutinovic, K.A. Brookhuis, M.P. Hagenzieker, V.A.W.J. Marchau, Behavioural effects of Advanced Cruise Control Use – a meta-analytic approach. EJTIR **5**(4), 267–280 (2005)

ETSC, *In-Car Enforcement Technologies Today* (European Transport Safety Council ETSC, Brussels, Belgium, 2005)

M. Fowkes, Future speed limiting technologies. Advanced Engineering, CILT Forum: 9 Oct 2007, MIRA

C. Gelau, M. Penttinen, I. Spyropoulou, T. Vaa, ITS and effects on road traffic accidents, in *Traffic Technology International Annual 2007 German Luxury Car Makers are Racing to be the First to Introduce Technology That Can Read Speed Limit Signs*, By STEVE COLQUHOUN, 2007, http://www.drive.com.au/Editorial/ArticleDetail.aspx?ArticleID=63948. Accessed 26 June 2009

M. Hoedemaeker, K.A. Brookhuis, Behavioural adaptation to driving with an adaptive cruise control (ACC). Transp. Res. F **1F**(2), 95–106 (1998)

M. Hoedemaeker, *Driving with Intelligent Vehicles; Driving Behaviour with Adaptive Cruise Control and the Acceptance by Individual Drivers*. TRAIL Thesis series T99/6, Delft University Press, Delft, The Netherlands, 1999

A.E. Hoetink, Advanced Cruise Control in the Netherlands; a critical review, in *Solutions for Today... and Tomorrow; Proceedings of the 10th World Congress and Exhibition on Intelligent Transportation Systems and Services ITS*, Madrid, Spain, 16–20 Nov 2003. ERTICO – ITS Europe, Brussels, Paper nr. 4082, Based on SWOV report R-2003-4, 2003.

J.H. Hogema, A.R.A. Horst, W.H. Janssen, A simulator evaluation of different forms of intelligent cruise control. TNO report TNO-TM 1994 C-30, TNO Human Factors Research Institute IZf TM, Soesterberg, The Netherlands, 1994

J.H. Hogema, W.H. Janssen, Effects of intelligent cruise control on driving behaviour: a simulator study. TNO report C-12, TNO Human Factors Research Institute TM, Soesterberg, The Netherlands, 1996

J.H. Hogema, A.M. Rook, Intelligent speed adaptation: the effects of an active gas pedal on driver behaviour and acceptance. On request of the Dutch Ministry of Transport, Public Works and

Water Management, Transport Research Centre AVV and the European Commission, TNO report TM-04-D011, 2004
E. Hollnagel, J.E. Kallhammer, Effects of a night vision enhancement system (NVES) on driving: results from a simulator study, Proceedings of Driving Assessment 2003. Park City, Utah
S. Hoogendoorn, M. Doelman, A. Kesting, S. Ossen, F. Viti, F. Zijderhand, "Full Traffic- WP Dataloggers & WP Verkeersimpact". Delft, April (2007)
http://www.prosper-eu.nl. Accessed 18 Sept 2009
http://www.veiligverkeernederland.nl. Accessed 15 Sept 2009
http://www.wayfinder.com/?id=3764. Accessed 20 Nov 2009
H.M. Jagtman, V.A.W.J. Marchau, T. Heijer, Current knowledge on safety impacts of Collision Avoidance Systems (CAS), In: P.M. Herder, W.A.H. Thissen (eds.) Critical Infrastructures – 5th International Conference on Technology, Policy and Innovation, paper nr 1152, Delft, 2001
S. Jamson et al., Intelligent speed adaptation literature review and scoping study. The University of Leeds and MIRA Ltd, ISA-TfL D1, 2006
H. Katteler, "Acceptance of Lane Departure Warning Assistance Systems". Rotterdam, 2003
H. Lahrmann, J.R. Madsen, T. Boroch, Intelligent Speed Adaptation – development of GPS based System and Field Trial of the System with 24 test Drivers, in *Proceedings of the 8th World Congress on Intelligent Transport Systems*, Sidney, Australia, Sept 30–Oct 4 2001
R. Liu, J. Tate, R. Boddy, Simulation modelling on the network effects of EVSC. Final Deliverable (Del. 11.3) of the External Vehicle Speed Control research project, Institute for Transport Studies, University of Leeds, Leeds, UK, 1999
W.J.R. Louwerse, S.P. Hoogendoorn, ADAS safety impacts on rural and urban highways, in *Intelligent Vehicle Symposium 2004 IEEE*, Parma, Italy, 14–17 June 2004, pp. 887–890
W.J.R. Louwerse, ADAS safety impacts on rural and urban distributor roads, in *Young Researchers Seminar 2005*, TUDelft, 2005
K.M. Malone, B. van Arem, Traffic effects of inter-vehicle communication applications in CarTALK 2000, in *ITS for a Livable Society; Proceedings of the 11th World Congress on ITS*, 18–22 October 2004, Nagoya, Japan (ITS Japan/ITS America, Tokyo/Washington), 2004
K.M. Malone, P. Eijkelenbergh, T. Alkim, G. Bootsma, *Quantified effects of advanced driver assistance systems, the need for field operational trials*. TRB Annual Meeting CD-ROM, 2006
V.A.W.J. Marchau, W.E. Walker, Dealing with uncertainty in implementing advanced driver assistance systems (ADAS): an adaptive approach. Integr. Assess. **4**(1), 35–45 (2003)
M.M. Minderhoud, P.H.L. Bovy, Impact of intelligent cruise control strategies and equipment rate on road capacity, in *Towards the New Horizon Together; Proceedings of the 5th World Congress on Intelligent Transport Systems*, Seoul, Korea, 12–16 Oct 1998, Paper nr. 2145
M.M. Minderhoud, P.H.L. Bovy, Development of a microscopic simulation model for AICC, in *ITS: Smarter, Smoother, Safer, Sooner; Proceedings of 6th World Congress on Intelligent Transport Systems (ITS)*, Toronto, Canada, 8–12 Nov 1999
P. Morsink, Ch. Goldenbeld, N. Dragutinovic et al., Speed support through the intelligent vehicle; perspective, estimated effects and implementation aspects. R-2006-25, SWOV Institute for Road Safety Research, Leidschendam, The Netherlands, 2006
W.G. Najm, M.D. Stearns, H. Howarth et al., Evaluation of an automotive rear-end collision avoidance systems. NHTSA Technical report DOT-VNTSCNHTSA-06-01, 2006
G. Nilsson, The effects of speed limits on traffic accidents in Sweden, in *OECD Proceedings of the International Symposium on the Effects of Speed Limits on Traffic Accidents and Transport Energy Use*, Dublin, Ireland, 1981
L. Nilsson, Safety effects of adaptive cruise controls in critical traffic situations (Reprint from 'Steps Forward'; in *Proceedings of the Second World Congress on Intelligent Transport Systems*, Yokohama, Japan, 9–11 November 1995, pp. 1254–1259). Swedish Road and Transport Research Institute VTI, Linköping, Sweden, 1996
H.-I. Oei, *Veiligheidsconsequenties van Intelligente Snelheidsadaptatie ISA. R-2001-11* (Stichting Wetenschappelijk Onderzoek Verkeersveiligheid SWOV, Leidschendam, The Netherlands, 2001)

M. Paeaetalo, H. Peltola, M. Kallio, Intelligent speed adaptation – effects on driving behaviour, in *Proceedings of Traffic Safety on Three Continents Conference*, Moscow, Russia, 19–21 Sept 2001, pp. 772–783
H. Peltola, Weather related ISA – experience from a simulator, in *Proceedings 7th World Congress on Intelligent Transport Systems*, Turin, Italy, 6–9 Nov 2000, pp. 54–61
M. Pieve, L. Andreone, F. Visintainer, Vehicle-to-Vulnerable roAd user cooperative communication and sensing teCHnologies to imprOVE transpoRt safety. Final Deliverable (Del.6.1) of the WATCH-OVER research project, Brussels, Commission of the European Union-DG Information Society and Media (DG INFSO) (contract no. 027014), 2008
E. Portouli, V. Papakostopoulos, F. Lai et al., Long-term phase test and results. Final Deliverable (Del.1.2.4) of the AIDE IP research project, Brussels, Belgium, Commission of the European Union-DG Information Society and Media (DG INFSO) (contract no. 507674), 2006
M. Räsänen, Effects of a rumble strip barrier line on lane keeping in a curve. Accid. Anal. Prev. **37**, 575–581 (2005)
M. Regan, T. Triggs, K. Young, N. Tomasevic et al., On-road evaluation of intelligent speed adaptation, following distance warning and seatbelt reminder systems. Final results of the TAC SafeCar project, Monash University Accident research Centre, 2006
Report No. FMCSA-MCRR-05-005; Federal Motor Carrier Safety Administration, US Department of Transportation, http://www.fmcsa.dot.gov/facts-research/research-technology/report/lane-departure-warning-systems.htm. Accessed 15 Oct 2009
R.A. Retting, S.A. Ferguson, A.S. Hakkert, Effects of red light cameras on violations and crashes: a review of the international literature. Traffic Inj. Prev. **4**, 17–21 (2003)
R.A. Retting, S.A. Ferguson, C.M. Farmer, Reducing red light running through longer yellow signal timing and red light camera enforcement. Results of a field investigation, The Insurance Institute for Highway Safety, 2007
J. Krems, D. Rösler, S. Mahlke, Thüring, Evaluation of night vision enhancement systems: Driver needs and acceptance. *FAS2006, 4. Workshop Fahrerassistenzsysteme*, Löwenstein (2006)
K. Rumar, Night vision enhancement systems: What should they do and what more do we need to know? Report No. UMTRI-2002-12, The University of Michigan Transportation Research Institute, Ann Arbor, MI, 2002
SCANIA, Vulnerable road user detection system (varningssystem för oskyddade trafikanter). Reference: PU35, 2009
D. Shinar, E. Schechtman, Headway feedback improves intervehicular distance: a field study. Hum. Factors **44**, 474–481 (2002)
SPAL web site, http://www.spal.it. Accessed 20 Nov 2009
SWOV Fact sheet, Advanced Cruise Control (ACC), Leidschendam, The Netherlands, 2008
SWOV Fact sheet, Intelligent Speed Assistance (ISA), Leidschendam, The Netherlands, 2007
S. Talmadge, D. Dixon, R.S. Riney, *Development of performance specifications for collisions avoidance systems for lane change crashes (DOT HS 809 414)*. Washington, DC: US Department of Transportation, National Highway Traffic Safety Administration (1997)
Technical Information For The Collision Industry (2006), Night Vision Enhancement Systems, I-CAR Advantage Online TM, May 15, 2006, http://www.i-car.com
J. Törnros, L. Nilsson, J. Östlund et al., Effects of ACC on driver behaviour, workload and acceptance in relation to minimum time headway, in *ITS – Enriching Our Lives; Proceedings of the 9th World Congress on Intelligent Transportation Systems ITS*, Chicago, IL, 14–17 Oct 2002 (ITS America, Washington, DC)
University of Twente, Testrit met Citroen Testrit_Citroen_C5_met_LDWS, 2007, http://www.aida.utwente.nl/actueel/archief/Testrit_Citroen_C5_met_LDWS.doc/. Accessed 13 Nov 2007
B. van Arem, J. Hogema, S. Smulders, The impact of autonomous intelligent cruise control on traffic flow, in *Proceedings of the 3rd World Congress on ITS*, Orlando, FL, Paper nr. 2032, 1996
C.N. van Nes, I.N.L.G. van Schagen, M. Houtenbos et al., *De bijdrage van geloofwaardige limieten en ISA aan snelheidsbeheersing* (Stichting Wetenschappelijk Onderzoek Verkeersveiligheid SWOV, Leidschendam, The Netherlands, 2007)

J. Vanderwerf, S. Shladover, M. Miller et al., Effects of adaptive cruise control systems on highway traffic flow capacity, in *Intelligent transportation systems and vehicle-highway automation 2002; Transportation Research Record TRR 1800*, 2002, pp. 78–84

A. Várhelyi, *Dynamic Speed Adaptation Based on Information Technology; a Theoretical Background* (Lund Institute of Technology, Lund, Sweden, 1996)

A. Várhelyi, M. Hjälmdahl, C. Hydén et al., Effects of an active accelerator pedal on driver behaviour and traffic safety after long-term use in urban areas. Accid. Anal. Prev. **36**(5), 729–737 (2004)

A. Várhelyi, T. Mäkinen, Evaluation of in-car speed limiters: Field study, 1998

C. Visvikis, T.L. Smith, M. Pitcher, Study on lane departure warning and lane change assistant systems. PPR 374, TRL, 2008

S. Vlassenroot, S. Broekx, J. de Mol et al., Driving with intelligent speed adaptation: final results of the Belgian ISA-trial. Transport. Res. A **41**(3), 267–279 (2007)

WATCH-OVER research project web site, Brussels, Commission of the European Union-DG Information Society and Media (DG INFSO) (contract no. 027014), http://www.watchover-eu.org/. Accessed 15 Nov 2009

M. Weinberg, H. Winner, H. Bubb, Adaptive cruise control field operational test – the learning phase. JSAE Rev. **22**, 487–494 (2001)

What is speed alert? http://www.speedalert.com.au/. Accessed 20 Nov 2009

M. Wiethoff, *Technologieën voor snelheidsbeheersing. R-2003-12* (Stichting Wetenschappelijk Onderzoek Verkeersveiligheid SWOVC, Leidschendam, The Netherlands, 2003)

M. Wiethoff, Final Technical Report ADVISORS. GRD1 2000–10047. The European Community under the 'Competitive and Sustainable Growth' Programme, SWOV, The Hague, 108, 2003b

G.J.S. Wilde, *Target Risk 2: A New Psychology of Safety and Health*, 2001, PDE Publications, ISBN: 978-0-9699124-3-9, ISBN-10:0-9699124-3-9

Chapter 5
Drawing the Picture. Approach to Optimize Messages on Roads by Design

Stefan Egger

5.1 The Carriageway Narrows

The agreement on principles stated in the "Vienna Convention on Road Signs and Signals," done at Vienna on 8 November 1968, through which every contracting country declares to employ unified signs and signals on its territory, represents the first globally targeted attempt for the harmonization of traffic-related messages. Although considered a huge step at that time, the overall situation on roads has changed significantly during the last 40 years, as we are faced today with larger and internationally connected motorway networks, allowing for higher speeds, and matching vehicles able to take full advantage of the possibilities given. At the same time, traffic density is much higher than ever before. The interrelation of both aspects yields a potentially higher risk of accidents.

5.1.1 More Vehicles on Roads

For example, documentation provided on the online representation of the Austrian Chamber of Commerce (Statistik Austria 1997) shows that Traffic on roads was low in the 1960 compared to today; in Austria a mere 926,945 vehicles in 1960 against 5,796,973 in 2007, constituting only 16% of current vehicle count.

5.1.2 Larger, Interconnected Motorway Networks

Accordingly, the network of motorways expanded likewise, exemplified here by a comparison of the Swiss motorway network of 1960 and 2000 (Fig. 5.1) (Institut für Verkehrsplanung und Transportsysteme 2000) bold red lines represent motorways.

S. Egger
International Institute for Information Design (IIID), Vienna, Austria
e-mail: stefan.egger@iiid.net

E. Bekiaris et al. (eds.), *Infrastructure and Safety in a Collaborative World*,
DOI 10.1007/978-3-642-18372-0_5, © Springer-Verlag Berlin Heidelberg 2011

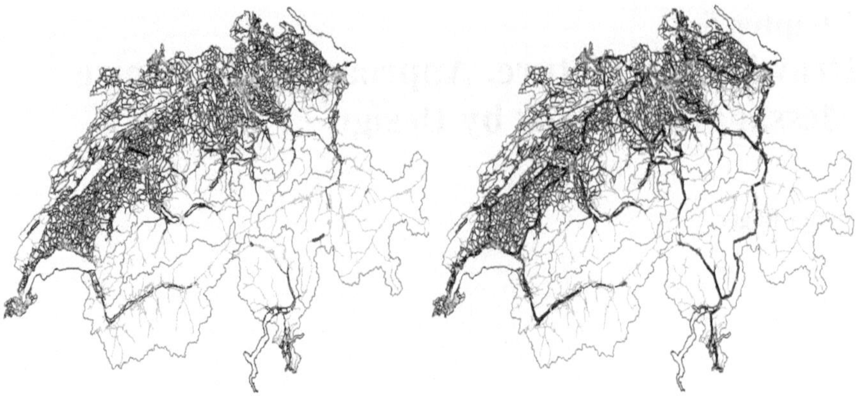

Fig. 5.1 Swiss motorway network of 1960 and 2000

Fig. 5.2 Three European scripts (Egger 2009)

Latin
Ελληνικά (Greek)
По-русски (Cyrillic)

5.1.3 Qualitative Development of Cars and Roads

The technical and – more subjcctively – tlie qualitative improvement of the network and its populating vehicles cater for the ability and willingness of individuals (not taking into account professionals, such as truck drivers) to drive ever-wider distances. This surfaces further factors, giving reason to rightfully question the validity of message signalisation principles currently in force.

5.1.4 Multi-Linguality

Considering the 23 official EU-languages, of which a driver is in command typically of one or two, long distance travellers are bound to encounter on their journey several unfamiliar languages in the form of text information. Different scripts (Latin, Greek or Cyrillic), very often do not even allow for "guessing" the meaning of a text-message. Even simplest content, like destination names, can prove impossible to be deciphered in time (Fig. 5.2).

The Vienna Convention states on multi-linguality (Economic Commission for Europe, Inland Transport Committee 1968/1995):

> The inscriptions referred to in paragraphs 3 and 4 of this Article shall be in the national language or in one or more of the national languages, and also, if the Contracting Party

concerned considers it advisable, in other languages, in particular official languages of the United Nations.

Conflicts arise here as Jamson et al. (2001) showed that, if a text is longer (" ... four lines ... "), due to a message being repeated in another language, driving behaviour is affected, prompting a reduced mean speed.

5.2 Disharmony of (Inter)national Road Signage Systems

5.2.1 Symbols

The Vienna Convention provides a large variety of road signs (symbols), of which the national interpretation of these symbols' appearance becomes part of the legislation of a signing country, prompting road signs to look tremendously different if compared country-wise. Quite regularly, the signs' depictions even differentiate from province to province within one country.

Presumably, the reason for this can be tracked back to (traditional) reproduction procedure. The Vienna Convention's annex depicting all signs (Annex 1: Road signs) gives indications on the visual appearance of symbols, but proves only good enough for giving an overview to the reader of the document. It does not provide samples to cater for precise dimensions, position or form of every graphical detail (crucial for exact reproduction), which is essential for the coherent appearance of symbols, if actually deployed on roads. Sample specifications are elaborated by each country individually, to create the needed prerequisites for actual application (reproduction), resulting in a national interpretation of the Vienna Convention's original symbols. Rarely countries share sample specifications for harmonization reasons.

One step further in the chain of production, hired companies fabricate road signs, according to the national sample specifications. Depending on the employed method (manual painting, silkscreen prints, foil laminated ...) and its inherent possibilities for precision, the resemblance of a finished road sign to the national sample specification and the original, as seen in the Vienna Convention, further decreases.

Key to a better approach towards a common appearance of signs should be the distribution of carefully crafted digital template files, while knowing that as soon as the sample specifications have legal status, a change of national laws concerned is needed, to allow for altered symbols. Here it needs to be emphasized that, if a country decides to take national effort in harmonizing its symbols by deploying digital templates to producers, a review of the samples specified by legislation is required to ascertain that requirements of today's traffic-improved long-distance image comprehension and legibility, as described in IN-SAFETY (Simlinger et al. 2008) – are met.

Examples for the diversity in appearance of the Vienna Convention symbols in several European countries (Egger 2005) follow in the figure below (Fig. 5.3).

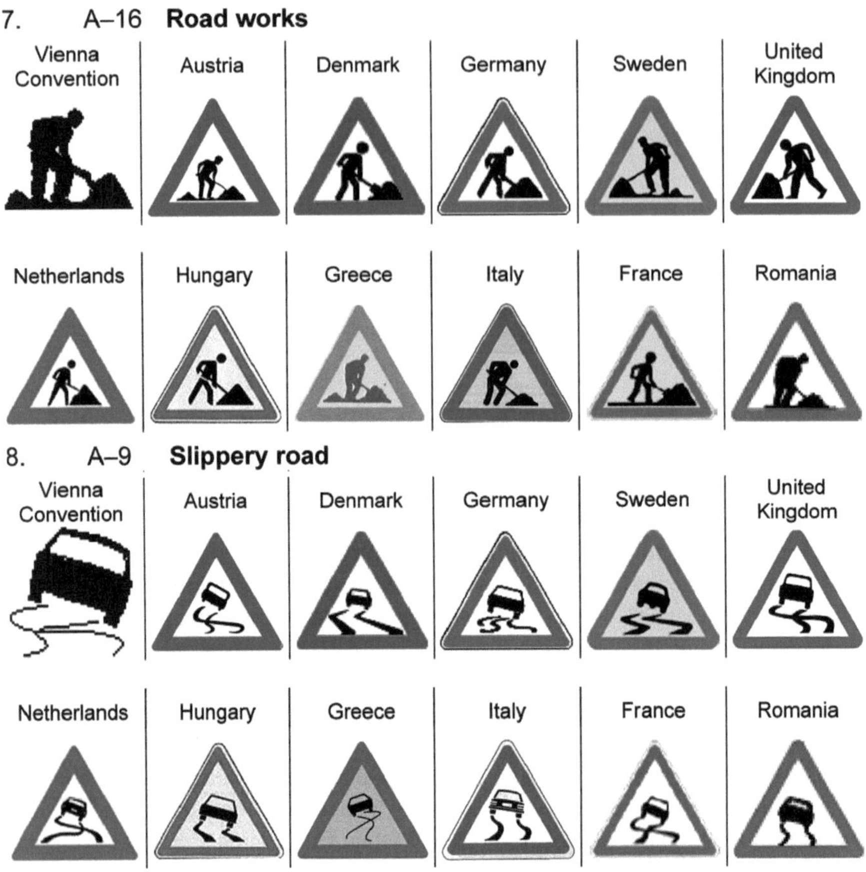

Fig. 5.3 Diversity of appearances of traffic symbols in several EU countries

Taking into account the difficulties involved in changing national – and even more so – international regulations to let improved designs of symbols enter the realm of road traffic, efforts for implementation should be focused on specific application fields, where improvement is needed most urgently, and where symptoms of traffic density, high speeds and international traffic collide: The motorways of the Trans European Road Network (TERN). Nevertheless, it is advisable to negotiate on the establishment of principles to let improvements have an effect on international traffic regulations in a timely manner.

5.2.2 *Typeface*

Focusing once more on language related issues, not only scripts governed by languages differ, but, just like the previously shown traffic symbols, typefaces

Fig. 5.4 Comparison of influential road traffic typefaces: RWS (The Netherlands), DIN Mittelschrift (Germany), Transport (UK) (Egger, 2008)

differ in terms of shape and appearance from country to country. The Vienna Convention does not prescribe specific typefaces to be used on roads – after all, requirements of scripts, depending on languages concerned, vary. Therefore, countries usually rely on their own typeface-developments, which are often considered as part of the (national) tradition and cultural heritage (Fig. 5.4).

Though, being forced to adjust to a different typeface after crossing a national border into a neighbouring country can increase the amount of time needed for a driver to read presented information, as a change of the appearance of text is bound to have an obstructive impact on legibility. This becomes imminent in potentially dangerous traffic situations, where attention strays away too long from traffic observation, in order to read messages. This rationale leads to the recommendation to internationally introduce one adequate typeface to cater for all the script requirements of every EU-language involved, while fulfilling qualitative prerequisites of early discrimination/legibility to support timely understanding and reaction, as laid out later on in Chap. 13.

5.3 VMS and Conventional Signage Harmonization

The Vienna Convention's statements related to variable message signs (VMS) are kept general (Economic Commission for Europe, Inland Transport Committee 1968/1995):

> 2. Nothing in this Convention shall prohibit the use, for conveying information, warnings or rules applying only at certain times or on certain days, of signs which are visible only when the information they convey is relevant.

In a way, motorways represent one of the spearheads of traffic development, considering infrastructural measures set to improve safety on the road. Among these, technology employing light emitting diodes (LED) used on gantries in VMS is especially important due to its capability of displaying critical on-the-spot/on-time information relating to e.g. danger warning, or current motorway status. In the observation of VMS development it becomes obvious that along with the refinement of the technological aspects, attempts are made to signal new messages which are not part of the Vienna Conventions Annex 1's set of messages, such as "dedicated lane for emergency vehicles", "danger warning- fog", or "danger warning- wrong way driver". For such yet unregulated messages, harmonization proves reachable, as agreement on national level suffices to be introduced country-wide. Countries participating in European Study 4 (ES-4, formerly "Mare Nostrum" (2006)), which is part of the

European-wide Easy Way project make use of this routine to reach international VMS harmonization, focused on the main corridors of the TERN

In addition, in the Vienna Convention, (page 9, article 8), it reads:

> 2. Contracting Parties wishing to adopt, in accordance with Article 3, paragraph 1 (a) (ii), of this Convention, any sign or symbol not prescribed in this Convention shall endeavour to secure regional agreement on such new sign or symbol.

Still, in this respect, harmonization of messages is only concerned with unification of signalisation among VMS and its current-practice technology. Until recently, no initiatives were involved in the harmonization of variable messages and conventional signage. The need for this is obvious: formal differences between messages on "static" and "dynamic" sign boards should be kept to a minimum, to make readjustment obsolete to a driver, to again help him/her keep the focus on traffic. As a side effect, e.g. harmonized typeface blends effortlessly into "mixed" signboards, where only inlayed partitions of VMS "windows" convey the variable part of the information.

Harmonising both VMS (employing coarse imagery due to the current low resolution display possibilities) and conventional signage (of high resolution), bears the unique chance to cater for the full range of VMS with increasingly higher resolutions emerging on current and future markets. In addition, these messages can be justly transferred to in-car displays in the same way, paving the road to a total harmonisation across all technological channels, which provide road messages and related information.

As VMS display information when needed, they evidently have the potential to lower the risk of information overflow posed to drivers by a high density of signposts. In addition, drivers consider messages signalled on VMS reliable, and react favourably upon. The ongoing improvement of the technological aspects, combined with declining prices, but also low energy consumption make it a candidate to become the future tool for signalisation. Therefore, work needs to be conducted to further develop the methodology for the display of combined messages on (emerging) freely programmable, "full matrix"-VMS (Fig 5.5), based on the proposed procedure vented in IN-SAFETY (Simlinger et al. 2008).

In particular, this consists of a structure for content arrangement, which allows for the display of combined messages in a systematic way, taking into account identified factors, such as prerequisites for message discrimination and comprehension, as well as the amount of information represented. In the latter topic, and in the technical requirements to host and use the proposed VMS content structure, efforts in terms of additional research are needed.

5.4 Requirements Towards Further Harmonisation

To summarize, following the previous rationales, (inter-)national harmonization (in its various aspects) of traffic related messages, it is duly important for maintaining and improving safety on the road. This should focus upon:

Fig. 5.5 Full matrix VMS (DRIP-) test bed in The Netherlands, performing a comparison of prospect VMS-typefaces (Remeijn 2008)

1. Prerequisites to secure early discrimination of messages at high speeds, allowing for prolonged viewing time,
2. The available methods to be employed to ascertain that a message's (symbol's) comprehensibility is sufficient to be used on roads,
3. Harmonisation must not be attempted without (re-)adjusting/(re-)developing messages to meet the requirements for discrimination and comprehension beforehand.

5.4.1 Requirements for Discrimination

A gap becomes obvious between the visual requirements for applicants for a driving license, and the actual sizes of displayed information. More specifically, Annex III of Council Directive 91/439/EEC of 29 July 1991 on driving licences requires:

> Group 1 (drivers of vehicles of categories A, B and B + E and subcategory A1 and B1): (6.1.) Applicants for a driving licence or for the renewal of such a licence shall have a binocular visual acuity, with corrective lenses if necessary, of at least 0.5 when using both eyes together.
>
> Group 2 (drivers of vehicles of categories C, C + E, D, D + E and of subcategory C1, C1 + E, D1 and D1 + E): (6.3.) Applicants for a driving licence or for the renewal of such a

licence must have a visual acuity, with corrective lenses if necessary, of at least 0.8 in the better eye and at least 0.5 in the worse eye.

Transferring this information into figures to determine the size of one minute of arc (MOA) needed for 0.5 visual acuity and a viewing time of 3.33 s (Road Standards Division (no year)), the size of information would require existing sign dimensions (as currently in practice) to be doubled. In other words, today drivers bearing visual acuity of 0.5 or less do not have the required time of 3.33 s to discriminate/understand/react, if they move at a speed close to the upper speed limit of motorways.

The smallest graphical detail (SGD) is the minimum dimension for any detail of a presented message (for instance, the strokewith of letters or linewidth of pictograms, the space between strokes or lines, etc.). By means of design, carefully adhering to the principle of the SGD (Egger 2009), which directly relates to 1 MOA, any road message can be improved towards better (earlier) discrimination for the benefit of drivers, providing a prolonged viewing time. If adhered to this, all details of a message are equally discriminable (Fig. 5.6).

The actual dimension of the SGD must be the same as MOA, which is governed by required viewing distance, which again depends on the viewing time needed to comprehend a presented message at a specific speed of travel, and the visual acuity of a driver. By adopting visual acuity of 0.5 (see Annex III of Council Directive 91/439/EEC of 29 July 1991) for this purpose, messages would cater for drivers with very weak eyesight, but prompt the VMS-panels to exceed the width of motorways. So, in order to achieve a workable dimension of the SGD, a higher visual acuity of 0.73 is proposed for speeds up to 100 km/h.

It should be noted that the Vienna Convention does not regulate the graphical appearance of its symbols in such detail, but the mere pictorial content (Vienna Convention):

Fig. 5.6 To the *left* – a current practice Vienna Convention symbol (F, 4), and to the *right* an equivalent symbol as proposed by ISO 7001 (PI CF 009), but rendered in conformity with the requirements for improved discrimination (Egger 2007)

> 1. In order to facilitate international understanding of signs, the system of signs and signals prescribed in this Convention is based on the use of shapes, and colours characteristic of each class of sign and, wherever possible, on the use of graphic symbols rather than inscriptions. Where Contracting Parties consider it necessary to modify the symbols prescribed, the modifications made shall not alter their essential characteristics.

Thus, discussions with the United Nations unit controlling the development of the Vienna Convention are urgently needed, in order to revise the depictions of traffic symbols according to the requirements of enhanced discrimination.

5.4.2 Requirements for Comprehension

As for the comprehension of messages – the International Organization for Standardization (ISO) provides through "ISO 9186:2001 Graphical symbols – Test methods for judged comprehensibility and for comprehension" valuable tools to verify the ease/difficulty of the comprehension of a symbol. As the methods were employed in IN-SAFETY, designers gained the ability to create and improve symbols, achieving highest possible comprehension scores, to secure fast and accurate understanding of an implied message. The procedure of ISO 9186:2001 can be summarized as follows: Examples for pictograms to represent a specific message are to be collected/designed. These examples undergo the Comprehensibility Judgement Test to come up with the most promising to be used in the following Comprehension Test. In the latter, the participants are presented a pictogram and are asked to state its meaning. From these answers, comprehension scores can be calculated, and insights gained for designers to improve pictograms to yield better scores. Optimally, the Comprehension Test is repeated after redesign.

Both described tests are to be carried out in at least three countries bearing distinctively different cultural heritage.

The procedures explained in Sects. 4.5.1 and 4.5.2 have been applied within IN-SAFETY and the relevant result are presented in Chaps. 13 and 14 of this book, respectively.

References

Statistik Austria, *Kfz-Bestand nach Bundesländern* (Wien/Vienna: Wirtschaftskammer Österreich, 1997)

Institut für Verkehrsplanung und Transportsysteme, *Poster Übersicht 1950–2000* (ETH, Zürich, 2000)

S.L. Jamson, F.N. Tate, A.H. Jamson, Evaluation of bilingual traffic signs. Leeds: University of Leeds (2001)

P. Simlinger, S. Egger, C. Galinski, *Proposal on Unified Pictograms, Keywords, Bilingual Verbal Messages and Typefaces for VMS in the TERN* (IIID, Wien/Vienna, 2008)

S. Egger, *Vienna Convention Signs for VMS* (IIID, Wien/Vienna, 2005)

S. Egger, (2007) Wien: IIID
S. Egger, (2008) Wien: IIID
Economic Commission for Europe, Inland Transport Committee, *Convention on Road Signs and Signals, Done at Vienna on 8 November 1968* (United Nations, 1968/1995)
Mare Nostrum (VMS) project/ES-4, Mare Nostrum: towards a European VMS contents harmonisation (Colmenar Impresores S.L., 2006)
Remeijn, H. (2008) Delft: Rijkswaterstaat
Road Standards Division, *Danish Technical Handbook for VMS* (Danish Road Directorate, København, no year)
S. Egger, *Legibility Criteria, the Smallest Graphical Detail and What it Means for Typeface and Pictogram Design* (IIID, Wien/Vienna, 2009)
ISO/TC145/SC1 (2006) ISO 7001. Graphical symbols—Public information symbols. Geneva: ISO
Council (1991) Council Directive 91/439/EEC of 29 July 1991 on driving licences
ISO, International Standardization Organization, *ISO 9186, Graphical Symbols – Test Methods for Judged Comprehensibility and for Comprehension* (ISO, Geneva, 2001)

Part II
New Developments in Modelling, Evaluating and Training

Chapter 6
Models on the Road

Thomas Benz, Evangelia Gaitanidou, Andreas Tapani, Silvana Toffolo, George Yannis, and Ioanna Spyropoulou

6.1 Traffic Simulation Modelling and Safety Aspects

Modelling has become a major part of all aspects in traffic engineering within the last decades. The models applied range from macroscopic models, treating network related facets of traffic, to microscopic models, which represent traffic flow by moving individual vehicles. The safety aspects can be integrated at several levels of modelling, targeting different parts of the driver behaviour.

The network effects of safety are typically handled by macroscopic models. They represent the supply, i.e. the road/street network, and the demand, i.e. the trips of people, and match both to create the traffic load on the network links. The basic idea of integrating safety in the microscopic models is to influence the routing behaviour of drivers in such a way, that either they observe the given safety levels on the links of the network, or to influence their route choice towards minimizing the (negative) safety effects of their trips. This has been a novelty introduced within IN-SAFETY project, which has gained significant momentum since.

The macroscopic approach adds safety as an additional parameter in the routing/ assignment. The information for the safety level comes either from actual traffic data for a specific network, or derives from known safety indicators for road/street

T. Benz (✉)
PTV AG, Karlsruhe, Germany
e-mail: Thomas.Benz@ptv.de

E. Gaitanidou
Centre for Research and Technology Hellas, Hellenic Institute of Transport (CERTH/HIT), Thessaloniki, Greece

A. Tapani
Swedish National Road and Transport Research Institute (VTI), Linköping, Sweden

S. Toffolo
IVECO, ER&C, Torino, Italy

G. Yannis and I. Spyropoulou
National Technical University of Athens (NTUA), Athens, Greece

E. Bekiaris et al. (eds.), *Infrastructure and Safety in a Collaborative World*,
DOI 10.1007/978-3-642-18372-0_6, © Springer-Verlag Berlin Heidelberg 2011

types. The algorithmic extension consists mostly of integrating the supplementary data into the objective function. This approach is, of course, relevant only to the drivers, not to other road users.

The microscopic approach has a similar basis, but tends to optimize individual trips on a network, in such a way that the overall safety is maximized. In this nature, the safety of other road users, e.g. pedestrians in a traffic calming zone, is implicitly included. Of course, the safety optimal routes do not necessarily lead to an optimal travel time distribution and may, thus, be different from a system optimum in the traditional sense.

Microscopic models are applied for all aspects that directly influence the task of driving a vehicle. Generally speaking, driving a vehicle, in this context, means the driver's control task in lateral and longitudinal direction. This task can be assisted by new Advanced Driver Assistance Systems (ADAS) or In-Vehicle Information Systems (IVIS). They provide warning or even take over part of the driving task continuously, like Adaptive Cruise Control (ACC), or only temporarily, like Collision Avoidance, under specific conditions. Such systems lead to changes in the trajectory of the vehicle and, thus, they may also lead to changes in the overall traffic flow. The analysis of the trajectories in various ways reveals changes in traffic flow as a whole, e.g., changes in the speed-flow-relationship, and also on safety relevant parameters, like time-to-collision (TTC) and its derivatives. Other indicators, like the shape of the headway distribution, can also be used for the estimation of safety consequences.

In the following sections, first an overview over the models applied within the IN-SAFETY project is provided, followed by the description of sample applications, which show the potential of the models for safety analyses. Then, possible extensions of the models which would improve them for safety indications are shown and, finally, an outlook onto the future of model applications for safety analyses is given.

6.2 Microscopic Models

Microscopic Models create the traffic flow from the movements of individual vehicles. Their difference lies in the way these movements are generated. While there exist a wide variety of approaches, the focus here is on such models that apply a rather detailed model for the driver-vehicle-environment interaction. In the context of "safety" these interactions are of major interest. It must be noted, however, that other approaches exist, with much less detailed movement description (e.g. agent-based models). Such approaches save run-time for the sake of complexity; they can be seen as a bridge between microscopic and macroscopic models. While they are able to handle larger scenarios at reasonable resource consumption (concerning time and memory), their results cannot be used for detailed analyses, like for safety or emission calculation purposes. The models we describe here are not only widely used in the traffic engineering community; they constitute the current state-of-the-art in microscopic modelling and incorporate a development background of many years.

6.2.1 RutSim

The Rural Traffic Simulator, RuTSim, (Tapani 2005) is a traffic simulation model developed for rural road environments. The model handles all common types of rural roads, including two-lane roads and roads with separated oncoming lanes.

Rural roads place different requirements on the simulation model than urban or highway networks. This difference is due to the fundamental differences in the interactions between vehicles and the infrastructure. The travel time delay in an urban or freeway network is dominated by vehicle-to-vehicle interactions, whereas the travel time delay on a rural road is also significantly affected by interactions between vehicles and the infrastructure. For example, speed adaptation with respect to the road geometry has a more prominent role on rural roads than it has on urban streets. A model describing traffic flows on rural roads must, therefore, consider the interaction between vehicles and the infrastructure in greater detail than models describing traffic flows in urban areas or on highways. Interactions between vehicles are nevertheless important on rural roads, particularly in overtaking situations. For modelling of two-lane roads, it is for example necessary to consider interactions between the oncoming traffic streams.

A rural road traffic simulation model was developed at VTI during the 1970s. This model has been continuously improved during the following decade. These model improvements also included large calibration and validation efforts. However, the original model applied simple rules for updating vehicle positions and speeds and was limited to simulation of uninterrupted flow on two-lane roads. RuTSim was developed based on this original model, to allow modelling of interrupted flows and new types of rural roads.

RuTSim is a micro-simulation model that consists of sub-models that handle specific tasks. The use of sub-models simplifies the future modification of RuTSim and increases its flexibility. The model is designed to handle one road stretch in each simulation run; i.e. rural road networks are not considered. The main road may incorporate intersections and roundabouts, and the traffic on the main road may be interrupted by vehicles entering and leaving the road at intersections located along the simulated stretch. The traffic flows entering the road at various origins may be time dependent. Turn percentages at intersections for each traffic flow are used to determine vehicle destinations.

RuTSim uses a time-based scanning simulation approach. The simulation clock is advanced with a user-defined step size, e.g., 0.1 s. The time-based simulation approach is chosen for RuTSim, because it allows more detailed modelling of an individual vehicle's interactions with the surrounding traffic and the infrastructure. With a shorter time step, the movement of vehicles from one time step to the next becomes smoother and therefore more realistic. Hence, a shorter time step may, given an adequate modelling logic, result in the driving course of events for an individual vehicle to be closer to the driving course of events found in real traffic. The use of a shorter time step does, however, increase the model run time. The model time step should therefore be chosen in relation to the current application. Outputs, in the form of aggregated traffic measures, do not require a time step as short as the one required if a driving course of events for a representative vehicle is desired.

The following steps are performed in every time step during a model run:

1. Add vehicles that are to enter the road during the time step to virtual queues, with one queue for each origin.
2. Load vehicles from the virtual queues to the road, if possible, i.e. if acceptable space is available on the main road.
3. Update the speed and the position for every vehicle on the road.
4. Remove vehicles that have arrived at their destination.
5. Update the state, i.e. free or car following, overtaking or passed, and acceleration rate, for every vehicle on the road.
6. Save the data.
7. If animation is enabled, update the graphical user interface (GUI).
8. If the stop time has been reached, terminate the simulation or else increment the simulation clock and go back to Step 1.

Before the simulation, the speed profile of the road and the traffic that is to enter the road are generated from the input road and traffic data, respectively.

The current version of RuTSim applies a car-following based on the "Intelligent driver model" (Treiber et al. 2000, 2006). This model accounts for driver limitations and anticipation to allow more detailed studies of traffic impacts of driver assistance systems. Details of this current car-following methodology applied in RuTSim are out of the scope of this chapter. Overtaking decisions on two-lane roads are controlled by a stochastic model depending on the current road characteristics and the distance to the oncoming vehicle.

Previous applications of the RuTSim model include quality-of-service studies of alternative rural road designs (Carlsson and Tapani 2006). RuTSim has also been utilized in a study of possibilities to conduct safety evaluations of driver assistance systems using traffic simulation (Lundgren and Tapani 2006).

6.2.2 S-Paramics

The application of S-Paramics focussed on the safety effects of route choice in a road network. Therefore, this description refers to the route choice characteristics of S-Paramics.

In S-Paramics, each vehicle tries to find the shortest route from the road section on which it is located, to its destination zone. The shortest route is the one for which the general journey costs are lowest. Each time a vehicle enters a new road section, the route is evaluated again, on the basis of the general journey costs that are 'stored' in route tables.

The road hierarchy in a network can be used to change the journey costs on special road sections for familiar and unfamiliar vehicles. The road hierarchy in a network is made up of major and minor road sections.

Major road sections are equipped with signs; the journey costs of familiar and unfamiliar vehicles are the same.

There are no signs on minor road sections and the familiar vehicles view the journey costs on minor road sections as being the same as the actual costs. Unfamiliar vehicles have a lower consciousness of minor road sections; they view the journey costs on these road sections as being twice the actual costs. These 'penalty costs' make it less likely that these unfamiliar vehicles will choose routes along minor road sections and they will therefore tend to stay on the signed road sections (i.e. the major road sections).

Familiarity with the road network has a fundamental influence on route choice in a hierarchical road network. If this directly influences the quantity of routes passing along routes with and without signs, it is important to properly calibrate the level of familiarity.

The standard familiarity value for all vehicles is 85%. This means that 85% of the vehicles make no distinction between the costs of major and minor road sections. The other 15%, the unfamiliar vehicles, view the costs on minor road sections as higher and will be more inclined to travel along major road sections.

The level of familiarity can be set separately for each vehicle type. For example, if a model includes taxis, it would be quite possible to set the familiarity at 100%, because taxi drivers usually know the road network well.

The general journey costs and the road category can be set for each individual road section.

The journey costs of an individual road section can be calculated using the general cost comparison (referred to hereinafter by its Dutch abbreviation, GVK). This represents a combination of factors that drivers take into consideration when choosing between various routes. The most important factors are time and distance. If a toll is charged for using certain parts of a road, these costs will also be taken into account.

The general journey costs GK of a road section are measured in time, distance and (if imposed) toll charges and can be weighted by means of coefficients, depending on the road category and the familiarity of the road users with the road network.

The general journey costs GK of a road section can be set to the same (generic) value for all vehicles, or they can be set by vehicle type.

In addition to calculating the general journey costs of an individual road section as described above, it is also possible to calculate the general journey costs for a road category. This determines the general journey costs for all road sections that fall into a certain road category. This is done in precisely the same way as described above.

If an individual road section falls into a category for which the general journey costs are 2 and, furthermore, it is allocated a specific value of 3 that applies only to this road section, then the final general journey costs are 6 (GK of the category multiplied by GK of the individual road section).

The route tables are filled in using the general journey costs of the road sections. The route costs are equal to the sum of the general journey costs of the road sections that form part of the route. Route tables give vehicles the opportunity to calculate the costs of a route choice at each junction along the route. When a vehicle approaches a junction, it consults the relevant route table and, after deciding whether to apply perturbation and/or dynamic feedback, the vehicle selects the route that has the lowest journey costs to the destination.

As standard, there are two route tables in a model in S-Paramics: one table contains the costs for vehicles that are familiar with the road network (familiar vehicles) and the other table contains the costs for vehicles that are unfamiliar with the road network (unfamiliar vehicles). Familiar vehicles have a different perception to unfamiliar vehicles of a route through the network. This is achieved by making use of a road hierarchy in the network and by calibrating familiarity.

In addition, a separate route table can be created for each type of vehicle, thereby producing a set of route tables. Each route table is calculated each time that a simulation is started.

The following allocation methods are possible in S-Paramics:

- All-or-nothing allocation
- Stochastic allocation
- Dynamic allocation
- Stochastic Dynamic allocation

6.2.3 VISSIM

VISSIM is a commercial micro-simulator that has been developed over the last two decades. It is based on a very detailed driver-vehicle model developed in the mid-1970s. Basically designed to re-create traffic flows on carriageways, like on motorways or urban arterials, it has recently been enhanced to integrate non-lane-bound vehicles, like two-wheelers and even pedestrians. In the context of safety applications, we focus here on the safety applications of Intelligent Transport Systems, namely Advanced Driver Assistance Systems, which pertain to passenger cars and trucks. The following description of VISSIM therefore concentrates only on the issues that are important for such applications.

The quality of the traffic flow model properties constitutes a major concern of their users: The traffic flow model used by VISSIM is a discrete, stochastic, time step based (1 s) microscopic model, with driver-vehicle-units (DVU) as single entities. The model contains a psycho-physical car following model for longitudinal vehicle movement and a rule-based algorithm for lateral movements (lane changing). The model is based on the continuous work of Wiedemann (1974, 1991).

Vehicles follow each other in an oscillating process. As a faster vehicle approaches a slower vehicle on a single lane it has to decelerate. The action point of conscious reaction depends on the speed difference, distance and driver dependent behaviour. On multi-lane links moved up vehicles check whether their speed improves by changing lanes. If so, they check the possibility of finding acceptable gaps on neighbouring lanes. Car following and lane changing together form the traffic flow model, being the kernel of VISSIM.

Figure 6.1 indicates the oscillating process of this approach. The thresholds of Fig. 6.1 are explained in an abbreviated form. Driver specific perception abilities and individual risk behaviour are modelled by adding random values to each of the parameters as shown for AX. For a complete listing of the random values the reader is referred to Wiedemann and Reiter (1992).

AX: Desired distance between the fronts of two successive vehicles in a standing queue.
$AX = VehL + MinGap + RND1 \cdot AXMult$ with RND1 normally distributed N(0.5, 0.15).

ABX: Desired minimum following distance, which is a function of AX, a safety delta distance BX and the speed v.
$ABX = AX + BX \cdot \mathrm{v}v$.

SDV: Action point where a driver consciously observes that he/she approaches a slower car in front. SDV increases with increasing speed differences (vΔv). In the original work of Wiedemann an additional threshold cldv (closing delta velocity) is applied to model additional deceleration by usage of the brakes with a larger variation than SDV.

OPDV: Action point where the following driver notices that he/she is slower than the leading vehicle and starts to accelerate again. The variation of OPDV is large (Todosiev 1963).

SDX: Perception threshold to model the maximum following distance, which is about 1.5–2.5 times ABX.

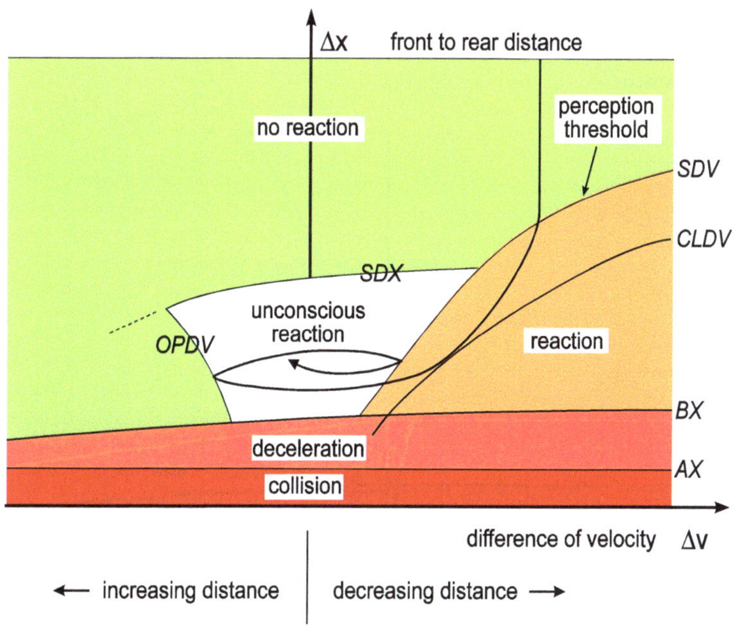

Fig. 6.1 Car-following model of Wiedemann – threshold and one vehicle trajectory

A following driver reacts to a leading vehicle up to a certain distance, which is about 150 m. The minimum acceleration and deceleration rate is set to be 0.2 m/s^2. Maximum rates of acceleration depend on technical features of vehicles, which are usually lower for trucks than the personal desire of its driver. The model includes a rule for exceeding the maximum deceleration rate in case of emergency. This happens if ABX is exceeded. The values of the thresholds depend on the present speed of the vehicle. Figure 6.2 denotes the values for two different speeds to display a current set of values.

In case of multi-lane roads, a hierarchical set of rules is used to model lane changes. First, a driver has a desire to change lane if he/she has to drive slower than his/her desired speed, due to a slow leading vehicle or in case of an upcoming junction with a special turning lane. Then, the driver checks whether he/she improves his/her present situation by changing lanes. Last, he/she checks whether he/she can change without generating a dangerous situation. In case of multi-lane approaches towards intersections, this method will lead to evenly used lanes, unless routing information forces vehicles to keep lanes.

The Network geometry is modelled using the graphical user interface of VISSIM. It is possible to load a scanned layout plan of the modelled network as a background for the network editor. Figure 6.3 show a layout plan and the resulting network model. VISSIM is using links and connectors between links. Each link has attributes, like number of lanes, gradient, free flow speed, etc. Nodes, like in transportation planning packages, are not required. Missing of nodes has the advantage that the full variety of lane allocations can be modelled. The network model of links and

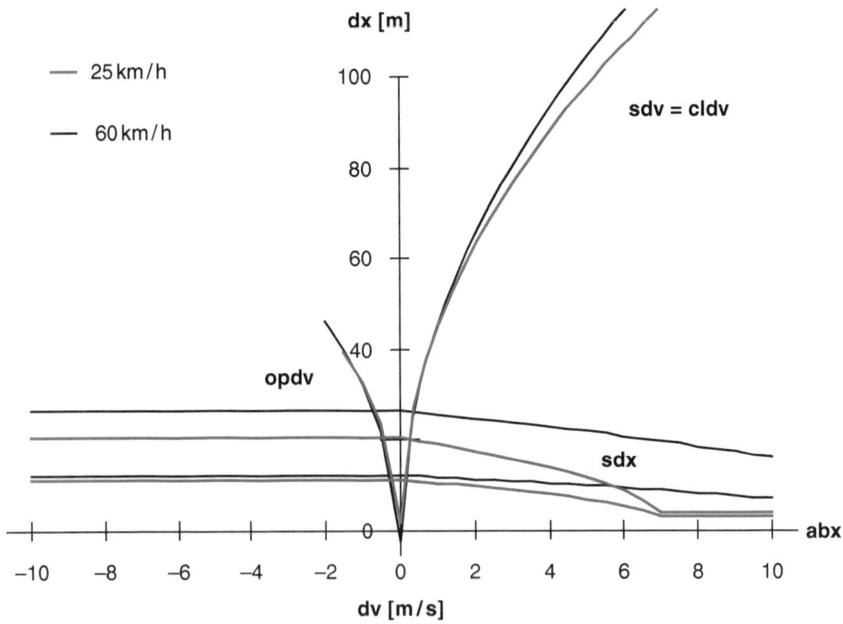

Fig. 6.2 Car-following threshold used in urban situations as a function of the speed

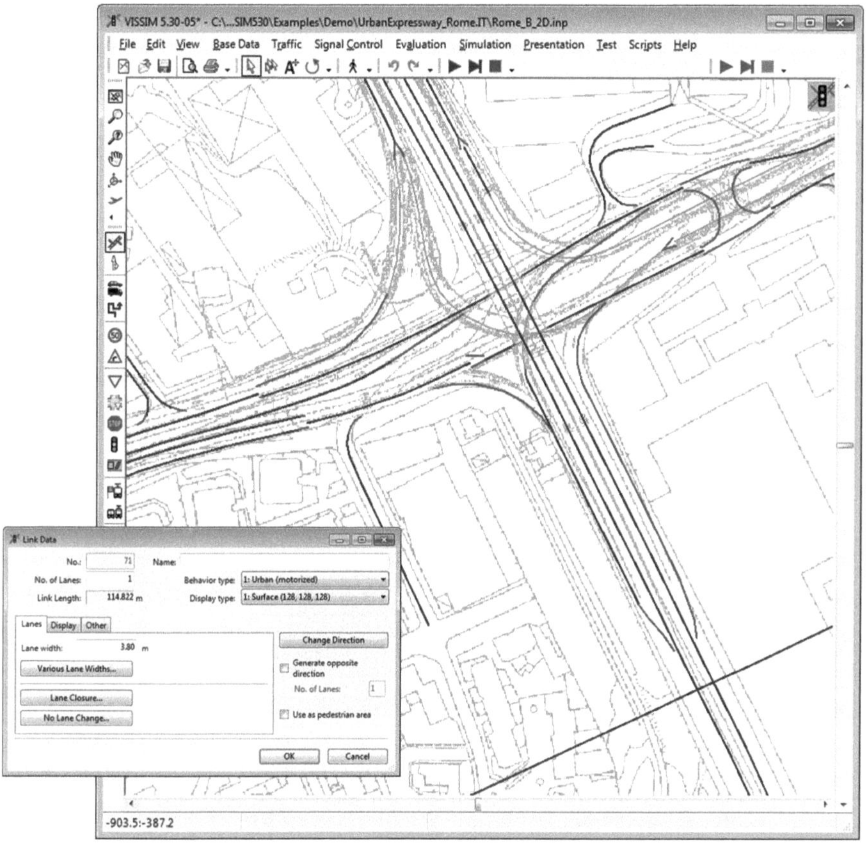

Fig. 6.3 Network model overlaid on junction layout plan in VISSIM model

connectors has been proven to be flexible enough to cover situations found in a variety of countries. Different driving habits between left-hand and right-hand driving are covered with the network model as well.

Additionally, traffic volumes and the vehicle fleet must be specified. It is possible to define different distributions of desired speeds, accelerations, vehicle lengths, and passenger boarding times. The road infrastructure, like signal heads, stop signs, yield signs, parking signs, speed signs, bus bays and tram stops are placed as particular objects allocated to links.

The following example shows how to model yield signs. Two tram tracks are displayed. The two trams are driving on sight instead of being signalized. One tram can only pass if the other has passed the conflicting area. Therefore, a time headway of 2.5 s plus a minimum spatial headway of 36 m (tram length plus reserve) has to be cleared (Fig. 6.4).

Since editing large networks may be time-consuming, VISSIM has now import filters from transport planning packages. The network model can be imported from the transportation planning model VISUM. VISUM is able to read

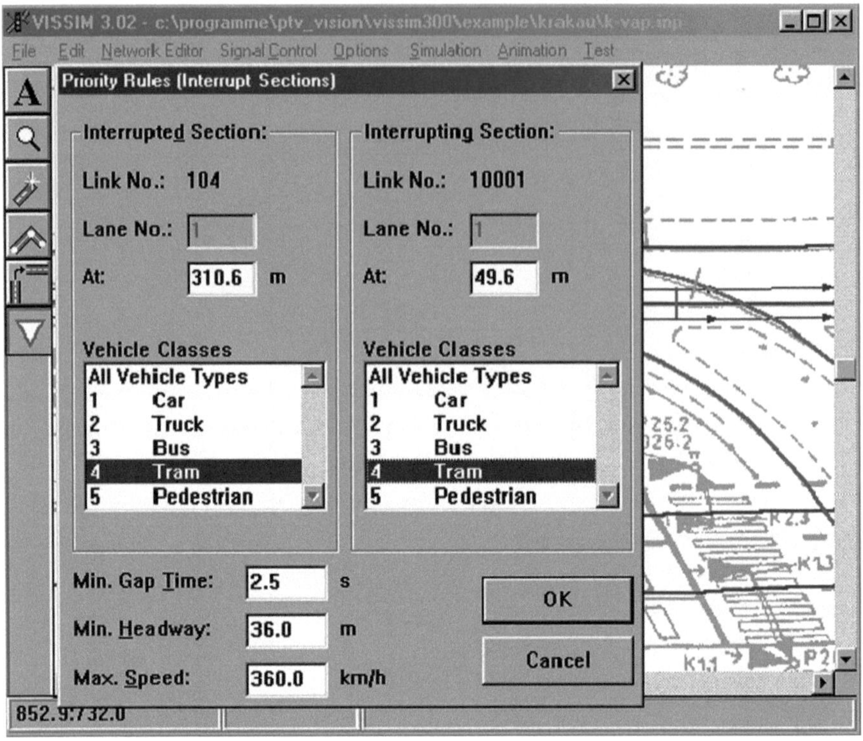

Fig. 6.4 Network model with yield signs for trams in VISSIM model

EMME/2-files. Therefore, VISSIM can read EMME/2 network and Origin-Destination data via the interface with VISUM. The largest networks currently modelled in VISSIM cover an area of about 400 km^2, including a little over 100 signalized intersections.

6.3 Macroscopic Models

Macroscopic models do not treat vehicles as individual entities but consider traffic as streams of vehicles. They apply macroscopic relationships between traffic volume, traffic density and average speed. Their original application is the assignment which matches the demand in terms of passenger (or goods) trips to the supply, i.e. the road or public transport network. This is indicated in Fig. 6.5. Their typical application lies in network-wide investigations that consider areas like cities or regions.

Over the years a number of such models have been developed – mostly for dedicated application; some, however, have become standard tools for transport planning, like SATURN or VISUM. In the following, we describe one of each category: SATURN as a standard tool and MT.MODEL as a dedicated tool.

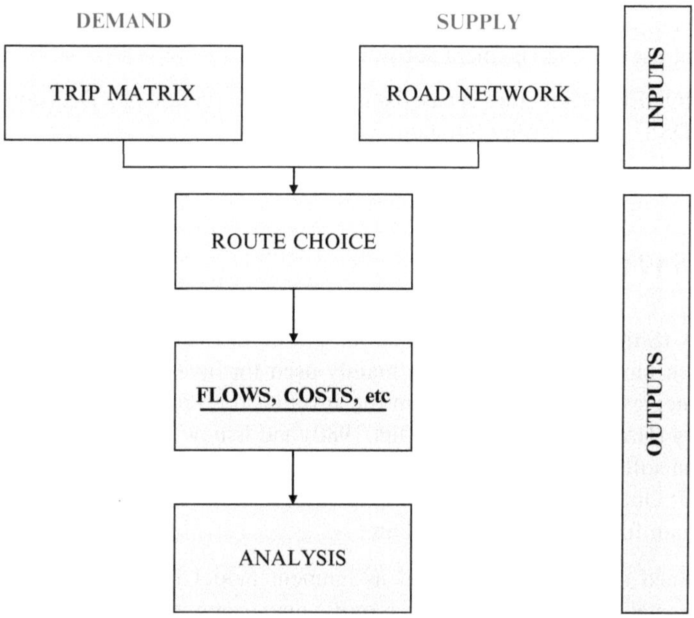

Fig. 6.5 Main structure of a typical assignment model

6.3.1 MT.MODEL

MT.MODEL is a user friendly and totally integrated system of mathematical models for decision support to the traffic and transport planning.

MT.MODEL allows to analyze the existing situation of a traffic system and, answering to questions as "what if?" (what would happen if...?), allows to estimate new suggestions of the area reorganization. The mathematical models that constitute its nucleus offer the opportunity to simulate the variations to the actual mobility and transport planning, previewing the effects that would derive from their realization.

In order to obtain a complete evaluation of the effects on the complete transport system, MT.MODEL allows:

- The analysis of the existing situation of demand, traffic supply and performances of the transport system
- The prediction of the mobility demand with regard to pre-assigned scenarios of socioeconomic and territorial evolution and pre-assigned configuration of traffic and transportation supply
- The valuation of the performances of transportation networks, according with these scenarios

The system is composed of:

- A data bank, that contains available information of demand and traffic supply
- Models of information management

- Models of demand and performance prediction of the transport system
- Current use and management software

MT.MODEL architecture is based on the general structure of a Decision Support System (DSS), proposed by Sprangue in 1986.

6.3.2 SATURN

SATURN (Simulation and Assignment of Traffic to Urban Road Networks) is an assignment model and as such it is mainly used for investigating traffic management strategies. It was developed in the Institute for Transport Studies of Leeds University (Hall et al. 1980; Van Vliet 1980) and is now widely used commercial simulation software for a variety of applications (Van Vliet et al. 1987; Matzoros et al 1987; Gulliver and Briggs 2005).

The main functions of SATURN are:

- Combined traffic simulation and assignment model for the analysis of road-investigated schemes ranging from traffic management over relatively localised to larger road networks.
- "Conventional" assignment model for the analysis of very large road networks.
- Simulation model for individual junctions.
- Network editor, database and analysis system.
- Matrix manipulation package.
- Trip matrix demand model covering basic elements of trip distribution modal split, etc.

There are two functions that are of interest within the framework of IN-SAFETY: the "conventional" assignment model and the network editor. The function of the assignment model assigns traffic, performing trips from an origin to a destination within the simulated road network, to different routes, based on a number of principles. The function of the network editor allows and can be applied for the analysis of network-based data which need not be in any way related to traffic assignment problems. As an example data related to accident rates per link, road resurfacings stored, etc., may be input and analysed.

SATURN offers a wide range of assignment methods, including generalised-cost, all-or-nothing, Wardrop equilibrium, etc. It employs the main structure of a typical assignment model which is illustrated in Fig. 6.5.

For the application of the traffic assignment, there are two main input elements that represent the demand and the supply of trips in a road network: the trip matrix Tij, which describes the number of trips from zone i to zone j that will take place, and the network, which describes the physical structure of roads, upon which the trips will be accommodated. The trip matrix and the road network are then input to a "route choice" model, which allocates trips to routes through the network. The result of this initial allocation is the development of traffic flows on the defined routes, based on which the corresponding network "costs" are estimated and the traffic assignment run by SATURN (according to pre-set user preferences) initiates. Last, within the analysis function the results of the assignment are estimated in the form that the user has defined and are provided to the user as an output of the program.

Part of the SATURN operation is the estimation of a cost for each route, based on which traffic is assigned into different routes. The cost is a function of the travel-time on the route and its distance (length of the route), and the corresponding formula for its estimation follows:

$$Cost = PPM * Time + PPK * Dis\tan ce + \sum_i PPU(i) * DATA(i), \tag{6.1}$$

where PPM and PPK are the weight factors of time and distance respectively and PPU (pence per unit) are those attributed to other data inputs (DATA). Hence, SATURN model allows the user to introduce further parameters in the cost estimation; for example DATA(1) might be a link route familiarity index and PPU(1) a weight to convert route familiarity value into monetary values. The required values of PPU(i) are provided by the user on a specific record, and may differ for different user classes. By definition, all DATA values are fixed, independent of flows.

6.4 Model Applications for Safety Assessment of Proposed Scenaria

The following sections give some insight into the application and the possible results that the models described above can provide. Of course, only few examples can be given here. However, they give insight into the complex issue of pre-evaluation of safety enhancement scenarios and new technologies. Such pre-evaluations are needed for further investigations, like cost-benefit analyses, for the planning of large-scale tests, e.g. field operational tests (FOTs) or also for system design. In the latter case, the models can be integrated in the feedback loop, when designing and evaluating a traffic safety system by yielding the results of a concrete design of a system.

6.4.1 Example Applications of a Microscopic Simulator

6.4.1.1 Adaptive Cruise Control (ACC)

This section gives insight on how to evaluate and estimate the impact of Advanced Driver Assistance Systems (ADAS) on the example of Adaptive Cruise Control (ACC). Such a system influences driver behaviour and is therefore a good example for the application of a microscopic simulator. The focus here is on the adaptation of the simulator to cover vehicles which are equipped with such a system. The results are only exemplary because a micro-simulator can produce many parameters of interest. Thus, any traffic related parameter, be it speed, volume or their relationships, can be easily obtained. Here we present results that require more detailed analyses and are of interest to a wider community, as they include both emissions and safety related results. In the beginning, a more detailed description of the implementation is given, in order to show the potential of a simulator, which allows for changes to the driver behaviour by varying parameters.

The ACC system functionality was modelled directly in VISSIM as a new "driver behaviour". VISSIM allows defining arbitrary parameter settings for the pre-defined behaviour in the state diagram as shown in Fig. 6.2. The available parameters to determine driver- or system-behaviour are according to the VISSIM manual:

CC0 (Standstill distance) defines the desired distance between stopped cars. It has no variation.

CC1 (Headway time) is the time (in s) that a driver wants to keep. The higher the value, the more cautious the driver is. Thus, at a given speed v [m/s], the safety distance dx_safe is computed to: dx_safe = CC0 + CC1 * v.

The safety distance is defined in the model as the minimum distance a driver will keep while following another car. In case of high volumes this distance becomes the value with the strongest influence on capacity.

CC2 ('Following' variation) restricts the longitudinal oscillation or how much more distance than the desired safety distance a driver allows before he/she intentionally moves closer to the car in front. If this value is set to e.g., 10 m, the following process results in distances between dx_safe and dx_safe +10 m. The default value is 4.0 m, which results in a quite stable following process.

CC3 (Threshold for entering 'Following') controls the start of the deceleration process, i.e. when a driver recognizes a preceding slower vehicle. In other words, it defines how many seconds before reaching the safety distance the driver starts to decelerate.

CC4 and **CC5** ('Following' thresholds) control the speed differences during the 'Following' state. Smaller values result in a more sensitive reaction of drivers to accelerations or decelerations of the preceding car, i.e. the vehicles are more tightly coupled. CC4 is used for negative and CC5 for positive speed differences. The default values result in a fairly tight restriction of the following process.

CC6 (Speed dependency of oscillation): Influence of distance on speed oscillation while in following process. If set to 0, the speed oscillation is independent of the distance to the preceding vehicle. Larger values lead to a greater speed oscillation with increasing distance.

CC7 (Oscillation acceleration): Actual acceleration during the oscillation process.

CC8 (Standstill acceleration): Desired acceleration when starting from standstill (limited by maximum acceleration defined within the acceleration curves).

CC9 (Acceleration at 80 km/h): Desired acceleration at 80 km/h (limited by maximum acceleration defined within the acceleration curves).

The parameters modified for simulating the ACC system were:

- Desired distance: it was assumed that the drivers keep a fairly large standstill distance of 2 s plus 1 m as the minimum headway time during following. Thus, the relevant VISSIM parameters were set to CC0 = 1.0 and CC1 = 2.00.
- Oscillations during following: the system can perform a much "tighter" following than a human driver. "Tighter" meaning that differences in relative speed are better perceived. The parameters CC4 and CC5 were set to 0.5. Furthermore, CC6 was set to 1.00.
- Acceleration during following: the system can keep a speed much better than a human driver. It was therefore assumed that the oscillations during following are performed at only CC7 = 0.1 m/s^2.

Other parameters were not changed. Especially the overtaking behaviour remained unchanged – which may not be so in reality. However, statistically representative data for the network, with pre-dominant weaving manoeuvres as causes for lane-changes were not available, so as to adapt this parameter too.

The network chosen was one that represents reality: a heavily loaded motorway junction in the Rhine-Main-Area, close to Wiesbaden, was chosen. Figure 6.6 shows the VISSIM representation of this network around the junction. The sections, especially north and south of the junction were much longer than displayed. The two motorways BAB A3 and BAB A66 intersect here, both carrying long-distance as well as commuter traffic. In this network weaving actions, which may possibly be dangerous and through traffic are combined. This was considered a very appropriate application for tests with ACC. The specific reaction of vehicles cutting in represents a demanding task for the system. Here, safety effects of this comfort application may arise.

Fig. 6.6 Simulation Network, Motorway Junction BAB A3 and BAB A66, near Wiesbaden, Germany

For the scenarios to be simulated, the traffic volumes that entered the network on both ends of the north–south directed A3 were each 1,800 (low), 3,000 (medium) and 5,000 (high) vehicles per hour; in all cases 10% of all vehicles were trucks. This traffic then splits up into the possible directions at the junction, according to shares derived from the real shares found in the morning peak period.

The share of equipped vehicles was 0 (base case), 10, 25, 50, 75 and 100% of all passenger cars. Each of these 18 cases, 6 penetration rates and 3 traffic volumes, was simulated 5 times with different random number seeds, to get a statistically sound basis for evaluation (Anund et al. 2007). As a first example for the results, the emissions of NOx are presented here in Fig. 6.7. The data shown relate to all vehicles, equipped and un-equipped. It becomes obvious, that the introduction of ACC vehicles has a positive effect on these emissions for all simulated traffic volumes. The three groups of bars reflect the three volumes simulated; within each group the bars of different colours indicate the rates of equipped vehicles between 0% (base case) and 100% (potential when all vehicles are equipped).

In order to indicate the possibilities to also evaluate safety related effects from microscopic simulation, results from a similar study by Benz (2008) are presented. Here, too, the effects of ACC were evaluated, however, in a different study design. The volumes were varied into more than three cases, in order to cover all possible situations. Especially, the range close to capacity was thoroughly modelled. By doing so, data for all volumes could be retrieved.

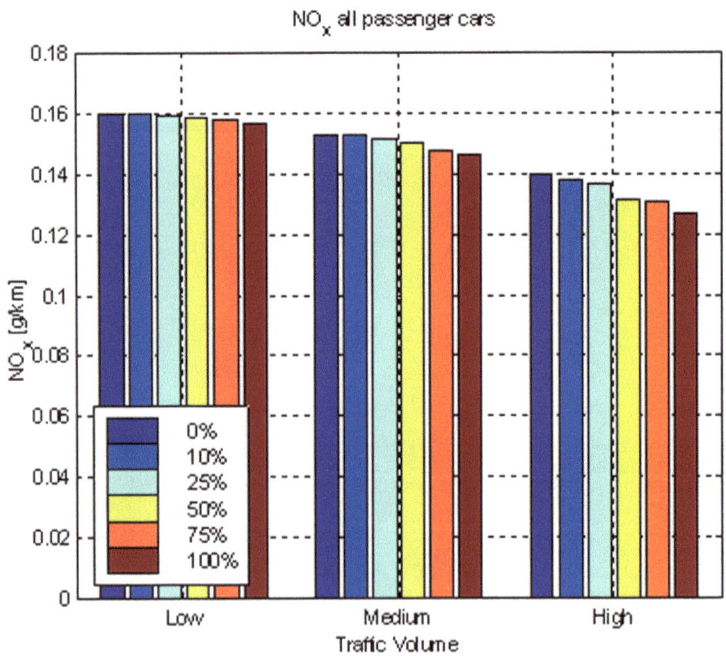

Fig. 6.7 NOx emissions

The safety effects were established by investigating time-to-collision (TTC) and the share of small headways. The two diagrams in Fig. 6.8 show the share of headway below 2 s (above) and below 1 s (below). These data were collected at a simulated cross-section. The lines in the diagrams relate to the base case (black), a low penetration rate (red line, 4% of passenger cars) and a high penetration rate (blue line, 13% of passenger cars).

It becomes obvious that the share of headways below 2 s is nearly independent of the presence of equipped vehicles; headways below 1 s, however, are less frequent with ACC vehicles in the network. Thus, ACC seems to reduce the very critical headways (Fig. 6.8).

6.4.1.2 Collision Avoidance System (CAS) and Lane Change Assistant (LCA)

Following the same scenarios design (Anund et al. 2007) (in terms of traffic volumes, penetration rates, etc.), another ADAS-related application was evaluated within IN-SAFETY. The aim of the application was to investigate the safety and traffic efficiency impacts of ADAS equipped vehicles, in several different penetration rates, on the same road and under the same circumstances. The network that has been simulated was a highway, including an intersection. The types of ADAS that were analysed were the Collision Avoidance System (CAS) and the Lane Change Assistant (LCA).

Following the structure of the VISSIM model, certain vehicle types and respective vehicle classes needed to be defined. Each of them represents a different group of vehicles with different characteristics. For the needs of the application in question, five different vehicle types/classes were defined, namely:

- PKW, including passenger cars, not equipped with any ADAS.
- LKW, including trucks, not equipped with any ADAS.
- ADASth, including passenger cars, equipped with the specific ADAS, following the theoretical behaviour parameters that the use of this equipment would imply (i.e. if the CAS warns the driver when TTC $\leq$ 2 s then we estimate that all drivers keep a min TTC of 2 s).
- ADASb, ADASc, including passenger cars equipped with the specific ADAS, following behaviour parameters, deriving from previous real tests with the ADAS in question (i.e. we consider different behavioural adaptations of drivers with CAS, such as different min TTC as measured in past tests with real users).

The driver behaviour parameters that were influenced in each category and their specific values are described in the relevant chapter.

In order to successfully simulate the behaviour of equipped vehicles in the network, certain default set parameters of driving behaviour needed to be changed, according to the expected effect of each ADAS. In VISSIM, there are several default parameters set, both for longitudinal and lane change behaviour, whose values determine the behaviour of the vehicle and whose differentiation could lead to different effects.

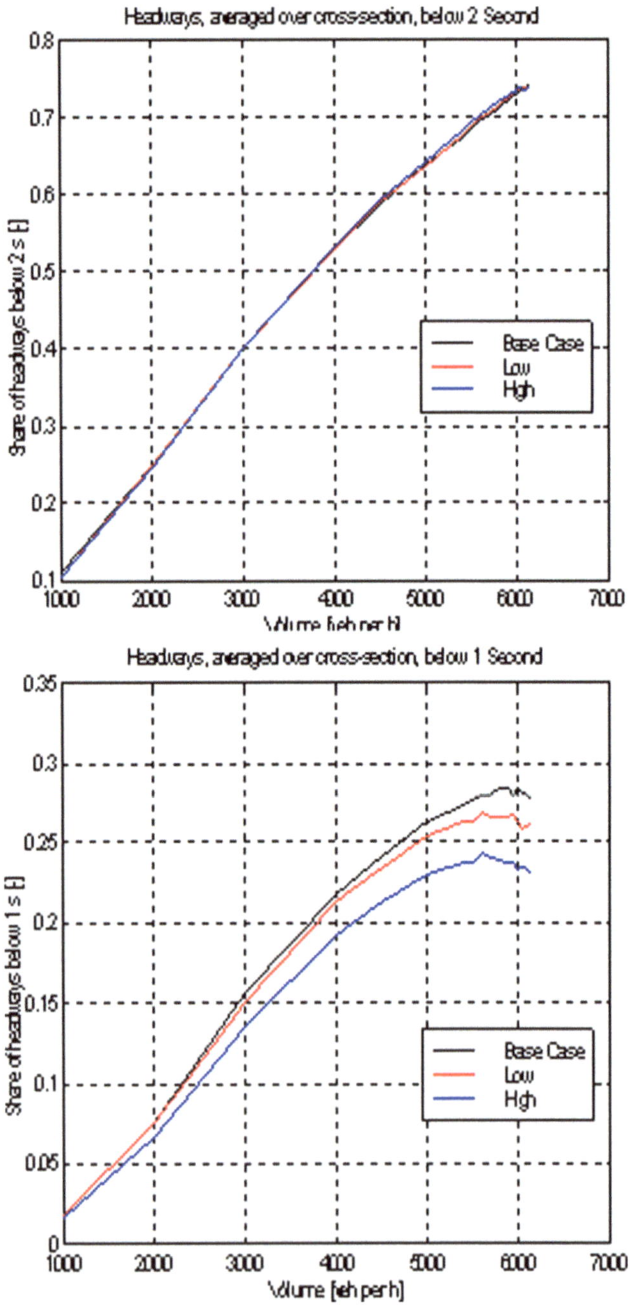

Fig. 6.8 Share of Small Headways depending on Volume
Source: Benz 2008

Longitudinal Behaviour Modelling of Selected ADAS

The longitudinal driving behaviour in the VISSIM micro-simulation traffic model is based on the "following" driving mode, as developed by (Wiedemann 1974, 1991). According to this approach, two different sets of parameters are included, defining the behaviour of the vehicle on the road. In terms of the study on the influence of the CAS, the parameter that has been influenced is the CC1, defining the time headway that the driver allows from the preceding vehicle. More specifically, in the case of CAS, the default value for time headway, as set in the Wiedemann 99 model, was 0.9 s. This value was changed to 1.0 s for the ADASth vehicle class, as is the theoretical value for the time headway used by the CAS. Moreover, in the case of ADASb and ADASc vehicle classes, which represent the behaviour from real tests, different values have been set for the CC1 parameter. More specifically, for the ADASb the value set was CC1 = 1.2 s and for ADASc, CC1 = 0.8 s, according to relevant on road tests results (Brouwer and Hoedemaeker 2006).

Lateral Behaviour Modelling of Selected ADAS

The VISSIM model includes a separate set of parameters ruling the lane change behaviour of the vehicles. Among them, the ones that were influenced during the performed study were:

- Min headway.
- Safety distance reduction factor.
- Max deceleration for cooperative braking.

The aim was to create a situation where the driver, influenced by the relevant ADAS, would be led to make more (or less) lane changes than in the default situation. However, as described in the results chapter, it was not possible to come up with a set of values that would create the desired effect, so as to simulate the behaviour of equipped vehicles' drivers.

The results indicated the influence of ADAS-equipped vehicles in the traffic composition over the total network.

Average Speed with Ideal (Theoretical) CAS

The overall average speed in the network changes in different terms as the CAS equipped vehicles penetration rate increases, depending on the traffic volume. At low traffic volume (1,800 veh/h) the average speed for all vehicles slightly decreases, with a max speed at 25% penetration rate and the min at 100%. The speed for the CAS equipped vehicle class shows a peak at 25% penetration rate but it generally decreases, with its minimum at 50%. At medium traffic volume (3,000 veh/h) the average speed for the whole network decreases at 10% penetration rate and then increases until 50%, where it has its max value, to decrease again until the minimum at 100%.The CAS equipped vehicle class gives two peaks at 25 and 75% penetration rates. On the other hand, the non-equipped vehicles have higher speeds at 25 until 50%, which decrease to reach the minimum at 75%. Finally, for the higher traffic volume (5,000 veh/h) all vehicle classes give lower speeds at 10, 50 and 75% penetration rates and the highest at 25%, while at 100% the speed is almost the same as at 0% (Fig. 6.9).

Noticing the high peaks at the diagram for high traffic volume (which however corresponds to very small absolute differences), one way analysis of variants (ANOVA) has been performed, to investigate the statistical significance of these differences. The result of the analysis ($F(2, 6) = 9.916$, $p = 0.06$) shows that the differences have no statistical significance.

Travel Time (Per Vehicle): With Ideal (Theoretical) CAS

As far as travel time is concerned, at low traffic volume the travel time for the whole of the network increases as the penetration rate of CAS equipped vehicles becomes higher. For CAS equipped vehicles the travel time is lower for penetration rates of

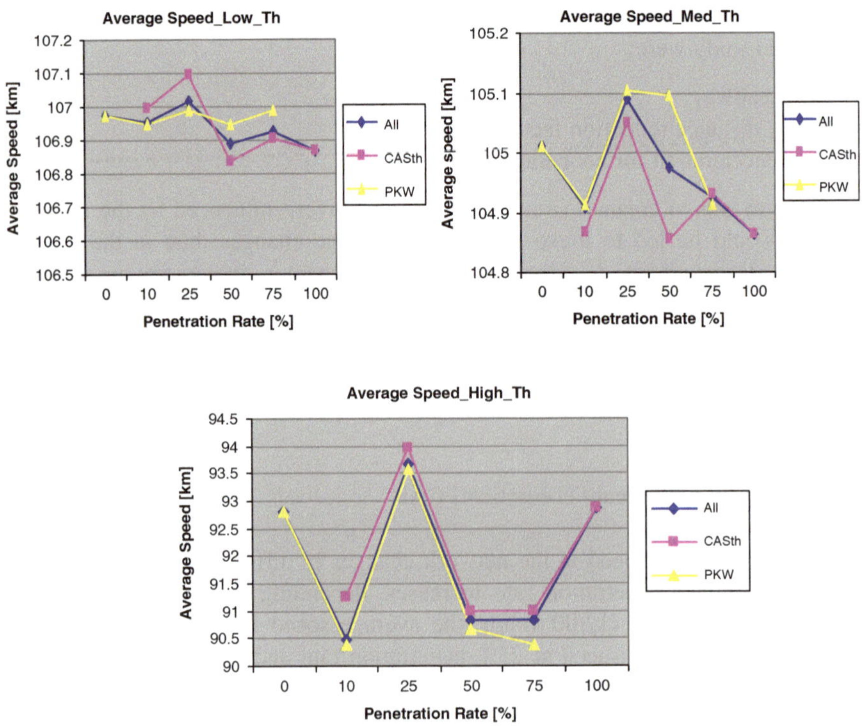

Fig. 6.9 Average speed at different traffic volumes with ideal CAS in different ADAS penetration rates

25% and 75% and higher for 50 and 100% penetration, while for the non-equipped vehicles it decreases until 50%, to increase at 75%. For medium traffic volume, travel time per vehicle at the network generally increases with penetration rate, only decreasing at 25%. The travel time for equipped vehicles is max at 10% penetration and min for 25%. For non-equipped vehicles the travel time generally decreases, having the minimum at 10% penetration. In the case of high traffic volume, travel time generally increases (max at 10%) with an exception at 25% penetration rate, where it decreases significantly and then increases again at 50%, to slightly decrease until 100% (Fig. 6.10).

Also in this case, ANOVA analysis has been performed to investigate the statistical significance of the variations noted in the "high" traffic volume diagram. The result ($F(2, 6) = 5.642$, $p = 0.056$) showed no significance.

Average Speed with Actual (Practical) CAS

A general remark on the average speed is that the speeds of the CASb and CASc classes are higher that the total, whereas the speed of the CASth and the PKW

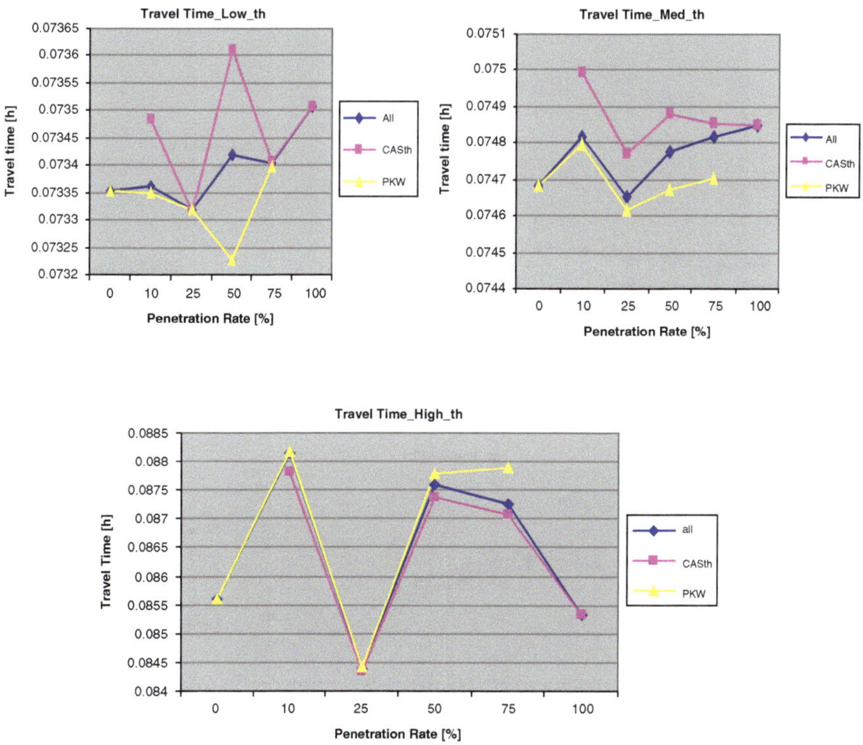

Fig. 6.10 Travel time per vehicle at different traffic volumes with Ideal (theoretical) CAS at different ADAS penetration rates

(non-equipped) vehicles are lower. The average speed on the network for low traffic volumes is slightly increasing. This is mainly due to the higher average speeds of the two classes of equipped vehicles (CASb and CASc), which are however decreasing as the penetration level is rising up. For the non-equipped vehicles the speed is decreasing, whereas for the third equipped vehicles' class (CASth) it is decreasing at 25% but then increases again until 75% penetration rate. For the medium traffic volume, the average speed for all vehicles is almost constant, with a slight increase. For all three equipped vehicles classes, as well as for the non-equipped, the speed is generally reducing. At high traffic volume the speed is decreasing in all cases, reaching the minimum at 75% penetration level (Fig. 6.11).

The results of the ANOVA analysis performed for the case of "high" traffic volume ($F(4,12) = 35.77$, $p < 0.001$) showed that there is statistical difference between some of the values. More specifically, CASb is significantly different from CASth and PKW, with CASb ($M = 99.57$) being significantly bigger than CASth ($M = 97.55$; $p = 0.022$) and PKW ($M = 97.44$; $p = 0.021$).

Also, in the case of "low" traffic volume, some statistically significant differences have been detected. ANOVA gave the result $F(4,12) = 18.88$, $p < 0.001$, which means that PKW is significantly different from CASb and CASc, with PKW ($M = 106.83$) being significantly smaller from CASb ($M = 107.534$; $p = 0.04$) and CASc ($M = 107.50$; $p = 0.05$).

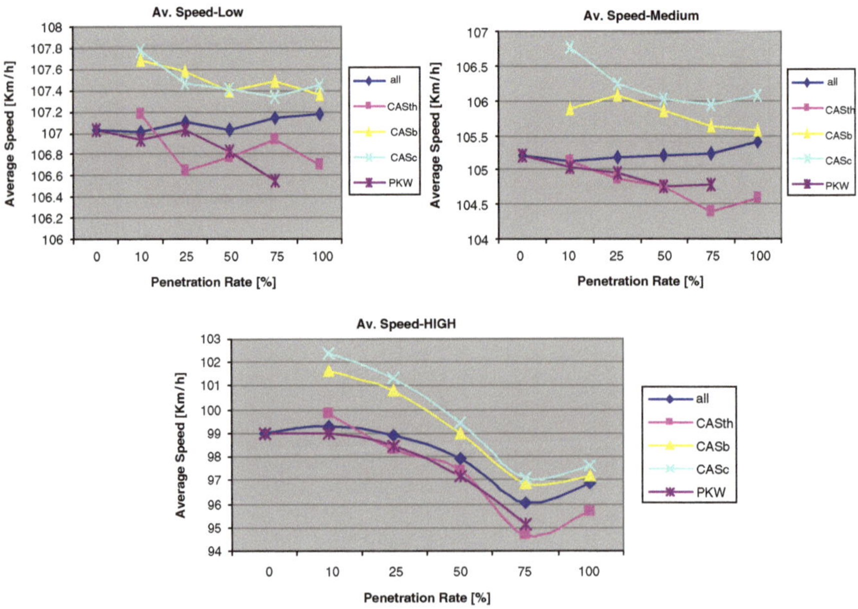

Fig. 6.11 Average speed (practical runs) at different traffic volumes with actual CAS in different ADAS penetration rates

Travel Time with Actual CAS

Regarding the travel time at the network per vehicle, at low traffic volume and for the total of vehicles, it is rather decreasing, except for the 50% penetration rate. The situation for the different vehicle classes is diverse, but generally the travel time tends to increase in all cases. In the case of medium traffic volumes, the travel time is generally not significantly changing for all vehicles. However, it is mostly increasing for all the separate vehicle classes. Finally, at high traffic volume, the travel time in all cases is clearly increasing, reaching its maximum at 75% penetration rate (Fig. 6.12).

In this case, ANOVA was also performed for the “low” and “high” traffic volumes. The result for the “low” was $F(4,12) = 6.03$, $p < 0.01$, indicating that the CASth ($M = 0.734$) values are significantly higher than the CASb ($M = 0.0731$) and CASc ($M = 0.0730$; $p < 0.05$ for both comparisons). Also, for the “high” traffic volume, the result ($F(4,12) = 44.12$, $p < 0.01$) indicates that CASth ($M = 0.0808$) is significantly different from CASb ($M = 0.0792$) and CASc ($M = 0.0788$; $p = 0.05$ for both comparisons). In addition, PKW ($M = 0.0809$) is significantly different from CASb ($M = 0.0788$) and CASc ($M = 0.0809$; $p < 0.05$ for both comparisons).

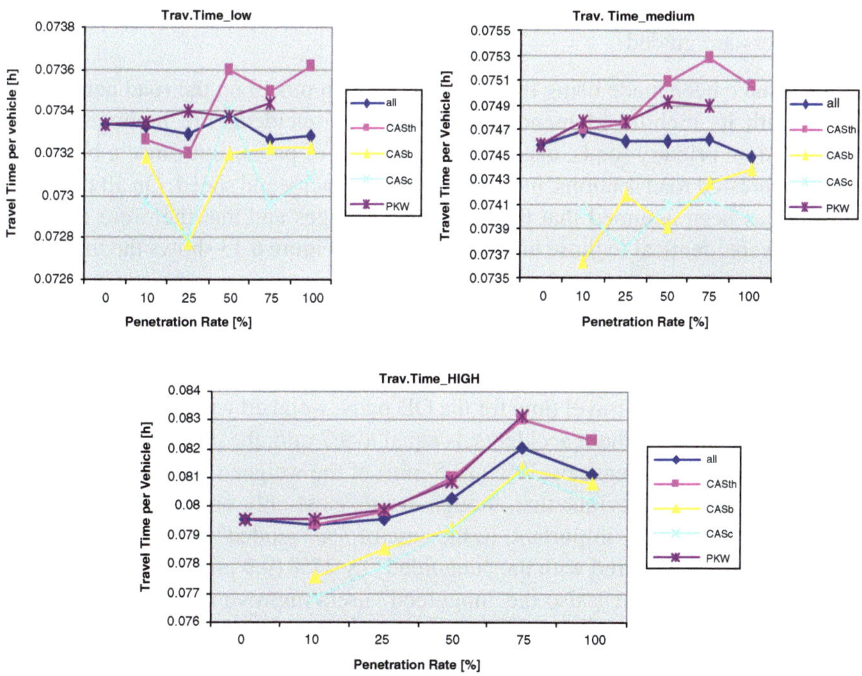

Fig. 6.12 Travel time per vehicle (practical runs) at different traffic volumes for actual CAS at different penetration rates

LCA Model Simulations

As stated above, different values were tested in the available lane change parameters, in order to investigate the influence of LCA in lane changing behaviour. However, no significant conclusion could be drawn from the results of the model, as the number of lane changes did not seem to be influenced. This fact was not in line with relevant results from real tests which have shown specific differentiation. Therefore, certain modifications should be considered for the lateral and lane change parameters of the model, in order for the model to be able to simulate reliably the actual driver behaviour, as affected in terms of lane changing.

6.4.2 *Example Application of a Macroscopic Simulator*

For the macroscopic simulation, an IVIS has been selected; namely route guidance. For its evaluation, a scenario of localized events has been examined, that expects local perturbations due to accidents. For this scenario three simulations have been made:

- All users are not guided
- 5% of Users are guided, 95% are unguided
- All users are guided

Tests have been made using the traffic simulation model on the road network of Turin, with its mobility demand. In the scenario, accidents are homogeneously positioned on primary roads, used by a big number of paths, and cause a big delay on the interested road sections, influencing both capacity and speed. On all the other links, it has been assumed that there were no changes and that therefore the road features were identical to those historical averages. Figure 6.13 shows the links with accidents.

The scenario considers the benefits which the "guided" users can obtain, if they avoid roads with accidents. In this case the total decrease has been evaluated on the hypothesis that all the users chose their usual path, using also the roads affected by accidents. The average travel time for the OD pairs, weighed with the volume of the OD pair, in the case without accidents, is equal to 9.7 min; the presence of unknown events involves an increase by 39% (13.5 min) of the weighed average travel time. In this situation it is obvious that, though events exist only on few links, to avoid such congestion is very important: in the case of 5% "guided" users, they improve their travel time compared with the "unguided" by 27% (9.9 min). In case these 5% of the users are "guided", also the "unguided" users improve their travel times: in fact the reduction of the 5% of congestion in the critical points improves the travel times of those who travel there. The total weighted average travel time for the "unguided" users becomes 12.9 min. Taking into account all the users together, the total average travel time on the network is equal to 12.8 min (Table 6.1).

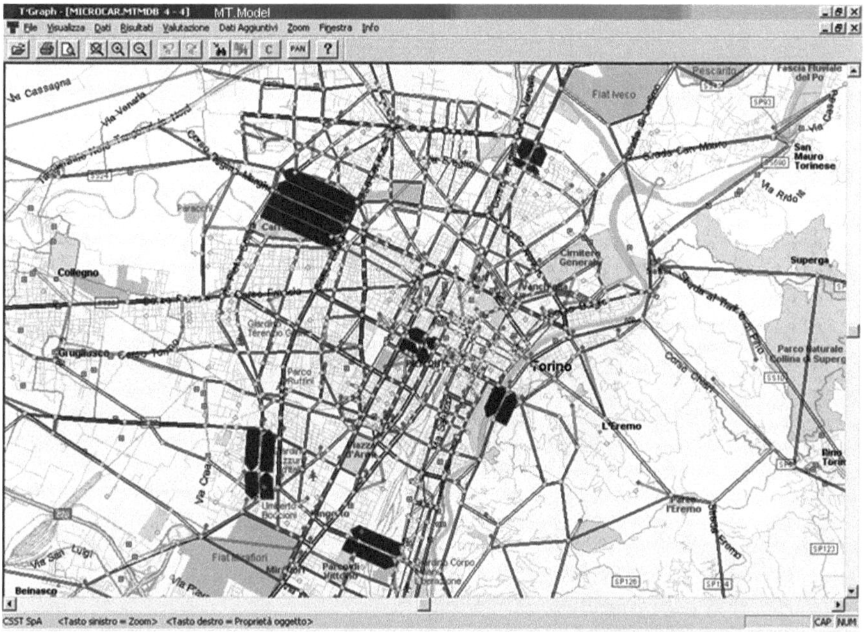

Fig. 6.13 Simulated accidents in the Torino network (Anund et al. 2007)

Table 6.1 Average travel time for the different scenario cases

Average travel time		Average travel time (weighted)
No accidents	All users are "unguided"	9.7
Localized accidents	All users are "unguided"	13.5
	5% of Users are "guided"	9.9
	95% of Users are "unguided"	12.9
	Scenario 5% "guided" + 95% "unguided"	12.8
	All users are "guided"	9.9

Figure 6.14 represents the flow distributions (the link thickness is proportional to the vehicular flow quantities on the link): the links with disturbance are represented with red colour. It is obvious that when the users are "guided", the roads with accidents are not used, while the parallel and neighbouring roads increase their flows. The "not-guided" traffic, instead, does not know about the accidents and therefore chooses paths blocked by accidents.

6.5 Into the Future

The use of simulation models in the traffic safety domain opens wide possibilities for different applications. In the current state of model development, it is already

Fig. 6.14 Flow distribution when 5% of users are "guided"

possible to model some applications of Intelligent Transport Systems and evaluate their effects, even if the models do not yet treat safety as such. Macroscopic models have been applied to determine the network-wide effects of driver information about safety levels on network parts; thus it was possible to establish the overall consequences for drivers with and without such a system.

Microscopic models are suitable to investigate such ITS applications that directly change the movement of vehicles. The ITS application is modelled either by adapting the driver behaviour or by including the system behaviour explicitly in the model. The results of a simulation run then reveal all changes to the traffic flow: macroscopic changes to the volume-speed-density relationships, changes in environmental aspects via a suitable environmental model and also changes to the overall safety level. Although current microscopic models do not include mechanisms for safety critical situations, they can provide indications about safety via so-called surrogate parameters, which allow an estimation of the safety level. It should, however, be noted that the reliability of results of the model depends highly on the driving behaviour parameters included in the model and the values selected for them by the researcher.

Further research will provide new insights into the processes that ultimately govern the occurrence of accidents. Including such processes into micro-simulations will allow investigating such critical situations in much more detail. This will not only lead to an improved ITS assessment but also to the design of improved safety relevant ITS measures. Especially in the case of LCA, the VISSIM microsimulation model, at its present state, does not seem to provide a reliable simulation of the effect of such a system in the lane change behaviour. Thus, there is need for the inclusion of adequate parameters, which would allow the model to effectively simulate such behaviours. As far as route choice is concerned (S-Paramics) the results of the different indicators do not all point in the same direction. The significance of the different indicators for research into route choice requires more attention. What is more, it is important to examine whether the indicators in the micro simulation provide a result that conforms to reality. The indicators for the safety of a route only comprise the safety of car drivers using a route. The indicators should be extended to the safety of all users (also cyclists and pedestrians) of a route. At the same time, a method should be developed for optimizing the safety of all (main) routes in the network. For planning applications the method should be integrated in existing planning models. For traffic management applications the safety criteria should be built into the choice algorithms of route planners.

Moreover, regarding the RUTSIM macrosimulation model, future research would include analysis of traffic effects of the individual driver results. These results include changes in driving performance due to driver fatigue and rumble strips on two-lane highways. In addition, further research on the relation between simulation-based safety indicators and accident risks is needed, in order to facilitate more accurate safety analysis using traffic simulation models.

References

A. Anund, Th. Benz, E. Gaitanidou, J. Spyropoulou, S. Toffolo, Improved micro and macro simulation models. IN-SAFETY, Deliverable D3.1, February 2007

R.F.T. Brouwer, D.M. Hoedemaeker (eds.), Driver support and information systems: experiments on learning, appropriation and effects of adaptiveness. AIDE IST-1-507674-IP, Deliverable D1.2.3, February 2006

A. Carlsson, A. Tapani, Rural highway design through traffic simulation, in *Proceedings of the 5th International Symposium of Highway Capacity and Quality of Service*, Yokohama, Japan, 2006

M.D. Hall, D. Vliet, L.G. van Willumsen, SATURN – A simulation-assignment model for the evaluation of traffic management schemes. Traffic Engineering and Control, **21**, 168–176. (1980)

J. Lundgren, A. Tapani, Evaluation of driver assistance systems through traffic simulation. Transport. Res. Rec. **1953**, 81–88 (2006)

Matzoros, A., Randle, J, Vliet, D. van Weston B., A validation of SATURN using before and after survey data from Manchester. Traffic Engineering and Control, **28**, 641–643 (1987)

J. Gulliver, D.J. Briggs, Time–space modeling of journey-time exposure to traffic-related air pollution using GIS. Environmental Research, **97**(1), 10–25 (2005)

A. Tapani, Versatile model for simulation of rural road traffic. Transport. Res. Rec. **1934**, 169–178 (2005)

Th. Benz, Evaluation of intelligent vehicle safety systems – a state-of-the-art example, in *15th World Congress on Intelligent Transport Systems*, New York, 2008

E.P. Todosiev, The Action Point Model of the Driver-Vehicle system, Engineering Experiment Station, Ohio State University Columbus, Ohio, Rep. Nr. 202A-1, 1963

M. Treiber, A. Hennecke, D. Helbing, Congested traffic states in empirical observations and microscopic simulations. Phys. Rev. E **62**(2), 1805–1824 (2000)

M. Treiber, A. Kesting, D. Helbing, Delays, inaccuracies and anticipation in microscopic traffic models. Phys. Stat. Mech. Appl. **360**, 71–88 (2006)

D. Van Vliet, SATURN, a modern assignment model. Traffic Engineering and Control, **23**, pp. 578–581 (1982)

Vliet, D. van, Vuren, T. van, Smith M. J., The interaction between signal setting optimisation and reassignment: Background and preliminary results. Transportation Research Record, **1142**, 16–21 (1987)

R. Wiedemann, *Simulation des Straßenverkehrsflusses* (Schriftenreihe des Instituts für Verkehrswesen der Universität Karlsruhe, Heft 8, Karlsruhe, 1974)

R. Wiedemann, Modeling of RTI-elements on multi-lane roads, in *Advanced Telematics in Road Transport edited by the Commission of the European Community*, DG XIII, Brussels, 1991

R. Wiedemann, U. Reiter, Microscopic traffic simulation: the simulation system MISSION, background and actual state. Project ICARUS (V1052) Final Report. Brussels, CEC. 2: Appendix A. (1992)

Chapter 7
Exploring Driver Behaviour Using Simulated Worlds

Andreas Tapani, Anna Anund, Nick Reed, and Alan Stevens

7.1 The Need

New measures to improve road safety need to be evaluated already at early stages of the development process to secure and maximise the proposed measures' benefits.

To assess impacts of already well-tried measures to improve the traffic system, one can conduct before and after studies or cross-sectional studies, based on field data. Road safety analysis of traditional safety measures can for example be conducted based on the actual crash turn out. New technologies to increase the forgiving or self-explanatory properties of the road traffic system can however not be reliably evaluated based only on field data. Even though some measures already have been introduced in the traffic system, they are not frequent enough for conclusions to be drawn. For example, for many types of recently introduced in-vehicle driver assistance systems, the proportion of equipped to un-equipped vehicles is still too small for conclusions to be drawn. Instead, evaluations of new measures to improve the forgiving or self-explanatory properties of the road traffic system have to be based on laboratory studies and modelling.

7.2 The Use of Simulators in Empirical Testing Methodologies

Driving behaviour is a complex, multifaceted information processing task that humans undertake with relative ease. However, mistakes or misjudgements at inappropriate times can result in tragic consequences for the driver, passengers, other road users and/or bystanders.

A. Tapani (✉) and A. Anund
Swedish National Road and Transport Research Institute (VTI) and Linköping University, Department of Science and Technology (ITN), Linköping, Sweden
e-mail: andreas.tapani@yti.se

N. Reed and A. Stevens
Transport Research Laboratory (TRL), London, UK

E. Bekiaris et al. (eds.), *Infrastructure and Safety in a Collaborative World*,
DOI 10.1007/978-3-642-18372-0_7, © Springer-Verlag Berlin Heidelberg 2011

Even though fully automated roads are possible to achieve using today's technology (Thorpe et al. 1997), it is still considered to be a utopia. The driver will, for the foreseeable future, remain an essential part of the driving process. There are several reasons for this; one non-negligible factor is that people are not willing to hand over the responsibility of driving to the vehicle. This conclusion can be drawn from the results of acceptance studies of driver assistance systems, which often show higher acceptance of purely information systems, than of systems that take over control of parts of the driving task (Brookhuis et al. 2001). Consequently, driver behaviour is, and will remain, crucial for successful introduction of measures to improve the forgiving or self-explanatory properties of the road traffic system. It is therefore appropriate to begin evaluations of measures for improvement with the measure's impact on driver behaviour.

The tools used for studying the measure's impact on individual driver behaviour have in common that they consider test drivers' behaviour in a laboratory situation. Although analysis of driver behaviour through observation of performance in the real world produces data with the greatest validity, it is also difficult to exert control over either the number or the type of vehicles involved or the demographics of the driving population. Moreover, since the measure under consideration can be assumed not to be widely available in the traffic system, it is not possible to measure data directly in the field. One alternative of implementing studies, on a dedicated test track using a suitably instrumented vehicle, allows highly detailed behavioural measurements, but one cannot expose participants to any risk of injury. An alternative approach is to implement the functionality of the measure in a driving simulator. This approach has the advantage that it is possible to control the traffic situation completely. Possible drawbacks of the driving simulator approach concern the realism and validity of the simulator. There are also other alternatives for studying driver behaviour, e.g. stated preference methods.

Knowledge of the impact of a measure on driver behaviour can be sufficient to enable measures design for improved driver comfort and acceptance. However, in order to evaluate the measures' potential to remedy road safety, traffic flow quality-of-service and environmental issues, it is necessary to aggregate the effects on individual driver behaviour to the traffic system level. This aggregation relies on modelling and estimation of the effects of the measure under different traffic conditions and on different road types.

Traffic simulation models, which describe conditions in a traffic network given the properties of the road network and the traffic demand, are useful for such analyses. Microscopic traffic simulation models consider individual vehicles in the traffic stream. It is therefore possible to include the characteristics of the measure and the driver behaviour associated with the measure in the driver/vehicle sub-models of the simulation. This makes it possible to estimate the effects on the traffic system through traffic simulation experiments, see Fig. 7.1. More on this can be found in Chap. 6. This new approach to bring methods for collecting data on individual driver behaviour and traffic simulation together in a unified empirical testing methodology was explored in the IN-SAFETY project.

Interactive driving simulation addresses both limitations and can provide detailed information about the behaviour of the driven vehicle, in relation to the environment and to other vehicles. These can be combined with physiological

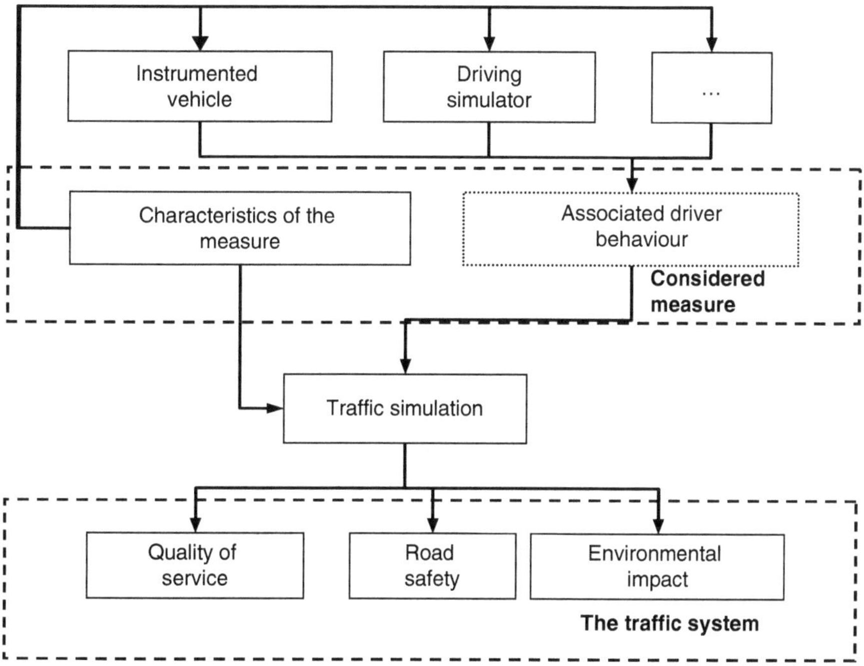

Fig. 7.1 Evaluation framework for measures to improve the forgiving and self-explanatory properties of the road traffic system

measures, such as electroencephalography (EEG), heart rate monitoring and eye tracking, to provide a detailed and comprehensive representation of behaviour and performance. Simulated scenarios can be created which present driving situations of different level of difficulty and/or danger, whilst the participant is at no risk of real harm. However, a participant suitably familiarised to driving in the simulated environment still perceives the element of risk and consequently produces behaviour that is representative of real driving (Törnros 1998).

A further advantage of simulation is the precise repeatability of scenarios. For example, an autonomous vehicle within a simulated scenario can be programmed to brake, achieving a precise deceleration rate when the driven vehicle is at a predetermined time headway value. The repeatability of trials and the precise measurement of behaviour are huge benefits for the researchers charged with analysing driver performance. Repeated-measure and matched-pair experimental designs can be exploited, allowing comparisons to be made between participants across trial conditions, with a high degree of confidence in the reliability of the results. Furthermore, detailed participant profile information and subjective opinions about the test conditions can be obtained through pre- and post-trial questionnaires; most of which would be almost impossible to apply in testing conducted on real roads.

TRL has successfully operated a driving simulator for more than 20 years and in that time the simulator has seen a number of different incarnations over time to keep pace with improvements in vehicle, projection, computing, and simulation technologies. The latest (Reed 2006) uses a standard family hatchback, a limited motion platform and realistic graphics and sound. Progress in computer graphics and 3D modelling now allows the creation of simulated environments that match real road schemes being constructed. The software enables full control of autonomous vehicles within the scene and an ability to make wide ranging changes to the performance characteristics of the driven vehicle (Fig. 7.2).

In the present chapter, four case studies that have been performed by TRL are presented, namely:

- Active Traffic Management
- Non-physical motorway segregation
- Actively illuminated road studs and Psychological traffic calming

Another case study has been implemented in the VTI simulator, during the IN-SAFETY project (Fig. 7.3).

Fig. 7.2 TRL Car Simulator during Red X trial

Fig. 7.3 The VTI moving base simulator

The objective of this driving simulator study was to study the effects of haptic in-vehicle HMI as a substitute for infrastructure elements installed to increase the forgiving and self-explanatory nature of rural road environments. The infrastructure elements considered are milled rumble strips. The effects of milled rumble strips and in-vehicle "virtual" rumble strips were studied for drivers being both not sleep deprived (alert) and sleep deprived, since there is a need for knowledge taking into account differences in driver status. Overtaking is a critical situation of interest in relation to rumble strips, which is difficult to study in real traffic. In such cases, a driving simulator is a useful tool, which allows the creation of a realistic scenario and collection of data on overtaking behaviour, while retaining a high degree of control over the experiment.

As a part of IN-SAFETY, new and improved traffic simulation models have been developed. The rural road traffic simulation model RuTSim (Tapani 2005) was improved to allow modelling of differences in overtaking behaviour. This allows driver behaviour data, from the driving simulator study, to be aggregated to the traffic system level using traffic simulation. It is a direct application of the relevant work on micro/macro simulators adaptation for safety impact analyses, as performed and described in Chap. 6. Thus, now it is possible to connect them to real data from experimental studies and evaluate the safety impact of the proposed measures.

As an example application of the new empirical testing methodology, the IN-SAFETY driving simulator pilot including milled rumble strips and the use of the RuTSim model to aggregate the individual driver behaviour observed in this driving simulator to the traffic system level, are also presented in this chapter.

7.3 Case Study 1: Active Traffic Management

Congestion brings many vehicles into close proximity, raising the probability of collisions such as rear-end shunts or sideswipes (Webb 1995). As well as reducing congestion, there is continuing pressure to make better use of infrastructure and reduce vehicle emissions (Stern 2006). One such scheme, being planned in 2004 as part of "Active Traffic Management" (ATM), was to implement Variable Speed Limits (VSLs) under conditions of congestion (3-lane VSL) and directing traffic to use the hard shoulder as an active traffic lane under conditions of heavy congestion (4-lane VSL).

ATM involves gantries at 500 m intervals with Advanced Motorway Indicator (AMI) signs above each lane (including the hard shoulder), to provide lane-specific information and a Variable Message Sign (VMS) for the provision of general safety guidance as well as information about accidents, delays and weather conditions. One option was to use a blank AMI above the hard shoulder (whilst all other AMIs display the VSL), indicating to traffic that normal motorway rules apply to the hard shoulder, i.e. it should be used for emergencies only. Alternatively, it had been proposed that a red X symbol should be used to give a definite signal to motorists that the hard shoulder is unavailable to traffic.

Prior to the implementation of hard shoulder running on the real motorway, it was possible to investigate the behaviour of drivers in response to these different signs using TRL's driving simulator (Thornton et al. 2005). Seventy-two participants were recruited and were assessed across experimental factors of Sign (Blank AMI vs. Red X AMI – to signal hard shoulder closure), Information (Informed vs. Uninformed about ATM), and Age (Younger vs. Older drivers).

During their drive, participants were instructed to hurry but then encountered clusters of simulated congestion. This was to encourage participants to make best progress along the route, using whatever road capacity they felt was open to them along the route. Analysis would then focus on the level of contravention and inappropriate use of the motorway. In one section there was no means by which a participant could overtake the congestion cluster, unless they used the hard shoulder whilst it was closed to normal traffic. In another section, the hard shoulder was opened to traffic (4-lane VSL) and the participant was thus able to overtake the simulated congestion traffic by travelling in the hard shoulder.

After completion of the trial, a questionnaire allowed assessment of the factors that were determinants in the decision by participants to use the hard shoulder, both at times when it was open and times when it was closed.

Participants who were aware of the operation of the ATM before taking part in the trial used the scheme more effectively than those who were uninformed. Informed participants used the hard shoulder more often when it was appropriate to do so and used it sooner and for longer than the Uninformed participants. It was also found that the four participants who misused the hard shoulder to overtake congestion were all in the Uninformed group. Questionnaire responses indicated that Informed participants were significantly more confident about using ATM and the effect it will have on motorway travel and safety than Uninformed participants. Once they had read the information leaflet post-trial, Uninformed participants recognised how useful it would have been in raising their awareness of the operational regimes of ATM before entering the scheme. These results were used to highlight that the information strategy must be comprehensive, to ensure that drivers both approach the scheme in the most positive frame of mind and, when using the scheme, do so as safely and as comfortably as possible.

Since completion of the simulator study, the M42 ATM scheme has been successfully rolled out and has enjoyed remarkable success, delivering improved traffic flow and travel times, while having no detrimental effect on safety (Department for Transport 2008). Wider implementation of the ATM measures is now being planned and its success is owed, in part, to the simulator testing prior to commencement of the scheme.

7.4 Case Study 2: Non-Physical Motorway Segregation

To improve traffic flow on the motorways just north of Manchester, one concept was to segregate longer distance 'Strategic' traffic from more local traffic by restricting them to the segregated outer two lanes of the four lane M60 motorway.

Fig. 7.4 Motorway segregation simulation

This was expected to result in increased journey times for local traffic and reduced journey times for strategic traffic. However, it would bring the advantage for all users of more reliable journey times. The proposed scheme segregated the lanes by non-physical means; a combination of specific line markings, road surfacing, signage, and operational regimes (Fig. 7.4).

Simulation offered an ideal way of assessing driver performance under the various operating conditions in a completely safe environment, before any commitments to infrastructure changes had been made. Seventy-two licensed drivers took part in the study, which employed a 2 (age) × 2 (route) × 2 (signage) between groups design. Participants fell into either the Younger (17–44) or Older (45+) age categories. They were assigned to drive either the Local route or the Strategic route. There were two signage schemes, Text destination and Symbol destination, tested in the trial and participants were presented with only one of the options. Therefore, there were eight experimental conditions and an equal number of participants were assigned to each.

As with the ATM simulator study, participants were instructed that they were late for an important meeting and then encountered heavy traffic in their designated lanes of travel. Of particular interest were the number of lane changes made within the segregation, the number of drivers who crossed the segregation, and the behaviour of those who crossed it. At the end of the trial, participants completed a questionnaire which explored their understanding of and opinions towards, the scheme. Driver behaviour was compared across conditions to determine the effects of age, route and signage scheme.

Results showed that most drivers joined their designated lanes well before the start of the segregation, after the gantry announcing the designated lanes and prior to the dashed hazard line indicating the start of the segregation. The questionnaire showed that most drivers understood where they were required to travel. Drivers on the local route tended to join their designated lanes of travel later than drivers on the strategic route. This may be because drivers feel more comfortable overtaking traffic on the right than undertaking traffic on the left.

A significant number of participants crossed the segregation. 21 out of 72 participants crossed at least once and there were a total of 33 crossing incidents in total. Younger drivers were more likely to cross than older drivers and this is consistent with the generally more aggressive driving styles displayed by younger drivers in the normal motorway sections at the beginning of the trial.

The average speed of participants when crossing into the incorrect region of the motorway was 40 mph; this was slower than the average speed of the traffic in this region which was approximately 50 mph. The average speed of drivers while travelling in the incorrect region was 57 mph. This exceeded the variable speed limit, which was set at 50 mph.

Many participants expressed frustration with the task and some participants commented that the signs or road markings were confusing.

It was concluded (Luke et al. 2006) that the trial demonstrated that significant numbers of drivers would cross the segregation under the conditions and scheme format presented in the trial. However, it should be noted that the results of this trial can be considered a worst case scenario for crossing incidences, as participants were placed under extreme time pressure in a situation where there was a large discrepancy in the traffic flow between the two segregated areas. In addition, Strategic route drivers were presented with conflicting signage, which is unlikely to be present in any on-road implementation of the scheme. The results of this study demonstrated the significant and undesirable impact of faulty signage.

7.5 Case Study 3: Actively Illuminated Road Studs

This study examined the potential improvement to road safety at night that may be achieved by illuminated road studs ('Active' studs) in place of standard ('Passive') retroreflective studs (Fig. 7.5).

Fig. 7.5 Active road studs

TRL's driving simulator was used to create a length of rural road and 36 participants were recruited from three age groups: Younger (17–25 years), Middle (26–54 years), and Older (55+ years). Each participant drove a 37.1 km trial route twice. The route had lead-in and run-out sections but the test section that was used for comparing across stud conditions was comprised of six repeats of a basic trial section. There were six critical corners in the basic section where the curve radius fell below 150 m and these were used for more detailed analyses.

In each drive, the participant experienced a simulated night-time environment and the road had sections with no studs and sections with studs. In one of their drives, the studded section had active studs; in the other drive it had passive studs. The studs were placed at varying intervals (based on the road characteristics) along the centreline of the road. Additional red studs (in both the active stud and passive stud versions) were placed on the nearside of the four sharpest bends in the repeat section used to create the trial route. The driven vehicle used dipped headlights throughout and no other traffic was present in the simulation.

As well as a pre-trial questionnaire, participants completed a post-trial questionnaire that recorded their subjective feelings towards each of the stud conditions once they had completed their two drives. Picture cue cards were used to remind participants of the environments that they had seen.

Results demonstrated that in each age group, participants' average speed when driving was significantly higher (by around 3 mph) in both studded conditions, relative to the no stud condition (Reed 2006). However, there were no significant differences between the active and passive stud conditions across the age groups in terms of overall speed. Assessment of how participants controlled their lateral position revealed that older participants spent significantly less time with the right edge across the centreline of the road with active studs than they did with passive studs.

More detailed analysis of braking results in the critical corners suggests that participants were better informed about how they needed to control the vehicle in order to negotiate the bends when the active studs were present. Similarly, analysis of drivers' lateral position in the corners revealed a marked difference between the passive and active stud conditions in right turns and suggests that enhanced delineation of the offside road edge may promote improvements in drivers' lateral control of their vehicle.

Broughton and Buckle (TRL Report 653 2006) reported that loss of control was the only precipitating factor in the causation of accidents (of all severities) that had shown a significant increase since 1999. The results from this trial suggest that the active stud installation that drivers observed in the simulator improved their control, particularly in right turns and for older drivers. It is, therefore, possible that the introduction of active road studs may help to reverse this trend.

Participants reported that active studs encouraged them to drive faster than they would normally. However, this is contradicted by the simulator data, which showed that there were only very slight increases in speed with active studs. This discrepancy between drivers' opinion and observed behaviour highlights the benefit that simulation can bring in allowing schemes to be tested by real drivers in a naturalistic environment. Participants also reported that they believed active studs would be highly beneficial to road transport and road safety.

Overall, it was concluded that active studs offer a significant safety advantage over standard passive retroreflective studs, since they appear to improve lane guidance in right turns without causing drivers to proceed at higher speeds.

7.6 Case Study 4: Psychological Traffic Calming

A reduction in traffic speed is associated with a reduction in accident frequency and severity (Taylor et al. 2000). However, traditional physical traffic calming measures (such as humps, speed cushions, speed tables and chicanes) can induce deceleration and acceleration which increase vehicle emissions (Cloke et al. 1997) as well as noise and vibration nuisance (Abbott et al. 1999). As well as construction costs, they may also damage vehicles, cause driver discomfort and be visually intrusive, particularly if warning signs are required.

The above concerns were sought to be addressed by developing traffic calming measures by evaluating interventions not involving construction, but that would be effective in reducing speeds. So-called "psychological traffic calming" has the potential to deliver speed reduction through interventions that raise perceived uncertainty or complexity.

To test the effectiveness of possible measures, a study (Kennedy et al. 2005) was instigated in which participants viewed photomontages of different road scenes into which different psychological traffic calming measures had been introduced. Participants were then asked to indicate how fast they would drive through such a road scene. The measures that caused participants to report the greatest reduction in speed were then taken forward into a study in the TRL Driving Simulator, where naturalistic changes in driving behaviour could be observed (Fig. 7.6).

Fig. 7.6 Simulated village traffic calming

The simulator trial demonstrated that continuous or repeated measures were required to sustain speed reductions, with a village gateway alone having little effect on speed within the following village. Although effective in the photomontage study, applying a coloured surface to the road did little to slow the speed of the driven vehicle in the simulator study. However, as found in the review, creating uncertainty was effective at speed reduction. Edge markings were introduced to create a visual narrowing of the road. This created a perceptual uncertainty about correct vehicle placement, leading to reduced speeds. This effect was enhanced when the edge markings were also given a texture. A similar measure tested was the use of red brick surfacing to narrow the road. This created uncertainty as it was not clear to motorists whether the brickwork was the footway or part of the road. The centreline was also removed and with the redbrick edging, drivers were concerned about meeting oncoming traffic, creating further uncertainty.

Where successful, the speed reduction measures caused the greatest speed reductions in drivers who drove at the fastest speeds in control sections (where no such psychological traffic calming measures were applied). The most effective measures, as found in the simulator study, were subsequently applied in the UK village of Latton. Traffic flow and speed detection techniques were applied, before and after the implementation of the psychological traffic calming measures, to test the efficacy of the new measures. In the village, two-way mean speeds fell by 12–50 kph and 85th percentile speeds fell by 13–16 to 60 kph. Although within the village over half of the vehicles still exceeded the new 30 mph (48 kph) speed limit during the 'after' survey, the proportion exceeding 40 mph (64 kph) fell from 50 to around 10%. A survey of the opinion of the residents of Latton found broad support for the new measures. The only concern raised was about removal of the centre white line from the road, but this reflects the uncertainty that the scheme is intended to create and from which speed reduction is a direct effect.

In addition to enabling the research team to test possible speed reducing interventions before implementing them in the real world, this study demonstrated the validity of the driving simulator for this type of work. The speed reductions observed in the simulator brought comparable speed reductions in the real world, suggesting that participants drive the simulator and respond to the interventions in a realistic manner.

7.7 Case Study 5: Virtual Rumble Strips

Effects of the different rumble strip conditions, no rumble strip, milled rumble strip and in-vehicle rumble strip, on individual driver behaviour were studied in the IN-SAFETY driving simulator study, performed by VTI. The road used for the driving simulator scenario was an approximately 9 km long uninterrupted 9 m wide Swedish two-lane highway. A repeated measures design including 20 test persons was adopted for the study. Each test person drove the simulator in both alert and sleep deprived condition. During each drive, the test persons drove on the same road without rumble strips, with visible milled rumble strips and with rumble strips presented as an in-vehicle assistance system with only sound and vibration. For each rumble strip condition, the test persons were given multiple opportunities to overtake a slower vehicle in front. The given overtaking opportunities differed with respect to the distance to the closest oncoming vehicle. Car-following and free driving situations were also included in the driving simulator scenario. The driving simulator views of the road with milled rumble strip and with in-vehicle "virtual" rumble strips are shown in Fig. 7.7.

The results of the driving simulator study indicated no significant differences in individual driver behaviour when equipped with in-vehicle rumble strips, compared to when driving on a road with visible milled rumble strips. There is consequently a potential to use in-vehicle rumble strips as a complement to milled infrastructure based rumble strips. There were however indications of differences between the rumble strip conditions and the two driver states that could influence performance on the traffic system level. These observations are presented below.

The test persons' driving speed in free driving conditions showed differences depending on driver state and rumble strip condition. Averages and standard deviations of the observed free driving speeds are shown in Table 7.1.

As can be seen in the table, there is an indication of slightly higher speeds for sleep deprived drivers. There are also larger variances in the speeds of sleep deprived drivers. Moreover, a tendency for higher speeds when supported by rumble strips was observed. This indication was stronger for sleep deprived drivers. Higher speeds in connection with rumble strips can possibly be attributed to behavioural compensation. The sense of increased safety when assisted by rumble strips might have stimulated the test persons to increased speeds.

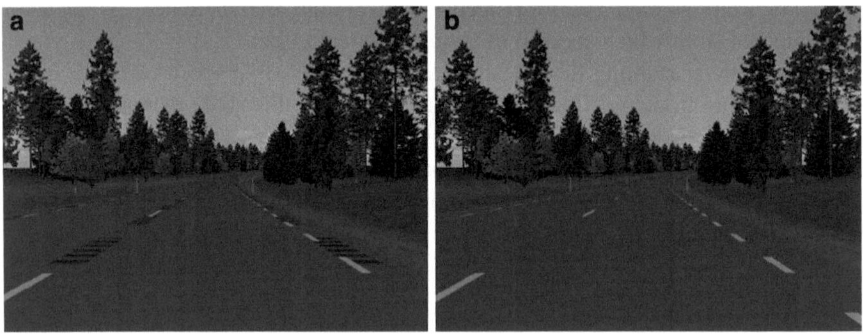

Fig. 7.7 Driving simulator views, (**a**) milled rumble strips and (**b**) in-vehicle "virtual" rumble strips

Table 7.1 Observed average free driving speeds (km/h)

	Alert		Sleep deprived	
	Average	Standard deviation	Average	Standard deviation
No rumble strip	95.9	8.5	96.1	10.7
Milled rumble strip	96.1	10.6	98.3	12.3
Virtual rumble strip	97.2	10.1	97.4	15.2

Car-following reaction times were also studied. A situation with a pre-programmed decelerating vehicle in-front of the driving simulator was used to estimate reaction times. The time from the start of the deceleration of the vehicle in front until brake force was applied in the simulator was used as an estimate of the reaction time. The average reaction time of alert drivers was 1.47 s and the corresponding standard deviation was 0.33 s. The observed reaction time of sleep deprived drivers was 1.63 s and the standard deviation was 0.29 s. There was no sufficient number of situations that could be used to estimate differences in reaction times for the different rumble strip modes.

The distance to the closest oncoming vehicle at the start of the overtaking manoeuvres was also considered. Since, it is known to be difficult to determine absolute distances in driving simulators, only relative differences were considered. Sleep deprived drivers were observed to accept overtaking with 4% shorter distance to the closest oncoming vehicle than alert drivers. There were too few overtaking situations with different rumble strip condition to estimate differences between the different rumble strip conditions.

7.7.1 Aggregating Impacts on Driver Behaviour to the Traffic System Level Using Traffic Simulation

The two-lane road used in the driving simulator study was modelled in RuTSim. The traffic flow on the road was set to 300 vehicles per hour and direction. This flow was chosen to represent typical traffic conditions on this type of road in Sweden.

Systems that give active support and thereby take over or actively interfere with parts of the driving process, e.g. adaptive cruise controls and speed limiters, will have an impact on both vehicle properties and driver behaviour. Assistance and information systems that do not give any active support can be assumed to only influence driver behaviour. Neither infrastructure based milled rumble strips nor in-vehicle virtual rumble strips give active support. Consequently only the observed driver behaviour needs to be considered in the traffic simulation modelling of rumble strips.

The observed differences in individual driver behaviour includes, as described above, differences in free driving speeds, reaction times and overtaking behaviour. These differences are to be taken into account in the traffic simulation modelling. Separate driver/vehicle classes, corresponding to each combination of driver state and rumble strip mode, were created to facilitate modelling of the different driver characteristics. All vehicle classes were based on the standard passenger car vehicle type in RuTSim. For each driver/vehicle class, the parameters of the RuTSim model was adjusted to take into account the observed differences in driver behaviour. Simulations with the RuTSim model, including the modelling of different rumble strip conditions and driver fatigue, can now be used to estimate traffic effects of driver fatigue and rumble strip support. The outcome of such simulations is presented below.

Average journey speeds over the simulated road are included as an indicator of the quality of service effects of the rumble strips. Safety effects are indicated by the time to collision based indicators introduced by Minderhoud and Bovy (2001). These measures use a critical time to collision threshold to distinguish safety critical situations from situations in which the driver remains in control. Time extended time to collision (TET) is the total time spent with sub-critical time to collision and therefore an indicator of the extension of possibly safety critical situations. The focus of the safety analysis in this study was on the interaction between oncoming traffic in overtaking situations. The safety indicators have therefore been calculated using the time to collision with respect to the closest oncoming vehicle during overtaking situations.

RuTSim is a stochastic simulation model, random numbers are used to assign vehicle properties and in decision processes during the simulation, e.g. in the assignment of basic desired speeds and for overtaking decisions. Simulations with different random number seeds will consequently give different results. Therefore, multiple simulation runs have to be conducted to estimate distributions of the results. The confidence intervals presented in this section are all based on ten replications with different random number seed and have been calculated by assuming normally distributed output from individual replications.

It is difficult to hypothesize on the combined effect of the rumble strips and driver states on journey speeds prior to the simulation. The resulting 95% confidence intervals for average journey speed for alert and sleep deprived drivers in connection with the three rumble strip conditions are displayed in Fig. 7.8. The confidence intervals corresponding to sleep deprived drivers are naturally wider than the confidence intervals for alert drivers due to the lower percentage of sleep deprived drivers in the simulated traffic. As can be seen in Fig. 7.8, the average journey speeds for alert drivers on the road with milled rumble strips are lower than the speeds for alert drivers on the two other rumble strips conditions. There is no distinguishable difference in the resulting average journey speeds for alert and sleep deprived drivers on the road with no rumble strip or on the road with a milled centre line rumble strip, see part (a) and (b) of Fig. 7.8. The average journey speed for sleep deprived drivers is however lower than the speed for alert drivers when equipped with an in-vehicle virtual rumble strip as can be seen in part (c) of Fig. 7.8. There is also an indication that the difference is decreasing with increasing percentage of sleep deprived drivers in the traffic. The lower speed for sleep deprived drivers equipped with in-vehicle rumble strips can most likely be attributed to the high standard deviation of free driving speeds for the individual drivers, see Table 7.1. As the percentage of sleep deprived drivers is increasing, more alert drivers will be constrained by slow sleep deprived drivers. The difference in journey speed between the two driver states will therefore decrease.

The difference in overtaking behaviour of sleep deprived drivers can be assumed to result in higher values of the TET and TIT indicators for sleep deprived drivers. Figure 7.9 display the resulting 95% confidence intervals for average TET with respect to oncoming vehicles during overtaking situations. There is an indication that sleep deprived drivers spend more time with a sub-critical time to collision. This observation can be directly attributed to sleep deprived drivers acceptance of 4% shorter distance to the closest oncoming vehicle at the start of overtaking

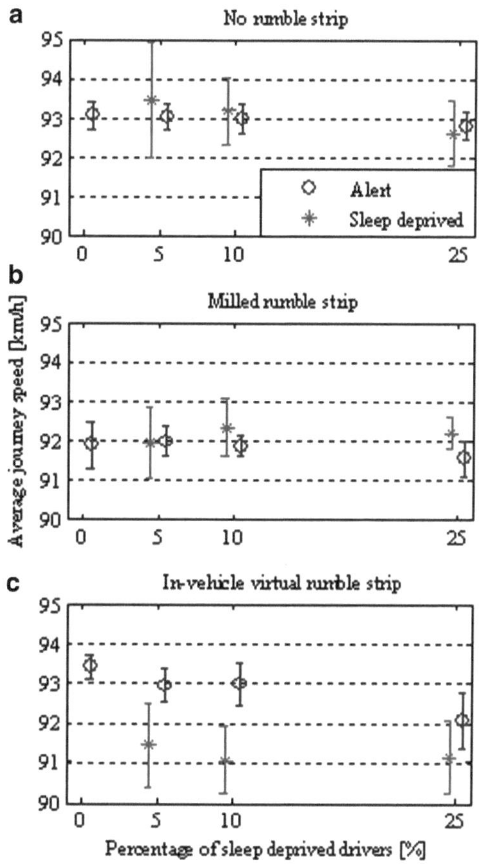

Fig. 7.8 Average journey speed for alert and sleep deprived drivers for (**a**) no rumble strip, (**b**) milled rumble strip and (**c**) in-vehicle rumble strip (95% confidence intervals)

manoeuvres. No difference in the results for different rumble strip modes can be observed. This is as expected, since no difference in overtaking behaviour between the different rumble strip conditions was observed.

In summary, the observed differences in individual driver behaviour give rise to observable indications of differences on the traffic system level indicators, average journey speed and TET.

7.8 Lessons Learned

These studies demonstrate that simulation can play a useful role in understanding changes in driver behaviour and anticipated safety outcomes as a result of modifications to the road infrastructure or introduction of new in-vehicle assistance system. They allow testing under a wide range of conditions whilst ensuring

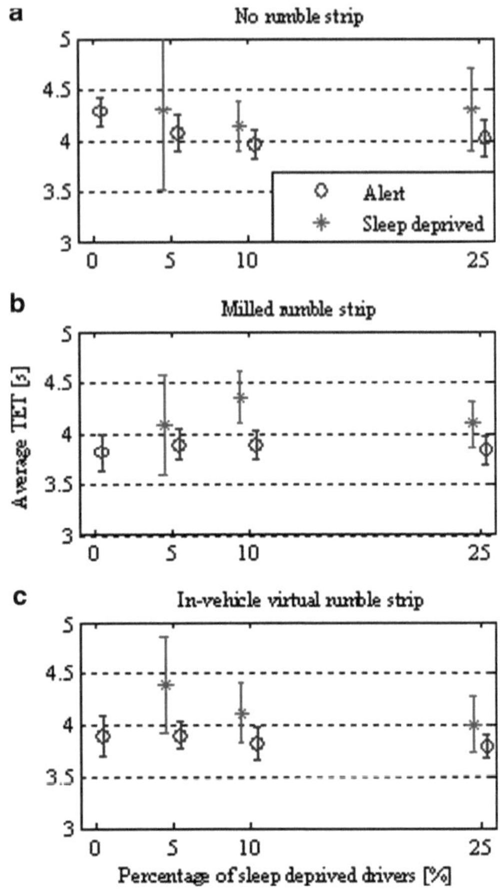

Fig. 7.9 Average TET for alert and sleep deprived drivers for (**a**) no rumble strip, (**b**) milled rumble strip and (**c**) in-vehicle rumble strip (95% confidence intervals)

participant safety, and enable evidence-based decisions to be made before innovations are actually applied.

Based on the simulator work and subsequent validation, a number of lessons can be drawn concerning aspects of SER and FOR enabling road design.

Studies of "Active traffic management" and "Non-physical motorway segregation" show that systems designed to ease congestion can also have implications for safety. However, any potential safety problems can be mitigated by informing drivers and helping them to understand the new designs e.g. through appropriate signage, thus actively contributing towards a more self-explanatory road environment (SER). The information strategy needs to be comprehensive, to ensure that drivers both approach the scheme in the most positive frame of mind and, when using schemes, do so as safely and as comfortably as possible.

Studies of interventions designed for rural roads specifically to improve safety and, more precisely, promote the forgiving nature of the road environment by vision enhancement and/or speed reduction, show that they may also have unanticipated consequences. For example, delineation of a road at night by "Actively illuminated road studs" offers a significant overall safety advantage compared with standard passive retroreflective studs but the greater visibility of the road ahead could be exploited by drivers choosing to drive at higher speeds. Also, where speed is a specific problem, such as in rural villages, creating uncertainty through psychological traffic calming can be effective in promoting speed reduction.

Application of the empirical testing methodology was exemplified by a study of the potential to replace milled centre line rumble strips with rumble strips presented to the driver as an in-vehicle assistance system. Individual driver behaviour data from a driving simulator study was used for traffic simulation using the RuTSim model. The simulation results display lower journey speeds for alert drivers on the road with milled rumble strip compared to on the road without rumble strip and the road with in-vehicle virtual rumble strip. There were also differences in the journey speeds for alert and sleep deprived drivers when equipped with in-vehicle virtual rumble strips, the journey speeds were lower for sleep deprived drivers. The derived safety measures indicated that sleep deprived drivers spent more time in safety critical situations due to changed overtaking gap-acceptance behaviour. The use of traffic simulation for the aggregation of individual driver behaviour made it possible to straightforwardly study the combined impact of changes in drivers' free driving speed, reaction time and overtaking behaviour.

As an overall conclusion, SER and FOR implementation scenaria need a careful and thorough evaluation (often employing driving simulators), to avert the risk of negative counter effects to traffic safety, as well as to optimize their application framework and, subsequently, their effectiveness.

On the other hand, driving simulator studies are commonly designed to study the subjects' reactions in relation to isolated critical situations. Application of driving simulator studies to collect driver behaviour to be used for traffic simulation place new requirements on the driving simulator scenario design. It becomes necessary to observe the subjects' continuous actions and reactions while driving to allow car-following and lane-changing/overtaking modelling for traffic simulation. A need for further research in this area has been identified through the IN-SAFETY project. Estimation of car-following and overtaking situations from the driving simulator data were found to be challenging tasks. Further research on driver behaviour modelling for traffic simulation and overall driver behaviour assessment, including vehicles equipped with driver assistance systems, is also needed. In addition, more reliable safety analysis using traffic simulation models require research on the relation between simulation-based safety indicators and accident risks.

References

K.A. Brookhuis, D. de Waard, W.H. Janssen, Behavioural impacts of advanced driver assistance systems – an overview. EJTIR **1**(3), 245–253 (2001)

J. Broughton, G. Buckle, Monitoring progress towards the 2010 casualty reduction target – 2004 data. TRL Report 653, Transport Research Laboratory, Crowthorne, 2006

Department for Transport, Advanced motorway signalling and traffic management feasibility study – a report to the Secretary of State for Transport, DfT, London, 2008
J. Kennedy, R. Gorell, L. Crinson, A. Wheeler, M. Elliott, 'Psychological' traffic calming. Published Project Report 641, Transport Research Laboratory, Crowthorne, 2005
T. Luke, N. Reed, A.M. Parkes, T. Thornton, Driver behaviour in response to non-physical motorway segregation. Presented at the international congress of applied psychology, Athens, Greece, 2006
M.M. Minderhoud, P.H.L. Bovy, Extended time-to-collision measures for road traffic safety assessment. Accid. Anal. Prev. **33**, 89–97 (2001)
N. Reed, Driver behaviour in response to actively illuminated road studs: a simulator study. Published Project Report 143, Transport Research Laboratory, Crowthorne, 2006
N. Stern, Review on the economics of climate change (H.M. Treasury, UK, October 2006), http://www.sternreview.org.uk
A. Tapani, Versatile model for simulation of rural road traffic. Transp. Res. Rec. **1934**, 169–178 (2005)
M.C. Taylor, D.A. Lynam, A. Baruya, The relationship between drivers' choice of speed and the frequency of road accidents. TRL Report 421, Transport Research Laboratory, Crowthorne, 2000
T. Thornton, N. Reed, N. Gordon, ATM – driver behaviour during operational regimes. Presented at smart moving 2005, Birmingham, England, 2005
C. Thorpe, T. Jochem, D. Pomerleau, Automated highway free agent demonstration, in *IEEE Conference on Intelligent Transportation Systems, Proceedings*, 1997, Boston, MA, pp. 496–501
J. Törnros, Driving behaviour in a real and a simulated road tunnel – a validation study. Accid. Anal. Prev. **30**(4), 497–503 (1998)
W.B. Webb, The cost of sleep-related accidents: a reanalysis (technical comments). Sleep **18**(4), 276–280 (1995)
J. Cloke, D. Webster, P. Boulter, G. Harris, R. Stait, P. Abbott, L. Chinn, Traffic calming: Environmental assessment of the Leigh Park Area Safety Scheme. TRL Report 397. TRL Limited, Crowthorne, 1999
P. Abbott, M. Taylor, R.E. Layfield, The effects of traffic calming measures on vehicle and traffic noise. Traffic Engineering and Control **38**(8), 1997

Chapter 8
Managing the Risks. Road Risk Analysis Tools

J. Stefan Bald, Katja Stumpf, Tim Wallrabenstein, and Le Thu Huyen

8.1 Why We Need Road Risk Analysis

Safety is an important criterion to be considered when designing road infrastructure, developing cars or organising road traffic. Road safety depends on numerous factors (e.g. human behaviour, infrastructure, natural influences, legal factors, etc.).

While introducing new elements to the system (e.g. Advanced Driver Assistance Systems (ADAS) in cars, dynamic warning signs on the road), it has to be verified that the safety of the system is not influenced negatively. Does this new element enhance safety? Does it introduce negative side effects, which (over) compensate its positive effect? Does it shift risks from one part of the system, where it is accepted, to another, where the new risks are considered unacceptable, even if the overall safety is enhanced?

To answer these questions, it is necessary to analyse safety in detail, preferably in a quantitative manner. Therefore, it is needed to evaluate all risks and to estimate their risk values in order to show, which parts of the system contribute to the risks, which parameters influence the risks and in which way. Prospective evaluations of this kind are called "risk analysis", which allow to systematically analyse risks, especially their values, their reasons, their consequences and to evaluate the effects of new policies and technical solutions.

Experience shows that absolute safety is impossible. In every system not all dangers can be avoided completely. Therefore, it is generally accepted to describe or quantify the residual risk. In this case, safety refers to the level of risk that is socially acceptable in these real-life situations. If the risk level is acceptable, the system is considered as safe.

Risk is a very general term. It may and should be defined more precisely by relating it to certain exposure groups or exposure periods of time. It may be given in

J.S. Bald (✉), K. Stumpf, T. Wallrabenstein, and L.T. Huyen
Technische Universität Darmstadt (TUDarm), Darmstadt, Germany
e-mail: jsbald@verkehr.tu-darmstadt.de

E. Bekiaris et al. (eds.), *Infrastructure and Safety in a Collaborative World*, 143
DOI 10.1007/978-3-642-18372-0_8, © Springer-Verlag Berlin Heidelberg 2011

objective scales (real risk) or in a more subjective way (perceived risk). It should be mentioned, that the opposite of risk is called "chance". The difference is that risks are connected to negative consequences, and chances are connected to positive consequences (Durth and Bald 1988).

In many existing systems, statistics of accidents may be taken as assessment indicators of risk. Common used parameters are accident frequency, accident severity, number of fatalities, number of injuries and amount of material damage. In effect, one describes the relative frequency of negative events as personal accident, injury or material damage.

In some situations, especially in the case of very severe potential consequences (e.g. chemical plants) or when planning totally innovative system components, this retrospective approach cannot be used. It is not possible to install such a system and await accidents to evaluate its risks and compensate for them. A prospective approach is needed, which allows to foresee the future and estimate the risks of not yet realised systems. The general approach is the same. Instead of the relative frequency (related to the past) of a certain (negative) consequence, one estimates its (uncertain but estimable) probability (related to the future). The combination of certain damage and its probability is called "risk" (ISO 12100-1; Mahalel 1986). The higher certain damage may be or the more likely the realisation of the damage is, the higher is the risk. Insurance companies "measure" the risk by multiplying the amount of damage by its probability and summing up the risk values for all possible damages. A system is considered safe if its overall risk is lower than an accepted value.

8.2 IN-SAFETY Risk Analysis Method

In the IN-SAFETY project, there was a need to evaluate the risks which appear when new ITS are introduced on the road system (Bald et al. 2008a).

The road system is very complex for different reasons, as it:

- Is a big and heterogeneous system.
- Has a high grade of individuality.
- Is strongly influenced by human behaviour.
- Encompasses many participating and influencing stakeholders with different interests and personal rights.

A systematic risk analysis methodology has thus been built within IN-SAFETY, to deal with this complexity, called INsafety Risk Analysis Method (INRAM).

Additionally, the new systems are shifting responsibility between different stakeholders or shifting general risk to a specific stakeholder. This may lead to the situation, that the installation of a new technology reduces the overall risk but burdens a specific stakeholder with additional risks, who is not willing to accept this. A systematic risk analysis methodology has to deal with these different aspects of risks, especially legal and economic risks.

The developed INRAM tries to address this complexity and variety by combining different approaches into one methodology.

As an overall framework, the method which was described in the EU project ADVISORS (Wiethoff 2003) is used. It allows the combination of technical, behavioural, legal and organisational issues.

To allow a more detailed analysis of the technical and behavioural aspects, the Darmstadt Risk Analysis Method (DRAM) was developed further and integrated into the above framework. It is a model orientated, modular approach, which allows to deal with high complexity, as well as the description of uncertainty.

Further applications will show whether it will be useful, or even necessary, to refine the legal and organisational parts as well.

There are also some other road safety related components of the IN-SAFETY project which are useful for risk management of roads (Bald et al. 2008a).

Road Safety Audits (RSA) is systematic investigations of road designs by specially trained experts. They discover conflicts of the road design with the current knowledge (especially the standards). They provide with the efficient assessment of the design of new roads, as well as the safety potential of existing roads. In fact, they are risk analysis procedures on a very low level. Tools like "DV-Test", which is a computer tool for the systematic determination of a "Sustainable-Safe Indicator", may be used to help in this process. High level risk analysis methods, like INRAM, may be used to provide information to such tools.

8.3 ADVISORS Framework

ADVISORS (Wiethoff 2003) was a research project (2000–2003) co-funded by the EC. Its aim was to enable the identification of all implementation risks and highlight possible mitigating strategies. A traditional FMEA analysis for technical risk was extended (Fig. 8.1) to incorporate other risks, such as the behavioural, legal and organisational risks.

The following methodology was integrated in the findings of the ADVISORS project (Bekiaris 2001) resulting in an encompassing Risk Analysis Method, to allow to assess all relevant risks, namely organisational, legal, behavioural and technical risks (Fig. 8.1).

The ADAS functions are described and the various risk numbers are determined by different analysis methods. Every risk (technical, behavioural, legal and/or organisational) can be calculated by the following formula:

$$\text{Risk number} = \text{Severity} \times \text{Probability} \times (\text{Detectability} + \text{Recoverability})/2 \tag{8.1}$$

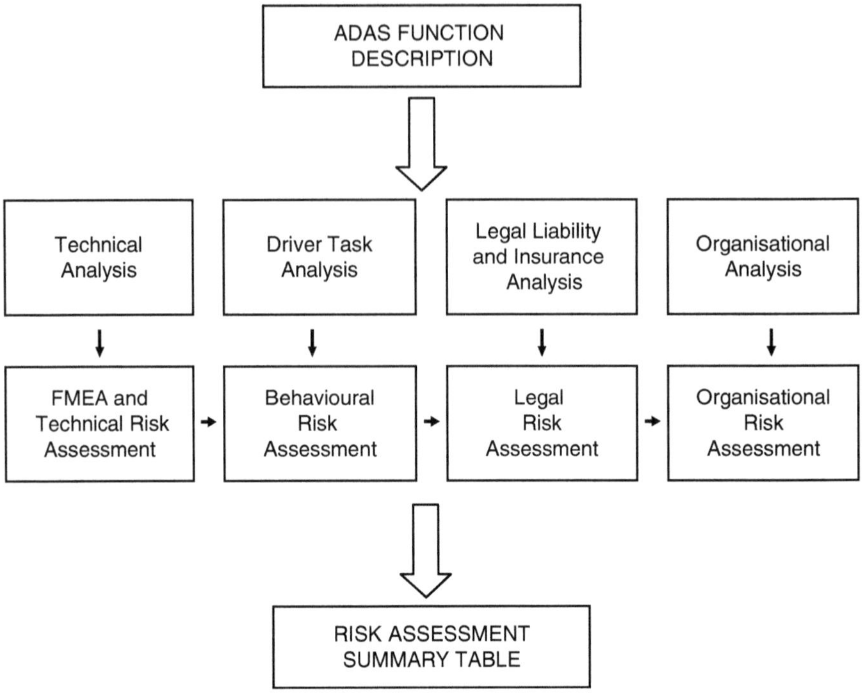

Fig. 8.1 Risk analysis process within ADVISORS (Bekiaris 2001)

At the end the overall risk number (ORN) is determined. It is given by the following equation:

$$ORN = \frac{RNT + RNB + \frac{RNL + RNO}{2}}{3} \tag{8.2}$$

- RNT = technical risk number
- RNB = behavioural risk number
- RNL = legal risk number
- RNO = organisational risk number

To obtain the RNT, a technical failure analysis considers hardware and software failures, and environmental events. It involves the evaluation of conceivable risks to the worker and public safety, and risk of damage to equipment, or the environment. Hazards can be identified by using a formal fault and hazard identification process, like the failure mode effect analysis (FMEA). Performing a FMEA starts with defining the system to be analysed, constructing a block diagram and, finally, identifying all potential items and interface failure modes.

To obtain the RNB, a Driver Behavioural Analysis is necessary. In the ADVISORS (Bekiaris 2001) project, its analysis method was the fault tree analysis, followed by the Success Likelihood Index Methodology (SLIM). The underlying idea was that the likelihood of an error occurring is dependent on a relatively small number of performance influencing factors (PIF), such as quality of training, procedures and time available for action. It is assumed that experts can judge how good or bad the PIFs are in specific situations. This rating is then multiplied with the relative importance of the PIF and all the outcomes are summed up to create the Success Likelihood Index. This Index predicts the probability of success of the specific situation. In the IN-SAFETY project, the DRAM method was introduced to obtain a more sophisticated objective approach.

For assessing the liability and insurance risk number (RNL), it is very important to understand the legal issues. As it is extremely difficult to obtain "general" legal opinions, it is helpful and necessary to subject specific circumstances to analysis. First, the relevant legal frameworks, in terms of traffic and product liability laws and insurance schemes, have to be described. The second step is to identify potential gaps and barriers. The last step is to evaluate and analyse the gaps through interviews, round table discussions and dissemination of preliminary results. As a result, recommendations for legislative action and insurance policies will be formulated and the overall risk can be assessed.

To obtain the RNO, familiarity and experience in the theory of Work Organisation and Management Structures is required for a detailed analysis. There is no single set of objective criteria for analysing the effectiveness of a particular organisation. To get the needed information about the organisation of the products, in this case ADAS, experts have been interviewed by questionnaires during workshops. It is possible to assess the risks and find their properties (i.e. severity or occurrence probability) for several systems.

8.4 Darmstadt Risk Analysis Method

The DRAM was introduced, aiming especially at analysing the technical and behavioural risk more systematically and in detail.

The first parts of DRAM were developed in 1988, as road safety research in Germany was facing the problem that pure accident analysis did not give enough information to derive valid conclusions on the actual risks, especially on the causes of dangerous situations. On the other hand, there was (and still is) much information available (e.g. on driver behaviour) which, with the existing methodologies, could not be systematically integrated into safety analysis. For this reason, research was initiated at the Technische Universität Darmstadt, to evaluate whether systematic risk analysis methods, which were used for risk analysis of complex industry installations, could be transformed and used in this domain (Durth and Bald 1998). Results of this first study were encouraging. The problem was to describe human behaviour in an appropriate way. For this reason, Bald, in his doctoral dissertation (1991), developed a predecessor of DRAM and used it for an early behavioural orientated road risk analysis of sharp bends.

Whether DRAM is suited to analyse legal and organisational risks as well, needs to be further investigated.

DRAM is a model orientated approach. The investigating team tries to simulate reality by describing its underlying or assumed cause and effect chain with a well structured model. If the outcome of the model does not contradict observations of the reality in the past, it is assumed that the model is valid. A valid model may be used to forecast outcomes of new systems or system responses in the future.

The main objectives of DRAM are:

- To be able to deal with uncertainty.
- To have the possibility to include all available information, even if it is highly scattered.
- To get access to the cause-and-effect chains of the road system.
- To have the possibility to improve and upgrade.
- To allow and encourage the cooperation of different research groups, even from different disciplines.

DRAM uses four different parts, called levels, and two tools to satisfy these objectives (Fig. 8.2).

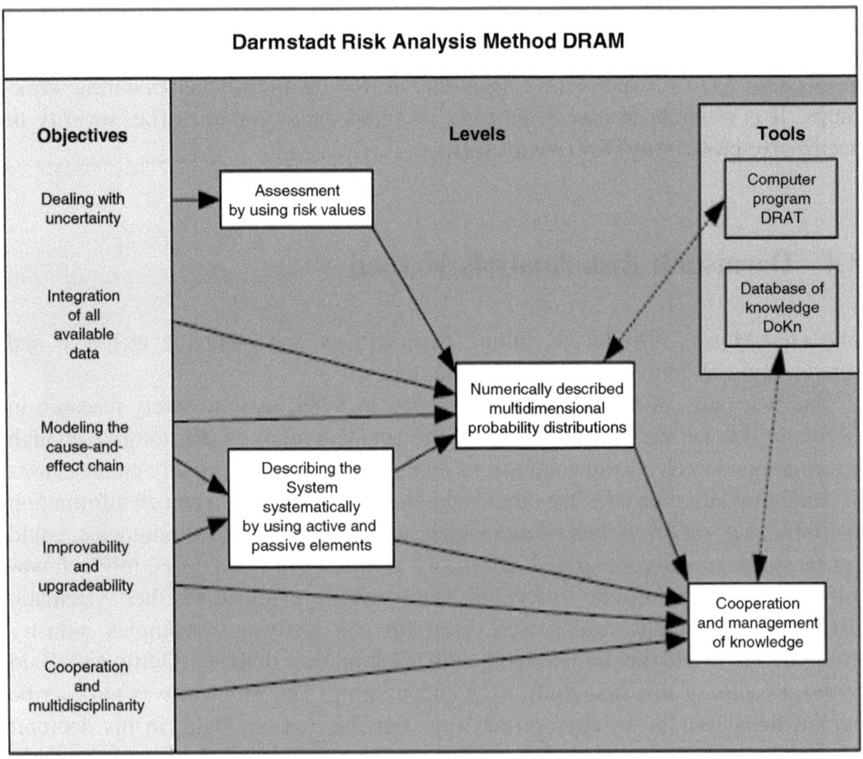

Fig. 8.2 Objectives, levels and tools of DRAM

It consistently uses probabilities to describe parameter values. As risk values are based on probabilities for certain damage values, DRAM considers risk numbers. This enables to deal with uncertainty.

DRAM uses numerically described multidimensional probability distributions (NDMPD) for describing its parameters and intermediate values. It uses distributions, so that the whole range of the parameter values and their related probabilities may be considered. Danger, as a consequence of many unfortunate effects, is not necessarily related to certain quantiles of the parameters. It uses numerically described distributions, as this enables to describe any shape of the distribution and to integrate all available data. It uses multidimensional distributions, as this enables to describe the complexity and the complicated dependencies of the system, which is necessary to model the underlying cause and effect chains. NDMPD may be a set of more than ten thousand numbers. To handle them, a computer tool, named Darmstadt Risk Analysis Tool (DRAT), is provided, which enables to combine and calculate with NDMPD without excessive effort.

DRAM describes the system, by using active and passive elements. This modular structure, where the active elements act as modules and the passive elements as interfaces enables:

- The cooperation across different research groups, even from different disciplines.
- To model the cause and effect chains of the process.
- To use intermediate, e.g. behavioural data and to evaluate parts of the model.
- To enhance, and even to refine single modules, to evolve critical parts of the model without the need for a complete rewrite (thus enhancing improvability and upgradeability).

Last but not least, DRAM facilitates cooperation and multidisciplinarity by its modular structure and its ability to build up a database of knowledge (DoKn), where system-related knowledge can be stored in the very general NDMPD format (considering information on copyright, etc.). The standards needed for collaboration, are organised very similar to the procedure to establish internet standards, the request for comments (RFC) procedure.

The components of DRAM are described in more detail in Bald et al. (2008b, c).

8.5 Getting Data for Risk Analysis

The real advantage is gained when the elements of a model are filled with data. As there is a tool provided, which enables to work with a huge amount of data, it is not necessary to reduce the data to more or less arbitrary characteristic values, which normally over- or underestimate risk systematically (Bald et al. 2008b). The possibility for quantitative description and evaluation (with NDMPD) principally enables any desired precision (only at the cost of a larger dataset and longer calculation time). The precision is only dependent on the availability and the gathering of data.

Generally, the question posed is where to get the data to fill all these NDMPD. The first steps of using NDMPD can and should be done with the data that is used by the traditional methods. Since the model structure is module orientated, it can be enhanced and refined on a module by module basis, when new information is available. One starts with quite simple partial models, which will be quite rough and constitute only a first step, but in fact, in most cases, will be more precise than the current approaches (which neglect many influences). These first models gradually evolve into a holistic and more and more detailed system model.

Using assumptions will give another chance to test new insights. The methodology enables to test the effects of assumptions against known results and – by varying the assumptions – testing their influence, enabling to estimate whether detailed analysis in that point will be worth the effort. Over time, more and more data will be available and – thanks to the flexible structure of NDMPD based models – may be integrated into the analysis.

The main source of knowledge are methods to analyse the past. With these methods, one tries to find correlations between accidents and their potential causes without bothering with the cause-and-effect chain. The statistical regression and correlation analysis tries to find the correlations between the variables by analysing the dispersion and approximating by finding formulas. If these methods are used to analyse the long process between parameters and accidents, they often fail. The failure occurs because the methods cannot model the complex functional relationships sufficiently. Furthermore, they cannot deal with the great dispersion of the data resulting from the very sparse number of accidents, if the parameters are very specific, or the large number of accidents, if broad parameter values are used. But if they are used to analyse specific parts of the system, they can be very efficient. So, they may be used to analyse certain system parts. Methods like INRAM are used to integrate the results of the parts into a system-wide model.

A sophisticated statistical approach is the quantitative causal chain (QCC) method and a comparative method, described by Lu (2006). This is a risk factor based statistical method, which deals with probabilities and consequences. The process between a measure and its safety effects is broken down into several links in a chain by identifying some controllable parameters (so-called determinants) in this chain, and the relationships between measures and determinants, whereas determinants and traffic safety factors are analysed respectively. In its practical application, the model can be used for comparative analysis of safety effects of different measures (e.g. infrastructure based versus autonomous, on-board related ITS).

Another approach is based upon the conflict analysing techniques. They try to open up a bigger reservoir of data by analysing conflicts and their reasons. This method is very effective but lacks the link between conflicts and accidents. In fact, certain situations with many conflicts have proved to be rather safe (e.g. roundabouts). So, Conflict Analysing Techniques are not useful as stand-alone methods. However, they are useful, if they are integrated into larger scale risk analysis approaches (Archer 2001).

8.6 Towards a Safer Future

The vision of INRAM and DRAM is to build a big model, which describes the whole driver – vehicle – infrastructure system. Each discipline uses it – very detailed in its own scope and more general in the other areas. The findings of each area are reflected to the general descriptions, so that the model is enhanced as a whole. This enables to do sophisticated analysis in an appropriate environment and to benefit from a system wide view and from the results of the surrounding research groups.

Which applications could result from the previous descriptions? In a first run, we could evaluate the risk component of the examples given in the preceding chapters more precise, especially considering human behaviour (e.g. self explanatory roads, VMS vs. ADAS), and to promote the cooperation and teamwork of researchers and government.

The modular structure of INRAM and DRAM enables to concentrate on individual parameters and effects of a traffic safety measure. It is no longer necessary to analyse the whole chain between parameters and (possible) accidents/damage. The modular structure enables to break this chain into smaller links: e.g. from (some) parameters to the speed – from the speed to the accident – from the accident to the damage. This facilitates to combine different methods of data gathering and to combine the findings from different research groups. For the first link behavioural analysis may be used (e.g. prediction models for speeds or conflict analysis), for the second link accident statistics, for the third link simulation programs or crash tests. The modular structure and the use of NDMPD enables to integrate the results of such different data sources into an overall system wide model.

The cooperation of different researchers and research groups even from different disciplines and the management of knowledge is supported by the expendable DoKn.

For the future and to approach the envisioned overall model, it is planned to build structures for specific fields, e.g. approaching sharp bends and tunnel safety. For that, it would be helpful to create general brick stones, e.g. modules for human behaviour (in cooperation with psychologists) and publish them via the DoKn.

The envisioned overall model is the foundation for analyzing and dealing more systematically and in detail with the technical and behavioural risks and chances, which will be given by new systems, methods and solutions.

References

J. Archer, *Traffic Conflict Technique (Historical to current State-of-the-Art)* (Institution for Infrastructure KTH, Stockholm, 2001)

St. J. Bald. Grundlagen für die Anwendung von Risikoanalysen im Straßenwesen (Basics for the Application of Risk Analysis in Road Engineering). Dissertation, Darmstadt, 1991

St. J. Bald, K. Stumpf, T. Wallrabenstein, Road risk analysis tools, Deliverable 3.2 of the 6th Framework EU Project IN-SAFETY (2008a) (http://www.insafety-eu.org)

St. J. Bald, K. Stumpf, T. Wallrabenstein, Modelling human behaviour by numerically described multidimensional probability distributions. Adv Transp Stud. **16** (2008b)

St. J. Bald, K. Stumpf, T. Wallrabenstein, Systematic risk analysis for safety assessments of road systems, in *Proceedings Transport Research Arena (TRA) Europe*, Ljubljana, 2008c

Bekiaris, E. (2001) Del D3/8.1 Compendium of existing Insurance schemes and Laws risk analysis of ADA systems and expected driver behavioural changes. User awareness enhancement, dissemination report and market Analysis and ADAS marketing strategy. Advisors GRD1 2000-10047

ISO 12100-1, *Safety of Machinery – Basic Concepts, General Principles for Design – Part 1: Basic Terminology, Methodology*, 2003

W. Durth, J. St. J. Bald, in *Risikoanalysen im Straßenwesen (Risk Analysis in Road Engineering) Darmstadt*. Reihe Forschung Straßenbau und Straßenverkehrstechnik (BMVBW, Bonn, 1998), p. 531.

M. Lu, Modelling the effects of road traffic safety measures. Accid Anal Prev **38**(3), 507–517 (2006)

D. Mahalel, A note on accident risk. Transp Res Rec. **1068**, 85–89 (1986)

M. Wiethoff, ADVISORS European Project (GRD1-2000-10047) Final Publishable Report and Annexes (2003)

Chapter 9
Back to School

Evangelia Gaitanidou, Evangelos Bekiaris, Maria Panou, Maria Gemou, Stella Nikolaou, and Martin Winkelbauer

9.1 Training as the Missing Link in ITS Application

The introduction of Intelligent Transportation Systems (ITS) as well as Advanced Drivers Assistance Systems (ADAS) and In-Vehicle Information Systems (IVIS) has boosted during the past two decades. Some of these systems have been gradually incorporated to the road infrastructure environment (e.g. the Variable Message Signs – VMS, etc.), while others have become optional vehicle equipment (e.g. Adaptive Cruise Control, Lane Departure Warning, etc.). In spite of recent advancements in the market penetration of these systems, their users, i.e. the drivers and the infrastructure operators, are not always aware of the actual needs and potential of these systems. The potentially important benefits of the use of these systems and technologies, such as the enhancement of road safety, the reduction of congestion by more efficient traffic management, the monitoring of critical infrastructure (e.g. tunnels), etc., may become useless if they are misused. Thus, there rises the need of training the users on the new technologies that have been introduced in the area of driving, traffic management, infrastructure equipment and more.

This need for training has been recognized by researchers and several attempts have been made, within the framework of related research projects. More specifically, in this chapter, training tools and/or curricula are presented, which have been developed within the EC co-funded projects IN-SAFETY, GOOD ROUTE, TRAIN-ALL, INFORMED and HUMANIST, addressing either drivers or infrastructure operators. The work has been undertaken by groups of researchers with a lot of experience on ITS technologies, as well as their applications and effects. Before developing the tools and/or curricula themselves, review of existing training schemes has been undertaken, so as to illustrate the existing situation and to identify

E. Gaitanidou (✉), E. Bekiaris, M. Panou, M. Gemou, and S. Nikolaou
Centre for Research and Technology Hellas, Hellenic Institute of Transport (CERTH/HIT), Thessaloniki, Greece
e-mail: lgait@certh.gr

M. Winkelbauer
Austrian Road Safety Board (KfV), Vienna, Austria

E. Bekiaris et al. (eds.), *Infrastructure and Safety in a Collaborative World*,
DOI 10.1007/978-3-642-18372-0_9, © Springer-Verlag Berlin Heidelberg 2011

the gaps and needs to be addressed. In fact, regarding drivers' training, a Multi-Media Tool (MMT) has been developed within HUMANIST Network of Excellence and relevant simulator training schemes within TRAIN-ALL FP6 STREP. Moreover, in view of training professional drivers and in particular drivers of dangerous goods vehicles, relevant curricula have been designed in GOOD ROUTE FP6 STREP, building on the experience gained in INFORMED Leonardo da Vinci project, where a series of ICT tools had been developed in addition; among them two multimedia tools presented herein. On the other hand, as far as the training of infrastructure operators on new technologies is concerned, IN-SAFETY FP6 STREP has developed a MMT for training road operators on new technologies in the area of telematics, road-based and in-vehicle systems while, within GOOD ROUTE a second training curriculum has been developed, this time aiming at the training of special infrastructure (e.g. tunnels) operators.

9.2 Drivers' Training in the Telematics Era

Many among the newly introduced technologies, such as the ones described in Chap. 4, interact with one or the other way with the driver. This is the case not only for in-vehicle devices, i.e. ADAS, IVIS, etc., but also for infrastructure-based equipment, which is providing information to the driver (e.g. VMS) and/or interacting with him/her or automatically with the vehicle, as in the case of cooperative systems. The drivers are usually interested in having state-of-the-art technological equipment in their vehicles (especially when this is provided as standard equipment) and may even pay more when purchasing their vehicle in order to include it. However, in most cases, the drivers are not fully aware of what the use of this equipment actually implies for their everyday driving routine and, in some cases, are either frustrated by its function or are misusing it. No matter how remarkable the design of an ITS, it will only be safe and beneficial if drivers are fully trained in its use. It is a common knowledge that user manuals are rarely read by the users and therefore do not constitute an integrated and effective driver's training method on such systems.

All the above mentioned risks are becoming even more important when it comes to drivers of dangerous goods vehicles, in which case the consequences of an accident are more significant, possibly affecting a wide area around the accident and their effects influence, not only the involved road users and the surrounding traffic, by means of damages and injuries, but also by means of environmental consequences.

The need for training the drivers in the use and functionalities of new technologies has been recognized and several initiatives have been undertaken towards this direction. However, despite the fact that several studies have been conducted on the issue of drivers' training, there are quite limited tools that are focused on ITS systems' training (Twisk and Nikolaou 2004).

The general idea behind driver training for ITS use is to address the drivers' needs for learning the use of the system, without negatively affecting safe driving behaviour. The main issue is for the drivers to understand the technical limitations of the systems, so as to be aware of the potential of the systems' performance, thus preventing from drivers' over-reliance or mistrust on the system. It is also very important to make the drivers aware of the capabilities of these systems and their effects on the driving behaviour, especially when it comes to system malfunction or to using vehicles with different equipment.

Increased automation of driving tasks currently performed by the driver can have several consequences, as their learned skills to avoid dangerous situations will be weakened, if the vehicle itself becomes increasingly capable of detecting and responding automatically to traffic hazards (Summala 1997). Thus, the training schemes should be accordingly adapted, so as to take this into account.

Training is also a way of minimising distraction due to the use of ITS. According to Young et al. (2002), it is important to:

- Make drivers aware of the risks involved in the use of some in-vehicle devices, such as mobile phones, whether hand-held or hands-free, as they are both distracting
- Educate and train drivers in the optimal manner to interact with existing and emerging in-vehicle technologies and services accessed through portable devices in order to minimise distraction
- Provide information on the way to operate in-vehicle devices, highlighting the most ergonomic and least distracting methods for doing so

With the appropriate training, driver trainees can understand the situations where each ADAS can apply and can increase their awareness and reduce their potential over-confidence on the reliability of these systems. Also, their knowledge on the ADAS functionalities and barriers will be increased and they will acquaint information on ADAS practical use, improving handling of effects while using them.

In conclusion, an ADAS can only be effective if (Simoes et al. 2006):

- It is used in the correct manner.
- Drivers don't rely too much on the system.
- It is not used for the wrong reasons.
- It is used by the driver for the purpose it was designed for.
- It is not used in situations where it can't work.
- It is used in the right conditions (e.g. road type).

In this context, several EC co-funded research initiatives have worked on the development of training schemes for drivers from different points of view. In this chapter, three indicative examples are being presented; the Multi-Media training Tool developed within HUMANIST NoE, the driving simulator based training schemes developed within TRAIN-ALL STREP and the training multimedia tools for professional drivers of dangerous goods vehicles, developed within the INFORMED Leodardo da Vinci project which have been embedded in the overall training schemes developed therein and further improved/enriched within GOOD ROUTE STREP.

9.3 The HUMANIST Multimedia Software Tool for Driver's Training in the Use of New ITS

9.3.1 Content

The HUMANIST Multimedia Training Tool is aiming at driver's training on new in-vehicle technologies that enhance driver comfort and safety. The MMT is provided as a software prototype in a CD-ROM, but can also be used as a web-based on-line training tool (Papamathaiakis and Nikolaou 2004).

The MMT includes in total 18 systems, which are clustered in three main categories; ADAS (9 systems), IVIS (7 systems) and DSuSy (2 systems):

- **Advanced Driver Assistance Systems (ADAS)**: Cruise Control, Advanced Cruise Control and Collision Avoidance, Lane Keeping and Lane Change Support, Parking/Reversing Aids, Vision Enhancement, Driver Monitoring, Vehicle Status Monitoring, Blind Spot Detection, Intelligent Speed Adaptation.
- **In-Vehicle Information Systems (IVIS)**: Navigation and Route Guidance, Automatic Comfort Settings, Web Services and Telecommunication, Traffic, Weather, Accident and Parking Information, Location-based Services, Emergency-Related Information.
- **Driving Support Systems (DSuSy)**: Anti-lock Braking System, Electronic Stability Program.

9.3.2 Technical Framework

The MMT was developed in Adobe Flash CS3. This software is appropriate for multimedia applications development. The final product is platform-independent, which means it can be presented on any type of computer platform (Windows, Macintosh and Unix). Additionally, any change in the application can be performed easily and the content can be updated instantly. Also, additional content can be added, as should be in a constantly updated MMT. Furthermore, the application can be distributed either in CD-ROM or online.

9.3.3 Structure and User Interface of the MMT

The Multimedia Tool is divided in three main categories, as described above, ADAS, IVIS and DSuSy. The user can navigate among them in the home screen. In each category the most representative systems are included, as mentioned above.

In every system, six training sections are available: the system description, functionalities, benefits, limitations, indicative HMI and the application examples.

Additionally, a help function is present, in order to help users to navigate through the MMT. Finally, a quiz has been developed, in order to examine the knowledge that users have gained from the MMT.

The multimedia training tool structure is divided in two layers, the main layer and the system layer. The main layer includes the home screen and the description screen of every system presented in this MMT, as well as and the help and quiz sections. This layer is depicted in Fig. 9.1.

The system layer includes the training chapters of the systems, as shown in Fig. 9.2.

Special consideration has been taken on the user interface of the multimedia tool, in terms of different aspects:

- **Navigation**: This application has been designed in order to be easy-to-use and user-friendly, so that no need for further instructions is required. In Fig. 9.3, the home screen of the MMT is depicted.
- **Users performance tracking**: A multiple choice quiz is included in the MMT (Fig. 9.4).
- **Feedback messages**: Several feedback messages are provided to the MMT user for assistance and/or guidance.
- **Pictures and text**: Apart from theoretical text in the training sections, which is provided in page layout, also relevant pictures, diagrams, etc., that are relevant to the text, are included (Fig. 9.5).
- **Videos**: A number of videos have been included in the MMT to visualize the available text and explain in detail the systems (Fig. 9.6).

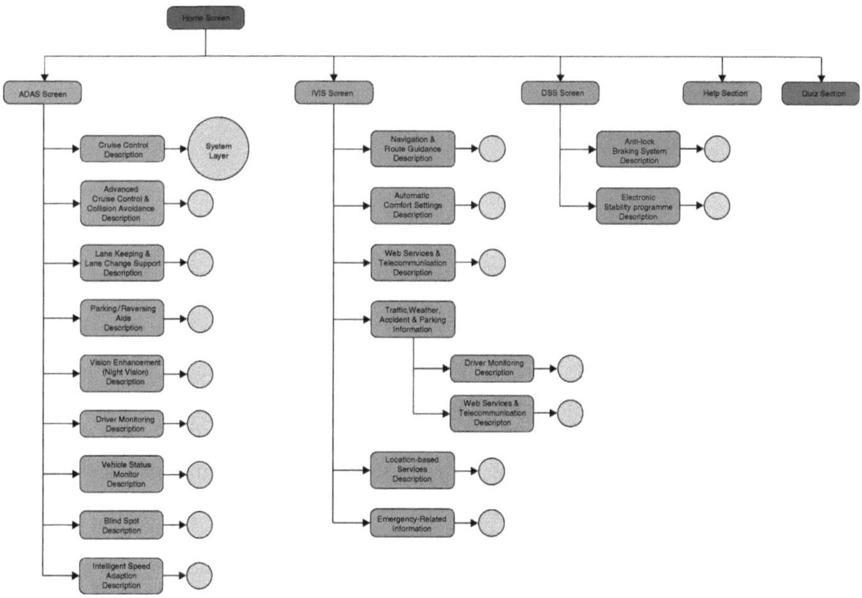

Fig. 9.1 The main layer of the HUMANIST MMT structure

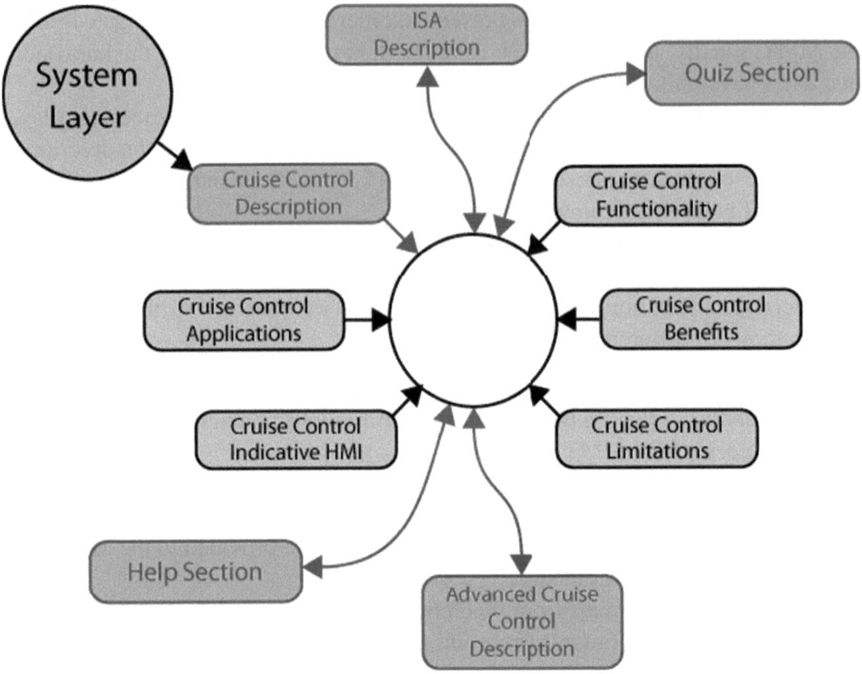

Fig. 9.2 The system layer of the HUMANIST MMT structure

Fig. 9.3 The HUMANIST MMT home screen

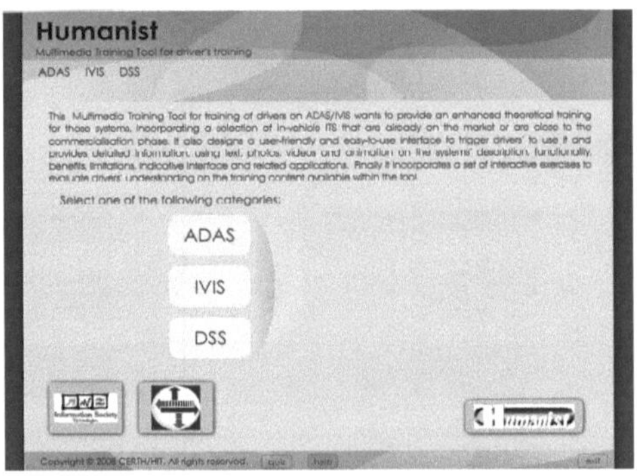

- **Simulations-animations**: Simulations and animations are videos developed in simulators and computer graphic applications. The main advantage of the simulations and the animations is that they present precise information, designed to provide information to the user that cannot be provided with real-life videos (Fig. 9.7).
- **Control buttons and menu**: This MMT is designed in order to provide users the ability to navigate through the application easily. Users can navigate through the screens by clicking the buttons available on them.

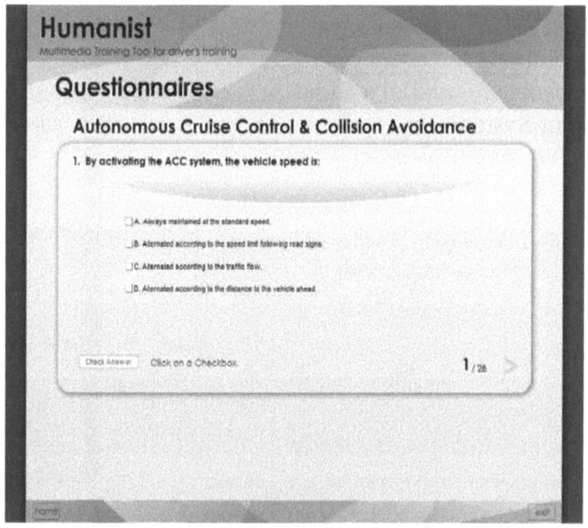

Fig. 9.4 The multiple choice questions available of the HUMANIST MMT Quiz

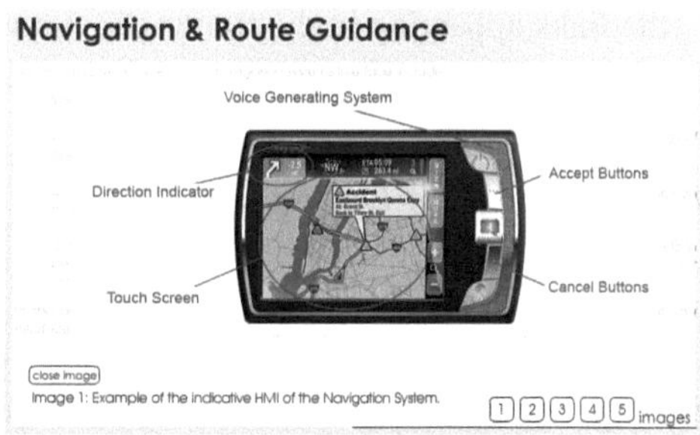

Fig. 9.5 A picture appearing in the HUMANIST MMT with its description

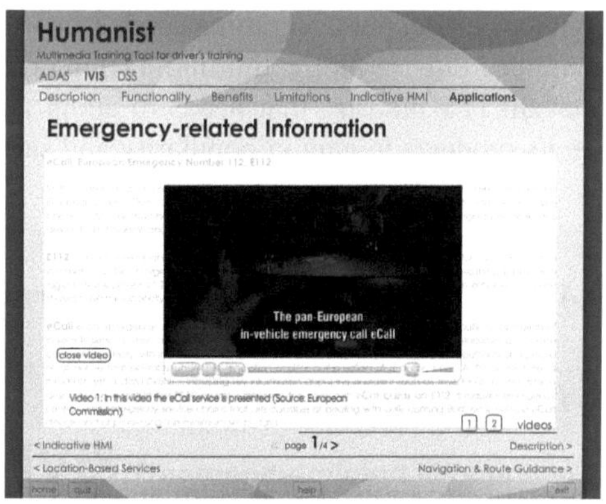

Fig. 9.6 A video appearing in the HUMANIST MMT

- **Help and support functions**: This multimedia training tool supports its autonomous operation by the user, providing help for the main functions of the application. The user can enter to the help screen by clicking the help button

Fig. 9.7 An animation presenting the areas where sensors are used in a Blind Spot System

on any screen. In the help screen, the user can navigate through the buttons and the links appearing in the training sections' screens.

9.3.4 Ready to Get Started

The HUMANIST MMT is a holistic and open tool, that aims to bring ITS to the average (non-technical) driver, explaining him/her their functionalities and highlighting relevant limitations of use. It is suggested to be integrated in driving schools theoretical training curricula or even at ordinary school lessons on traffic safety. However, the rather limited ITS introduction in the vehicle market currently prevents it from becoming a priority for driving schools. On the other hand, in a vicious cycle, not using it means that driver's unawareness on ITS functions will continue, thus severely limiting ITS market penetration. Most likely it is the relevant ITS industry that needs to adopt and sponsor such tools, if it wishes to see its products to gain in popularity and sales.

9.4 TRAIN ALL ADAS/IVIS Training Simulation Module

9.4.1 Simulation for Training

Theoretical training has its limitations. To really understand and appreciate ITS functionalities, drivers would optimally need to try them out. As it is however not feasible to equip several driver training school cars with such equipment, the best alternative is to use training driving simulators for this task.

The development of appropriate training schemes and scenarios for computer-based training in the use of new driver assistance and information systems (ADAS and IVICS), is one of the many objectives of the TRAIN-ALL EC co-funded project. In order to make drivers familiar and able to use the in-vehicle driver support systems, specific simulator scenarios have been built, along with the use of criteria for their safe use, so that the trainees can learn how to optimally use ADAS/IVICS without overestimating their functionality and knowing their limitations Such simulations would complement any theoretical knowledge gained on such tools, by multimedia tools, such as the HUMANIST one, presented in Sect. 9.3.

Two modules were developed for the simulation of ADAS/IVICS, which are complimentary. The one goes very deep in the analysis of the functionality and uses existing and new simulation elements and traffic participants, whereas the other is limited to the generic functionality (as different simulators have different simulation elements) but can be easily adapted to other simulators. This solution has been successfully implemented in five driving simulators (truck simulator, passenger simulators, motorcycle simulator and VR simulator) of different manufacturers.

9.4.2 Methodology and Functionality

As mentioned earlier, the specific ADAS to be simulated were initially selected and then, the use cases and the scenarios were specified (Poschadel et al. 2008), including all the necessary parameters and elements. The selected ADAS and IVIS are listed below. However, following the same methodology, any other ITS functionality can be simulated:

- Collision Avoidance System (CAS) – ADAS
- Lane Deviation Warning (LDW) – ADAS
- Communication while driving – IVIS

There are two training modules developed for each application:

1. Initially, a stand-alone module was programmed, with the simulated scenarios, showing the system functionality and limitations. The scenarios and software module were then installed at CERTH/HIT simulator (Smart vehicle-based dynamic simulator). The s/w was debugged and optimized.
2. In the meantime, the project architecture was finalized and the second module, with the ADAS module functionality has been programmed. It is a module (DLL based) that abides to the project's interoperable architecture (Huiskamp et al. 2008).

The two modules have some small differences in terms of content and functionality, the major being the scenarios on limitations which are included in the first module only. This cannot be easily integrated to a generic interoperable module, as they are based on very specific elements of the simulation, such as trailers, specific loads of trucks, etc.

For the IVIS, proprietary scenarios have been developed. The s/w determines the road type (according to specifications that are detailed in the use cases), as the scenario deals with the driver interaction with the mobile phone.

More details on the functionality of the developed modules follow below.

9.4.3 ADAS Module Description

As mentioned above, the focus of the ADAS module is twofold:

1. Training on the ADAS functionality.
2. Training on the ADAS limitations.

9.4.3.1 Functionalities

For the warning strategies, different algorithms can be selected by the driver trainer with different warning thresholds, in order to adjust the system functionality according to the specific needs of each trainee. In the proprietary version, a window appears before the start of the scenario and the trainer can select which algorithm will apply for the warning threshold of the CAS and the LDW. The windows are as below:

Criterion: $TTC \leq TTC_{limit}$ Criterion: $Theadway \leq Theadway_{limit}$ Criterion: $RT \leq RT_{limit}$	Criterion: $TLC \leq TLC_{limit}$ Criterion: $DLC \leq DLC_{limit}$

where TTC[1] *= time to collision, Theadway*[2] *= time leading, RT*[3] *= reaction time, TLC*[4] *= time to line crossing, DLC*[5] *= distance to line crossing.*

[1]TTC is the time until the collision of a vehicle with the leading one, given that the speed of both vehicles remains the same with that at the specific time instant. This time is infinite if the leading vehicle travels with a higher speed than the following one.

[2]Theadway is defined as the time until the collision of a vehicle with the leading one, given that its speed remains the same but the leading vehicle decelerates with infinite deceleration (i.e. it remains at the same position as in the specific time instant).

[3]According to literature (Lee et al. 2004; Warshawsky-Livne and Shinar 2002; Green 2000), the reaction time is composed of the mental processing time of the driver and the action time.

[4]TLC is defined as the time distance between the central axis of the vehicle and the side of the road lane in which it travels. There are two such times that exist, depending on which side of the lane we refer to, the right or left one.

[5]DLC is defined as the distance until the centre of the vehicle reaches the left or right lane marking of the lane where it is positioned. It is a more stable parameter the TLC, since it can be calculated directly by the CAS and not approximately, as the TTC.

For the limits of TTC, Theadway, RT, TLC and DLC, default values are used according to Panou (2008), however, the values of these parameters are open to be set by the trainer, following of course the minimum/maximum safety thresholds.

DLC limits may be provided as absolute or relevant values (percentage of lane width).

In the interoperable version, the choice of algorithms is for the moment set as default. Four of the above (except RT) have been implemented in the s/w code. The RT-based algorithm has not been implemented as the way it is calculated differs very much from simulator to simulator and is also simulator-scenario dependent.

9.4.4 Limitations

One of the simulation key aims is to avoid that the driver develops overconfidence to the system. For this reason, specific scenarios are designed, as described below.

9.4.4.1 CAS Limitations

Two scenarios have been developed to highlight the limitations of the CAS. Three different road types are included.

In the first scenario, a box falls from a truck in front of the driver, while at a critical distance from the driver's vehicle. This corresponds to a "leading vehicle" breaking with an infinitive deceleration, thus being outside the effective warning criteria of any CAS in the market. Figure 9.8 shows snapshot of this scenario.

In the second scenario, a vehicle in front is carrying a trailer in a road of upcoming slope, the trailer gets loose and slides backwards, towards the driver. When the time to headway is less than 1 s the system warns the driver, but the obstacle has negative relative speed, thus warning comes too late to the driver. This is depicted in Fig. 9.9.

Also system malfunctions are simulated, which do not require special modules/ scenarios, as the trainer can simply turn on/off the system randomly, to see the reaction of the driver while driving.

Fig. 9.8 Scenario of a box falling from a truck in front

Fig. 9.9 Scenario of a trailer that gets loose from the vehicle that carries it, in an upcoming slope road

9.4.4.2 LDW Limitations

For the demonstration of the LDW system limitations, two specific scenarios have been developed in a highway environment. In the first one, there are two lane markings on the lane (one is an old marking which is still quite visible) and the system warns the drivers arbitrary based on the old and the new one. In the second scenario, there is snow that has covered the lane marking in several parts of the road and the system does not provide any warnings in those parts.

Apart from the above scenarios, the trainer has to switch on/off the system at random periods of time, as in the CAS case, to simulate system malfunction.

9.4.5 IVIS Module

9.4.5.1 Functionality

The simulated functionality is aimed to train drivers on how to restrict communications (esp. mobile phone use; even hands-free) and how to optimally use adapted communication aids while driving.

For this scenario, a challenging driving environment has been created, with complex driving situations, where the full attention or the imminent reaction of the driver are needed, e.g. roads with sharp curves, close interaction with other vehicles, slow-driving car ahead to be overtaken, etc. Three different road topologies have been created with combined road types (corresponding to highway, rural and urban roads).

While driving in this environment, the driver is asked to use the mobile phone and read an SMS that has arrived. The system records the vehicle behaviour (position in the lane, TTC, etc.), both before and during the reading of the SMS. At the end of the journey, the vehicle behaviour is shown to the driver at a bird-eye/helicopter view.

As a second step, the system keeps on hold the SMS/call while performing a manoeuvre (e.g. overtaking, changing lane) or driving over 100 km/h, etc. and sends them when the critical event is over. The realisation of this second step however, depends on the capabilities of each simulator's simulation environment.

IVIS Limitations

For the system limitation and malfunction effects scenario, the aim is to prove that overconfidence and trust in the system may cause accidents or driver annoyance when the system does not act as expected. In this scenario only one road topology with combined road types is used (consisting of highway, rural and urban roads).

The system keeps on hold the received messages and calls until the driver finishes the critical driving action. The next steps is that the system stops working (i.e. does not delay SMS/calls) or works only periodically. For these steps, the trainer has to turn off/on the system manually.

9.4.5.2 Text-Bed Scenario

To demonstrate the above functionality, the following example scenario has been implemented:

The scenario has a highway road with two lanes. A message appears on the screen, which directs the driver to change lanes randomly (e.g. from lane 1 to move to lane 2 and then to lane 1, etc.), about twice to three times per minute. The road is ~6 km long, with no traffic (only on-coming traffic). The VMS has small text 'Change lane' and there is a flashing arrow above the lane indicating in which lane the driver has to go. See the example in Fig. 9.10, where the VMS shows that driver has to go to lane 1.

The driver will drive initially without using the mobile phone, and then using it. The TLC is measured in each ride. Afterwards, the system shows the driving in both cases (we expect more and wider lane deviations with the use of the mobile phone).

This test is in accordance to the emerging ISO standard ISO/DIS 26022 'Simulated lane change test to assess in-vehicle secondary task demand' (ISO/DIS 26022 2011).

Fig. 9.10 IVIS limitation screenshot

9.4.6 HMI Modes

The developed module is comprised of various training scenarios, with the possibility for the instructor to alter key characteristics regarding the functionality of ADAS/IVIS. Four possibilities of HMI (i.e. of giving the system message to the driver) have been specified and proposed: optical, acoustical (voice message or directional sound indicating the origin of the danger, i.e. left or right side, continuous beep, repeating beep, Wierwille alarm (a rectangular type of warning tone with frequency 1,000 Hz, that has been proposed for crash avoidance systems (Fahey and Wierwille 1995)) of increased volume or frequency as the distance to the obstacle is being reduced, etc.), haptic (brake pulses, seatbelt vibration, 'rumble strips' emulation, etc.) and combinatorial (visual and acoustic feedback for imminent situations, and also possibly tactile). Of them, only the optical and acoustical ones have been implemented so far, because of simulator limitations.

There are different acoustic warnings that the trainer can choose for warning on a risk in the longitudinal axis of the road (frontal collision risk) and a risk at the lateral axis (lane deviation). As default, the Wierville alarm sound is used for the frontal collision warning, while for the lane deviation warning the sound provided is a simulation of rumble strips.

Furthermore, the trainer is able to select various sub-modalities, such as different optical warnings (icon on screen, flashing or steady LED on the inside or side mirror, or small icon, etc.). Figure 9.11 shows an example of alternative ways of presenting a warning message coming from a LDW system, by providing icons or a light on the side mirror, in order to warn the driver when there is a risk while attempting to overtake (Bekiaris et al. 2007).

The possibility to choose among different HMIs or system logics/functionalities provides further learning opportunities.

9.4.7 Into the Action

The developed simulator scenarios can be easily adapted and installed in a wide range of simulators with minimal need for reprogramming. Thus, they can very well supplement other knowledge (i.e. given by theoretical courses or Multimedia tools)

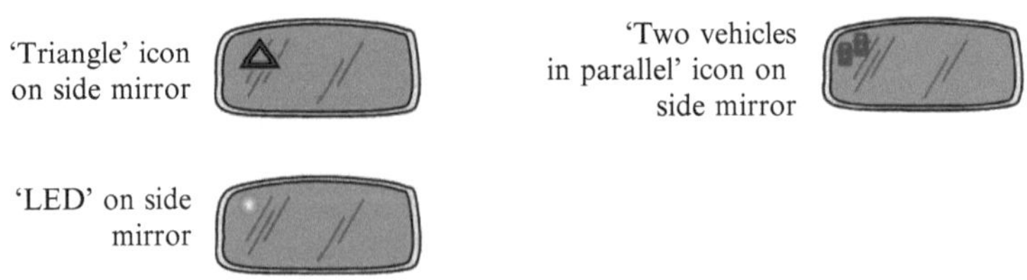

Fig. 9.11 Alternative ways of presenting a LDW warning message

and is the next best option after driving an ITS-equipped vehicle. Also, the established principles may be followed to model and simulate more ITS functions.

As several hundreds of training simulators exist in driving schools Europewide, it is becoming more and more feasible and practical to use such simulations for training novice drivers on ITS functionalities. Furthermore, ITS vendors could use such simulations in dealer rooms and motor shows to familiarize and even train their customers, thus enhancing public awareness, while also promoting the safe use of their products.

9.5 INFORMED Training Tools for Dangerous Goods Drivers

The transportation of Dangerous Goods involves high risks. If these substances are mishandled, injury and property damage risks are increased. Those at risk include drivers, depot workers, station attendants, infrastructure and emergency units workforce, other road users and third party populations. Especially drivers are the ones first recalled to abide certain specifications in order to decrease inherent risks in the Dangerous Goods transportation.

The GOOD ROUTE project developed a cooperative system targeting at the monitoring of the Dangerous Goods (DG) fleets, as well as routing, re-routing and enforcement, whenever needed, in order to minimise the Societal Risks related to their movements, whereas still generating the most cost-efficient solution for all actors in their logistic chain.

Within this concept, GOOD ROUTE developed also supporting training curricula for both the drivers of the dangerous goods vehicles and the road and TMIC operators. Those training curricula have been based on the pre-existing outcomes of the INFORMED Leonardo da Vinci project (INtegrated system FOR an Advanced and Life-LongTraining MEthodology of Dangerous Goods Drivers and Trainers. EL2002/B/F/PP-114010), which aimed at the development of a new training curriculum and innovative multimedia tools, to support the training of drivers and their instructors on transporting dangerous goods and handling them efficiently in case of an accident. In this context, and among other ICT tools, two Multimedia tools were developed and embedded in the suggested training curriculum. These were namely a *Multimedia tool for ADR training* and a *Multimedia tool for the advanced driving of vehicles carrying dangerous goods.*

9.5.1 Current Legislation and Existing Training Schemes

As a first step, existing training schemes for DG and their needs, as well as the current legislative framework were investigated. Currently, the main European legislative framework is the so called "The European convention for international road transport of dangerous goods" – often mentioned as "The ADR convention" or just "ADR". EU has adopted the ADR legislation as the foundation for the practical carrying out of road transport of dangerous goods within the boundaries of the Union as well as for the entire content of the ADR training courses met in its member countries. As a supplement to the ADR regulation, the road transport of dangerous goods is further regulated via a number of directives (e.g. 94/55/EU, 2000/61/EU, 2001/7/EU, 95/50/EU, etc.).

Despite the above legislative framework, there is no homogeneity detected in the application of ADR in the various EU members in terms of training.

It is worth noted that there are no standard requirements for the trainees and the trainers to be accepted at the training institutes. In most cases, trainees are obliged to own a driving license, whereas illiterate persons are not accepted. On the other hand, only few European countries (i.e. France, Greece), trainers must have a high educational level and in most cases they have been recognised as Safety Advisors according to the ADR agreement and/or are certified by a qualified EU institution.

The license is renewed every 5 years, whereas in some countries, as Denmark, training courses are conducted annually, preparing trainers for ADR examinations.

The content mainly taught is the ADR content and only in few cases, some additional sections are included, like in case of the Netherlands, mostly dealing with local regulations. There is also no homogeneity regarding the way the training and assessment are performed.

The ADR training organisations are officially supervised by the corresponding Ministries (e.g. Ministry of Transport). The examinations take place at training centres or in fewer cases at independent examination centres, as in the case of Belgium (Bekiaris and Gemou 2009).

9.5.2 *Identified Training Needs and Gaps*

Within the context of INFORMED project, a thorough accident analysis and interview survey was held in September 2003, aiming to identify the training needs of the drivers and trainers in the Dangerous Goods transportation segment (Bekiaris et al. 2003).

The most interesting and relevant outcomes of this survey, related to the DG drivers training needs are summarised below:

1. The drivers should fulfil some primary requirements, before they start attending the training courses.
2. As regards the training course itself, the accident statistics and the employers' interviews have proven that the ADR training is not enough, even if it is performed in the right way.
3. Training courses and scenarios should also focus on those issues that accident statistics admonish to.
4. The training environment and the equipment used needs a series of improvements.

5. The preferable duration of each training session is, according to the trainees, 1–1½ day, while, according to the trainers 2–3 to 4–5 days.
6. Regarding distance training schemes, trainees would desire also the voice and the video in such a course, which should be taken into consideration.
7. For the life-long training issues, 58% of the trainees agree with their re-evaluation every 5 years and they prefer printed updates, perhaps because they are not accustomed with Internet use.
8. Finally, it is clear that the trainees training effectiveness strongly depends on the trainers qualifications and abilities.

9.5.3 DG Drivers Training Tools

As aforementioned, there were two Multimedia tools (*Multimedia tool for ADR training* and a *Multimedia tool for the advanced driving of vehicles carrying dangerous goods.*) developed in the context of INFORMED, in order to support the overall training framework developed for the DG drivers but also for their trainers. They are described in short in the following sections.

9.5.3.1 Content

The *Multimedia tool for drivers' ADR training* is a software tool, which can be used by all drivers who wish to be employed in the ADR sector (Mousadakou and Gemou 2004a). In addition to this, the *Multimedia Tool for the Advanced Driving Carrying Dangerous Goods* software tool has been developed which aims to train those drivers coming from all European countries, wishing to be involved in the dangerous goods haulage sector as well as all drivers, not exclusively professional ones, who wish to be informed on the risks of driving, on ways to effectively prevent their and other people's life from road accidents and mitigation strategies in case a road accident cannot be avoided (Mousadakou and Gemou 2004b).

The feedback for the realisation of the MMT has been provided by the conclusions and results emerging from an extensive State of the Art conducted in INFORMED (Gaumet et al. 2004), the training material gathered by all INFORMED Partners especially adapted by the INFORMED experts and the additional training material developed specifically for this tool. Copyrights for each picture, video, animation, etc. embedded in the MMTs have been preserved.

The MMTs have been produced in executable format in CD-ROMs, which have enclosed the instructions for MMT use and maintenance, but can also be used as web-based on-line training tools. They are available in Danish, Dutch, English, French and Greek.

Both MMTs cover the most significant training issues which are to be trained within the INFORMED curriculum for Drivers training course. However, the self-instructive format of the MMTs enables their use in the self-training course of the INFORMED curriculum (or beyond) as well.

The training issues covered in the *Multimedia tool for ADR training* succeed to fulfill the training needs of the ADR drivers in the issues considered by the INFORMED Consortium as the most significant ones (Christiansen et al. 2003).

The *Multimedia tool for ADR* training includes also an Introductory course, entitled "Introductory Safe Driving Course" for the *Multimedia tool for the advanced driving of vehicles carrying dangerous goods*, which provides an accomplished training in safe driving of ADR drivers.

The *Multimedia tool for ADR training* consists of 2 modes (Training and Test mode) and encompasses 90 scenarios, which are subdivided in the following main MMT sessions:

- Training Mode
  - Dangerous Goods
  - Means of Transportation
  - Legislation/Documents
  - Labeling/Packaging
  - Tank Specialisation Course
  - Health Risks Awareness
  - Personal Protective Equipment
  - First Aid
  - Advanced Fire Fighting
  - Emergencies
  - Driver's Responsibilities
  - Safe Work Practices
  - Technical Fundamentals
  - Introductory Safe Driving Course
- Test Mode
  - "Basic Course" Test
  - "Tank Specialisation Course" Test
  - "Class 1 Specialisation Course" Test
  - "Class 7 Specialisation Course" Test
  - "All Classes" Test

The subjects constituting the "Training Mode" of the tool have been determined with the agreement of all INFORMED Partners and the material composing them have been provided by all of them. In some cases the provided material, already used by the INFORMED Partners training institutes for internal training purposes, has been translated from their native languages, like Dutch or Danish to English. Eventually, the multimedia tool emerged is a compendium of the material available at several European training institutes, filtered and adapted in order to be applicable for all European countries drivers as well as of new material, specifically developed for this tool. The INFORMED Consortium Partners experience and knowledge has significantly contributed to the integration of various material gathered and finally to the composition of an innovative and effective multimedia training tool, that addresses all drivers throughout Europe who wish to be involved in the ADR sector.

The "Test Mode" of the MMT, which should be normally run by the trainee after being trained with the "Training Mode" scenarios, consists of assessment scenarios, including multiple choice questions for the most significant training sessions.

The *Multimedia Tool for the Advanced Driving Carrying Dangerous Goods* consists of 44 training scenarios. This is a really innovative tool, since there no other relevant software tool is found, at least in terms of the INFORMED survey, providing training in advanced driving techniques for drivers involved in the dangerous goods haulage sector. The training provided with this MMT is being supplemented by the on-the-road training scheme developed in INFORMED (Gaumet et al. 2004). The *Multimedia Tool for the Advanced Driving Carrying Dangerous Goods* consists also of a training and test mode, which follow the same concept as in the *Multimedia tool for ADR training*. The respective training and test mode sessions of this MMT are as follows:

- Training Mode
  - Defensive Driving
  - Antiskid
  - Antirollover
  - Fatigue Management
  - Eco Driving
- Test Mode
  - "Defensive Driving" Test
  - "Visibility Issues" Test
  - "Road Obstacles" Test
  - "Gap Acceptance" Test
  - "Safety Distance" Test
  - "Hazard Perception" Test
  - "Tracking of Unexpected Behaviour of Road Users" Test
  - "Lane Change" Test
  - "Speed Choice" Test
  - "Fatigue Management" Test

In general, the content of both MMTs (training and test part) addresses all drivers throughout Europe, providing information and regulations that are valid and common Europe-wide in this sector and does not focus on potential exceptions and additional special regulations valid in each European country.

9.5.3.2 Technical Framework

The software tool, which has been used for the development of both MMTs is the Macromedia Authorware V6. It has been selected as the most appropriate one among others, since it enables developers to create highly interactive, rich-media learning applications, which can be delivered to customers on the Web, LANs and CD-ROM and additionally track end-users progress and performance results.

9.5.3.3 Structure and User Interface of the MMTs

As aforementioned, both MMTs consist of two basic modes, namely "Training Mode" and "Test Mode" (Fig. 9.12). The "Training Mode" encompasses theory and exercises for the effective training of the drivers and the "Test Mode" includes a series of multiple choice questions, which aim to detect the level of the trainees' acquired knowledge around the training subjects of the "Training Mode". Within the "Training Mode" a series of videos, animations, descriptive pictures and texts are provided in order to enhance the training procedure.

Figures 9.13 and 9.14 depict the structure of the MMTs.

Special consideration has been taken on the user interface of the multimedia tools, in terms of different aspects:

Navigation: Both MMTs are designed in order to be easy to navigate without further instructions. The first screenshot presented to the MMT user is the "Welcome Page" of the MMTs (Fig. 9.15).

After that, the trainee has to select between the "Training Mode" and the "Test Mode". Normally, the trainee has to be trained first with the scenarios of the "Training Mode" before s/he proceeds with the assessment scenarios of the "Test Mode". However, there is no restriction for that on behalf of the software; thus the trainee is free to select whichever mode s/he wishes to. In case s/he selects the "Training Mode", the main training sessions (of each MMT respectively) will be presented to him/her, which in turn lead to respective submenus and finally scenarios they consist of (Fig. 9.16).

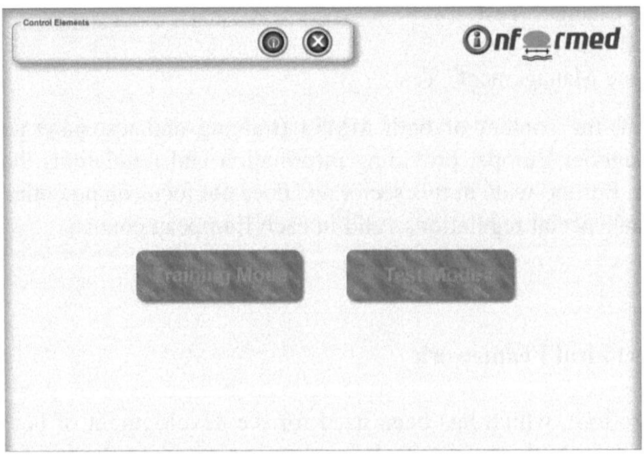

Fig. 9.12 INFORMED MMTs – "Training Mode–Test Mode selection"

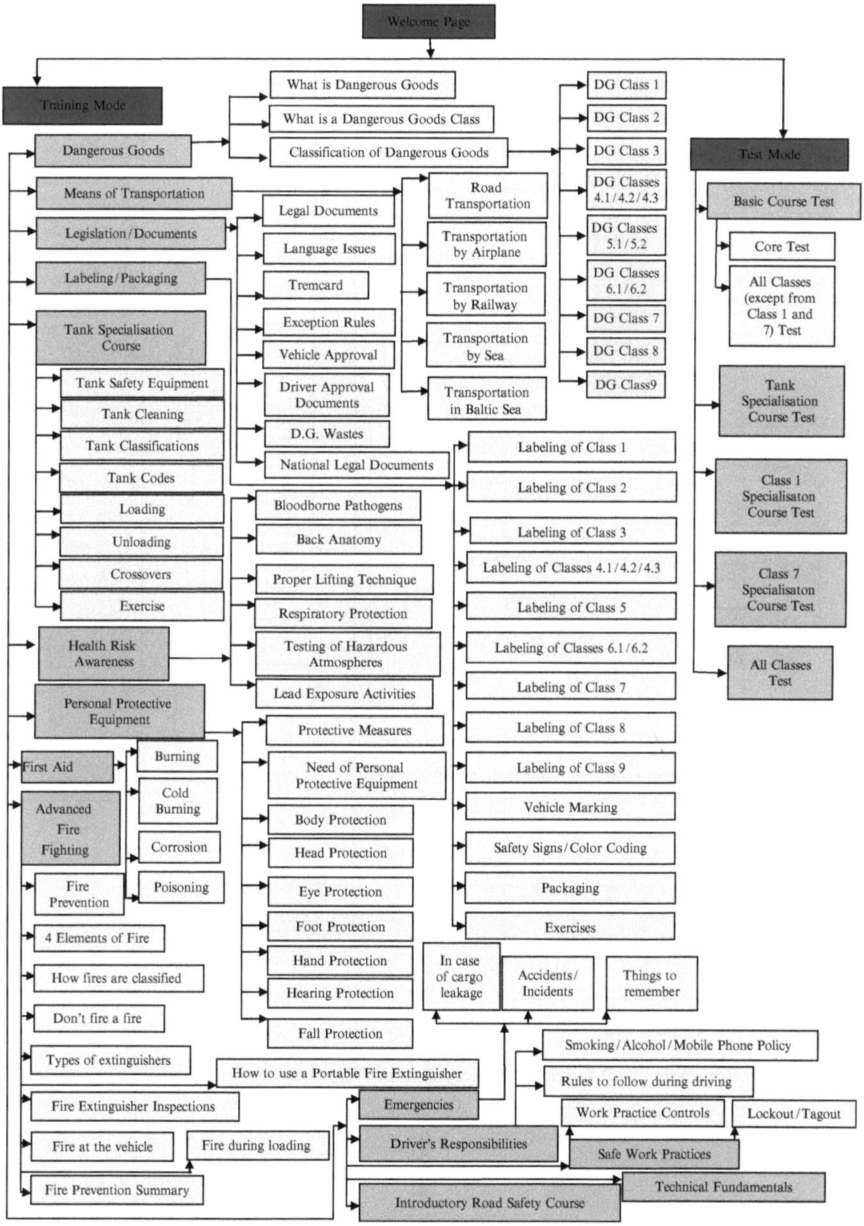

Fig. 9.13 "Multimedia tool for ADR training" structure

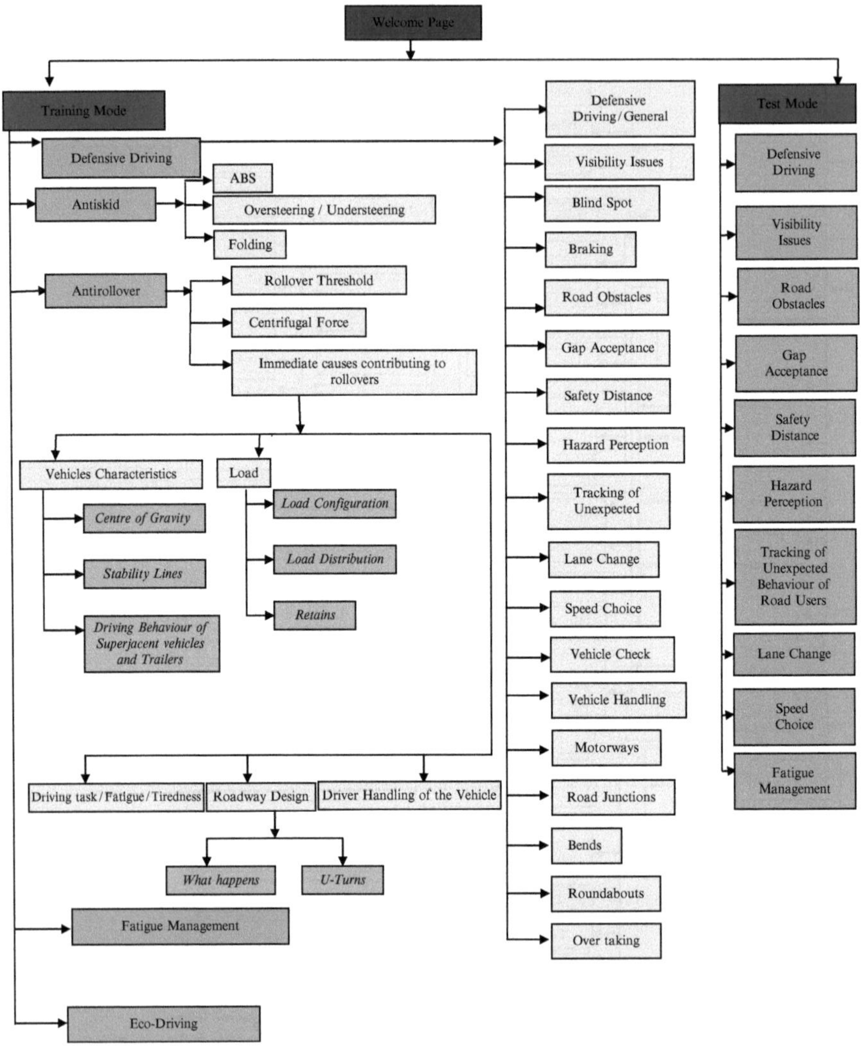

Fig. 9.14 "Multimedia Tool for the Advanced Driving Carrying Dangerous Goods" structure

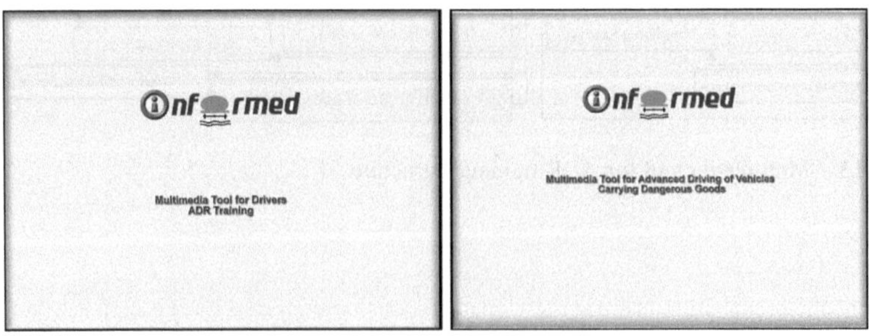

Fig. 9.15 MMTs "Welcome page"

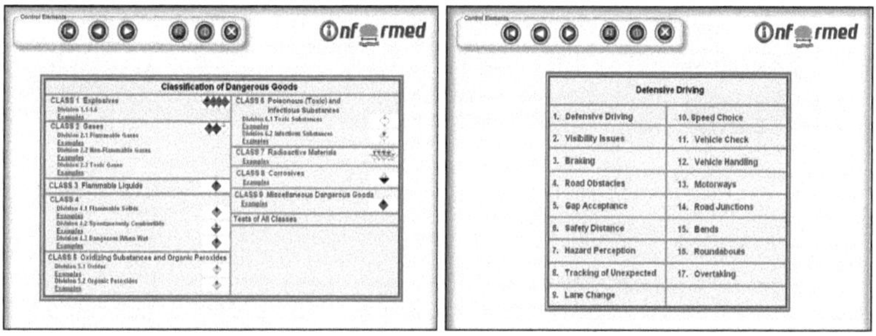

Fig. 9.16 INFORMED MMTs (*left*: "Multimedia tool for ADR training" – "Dangerous Goods" session–"Classification of Dangerous Goods" session submenu; *right*: "Multimedia for advanced driving carrying dangerous goods" – "Defensive Driving" session submenu)

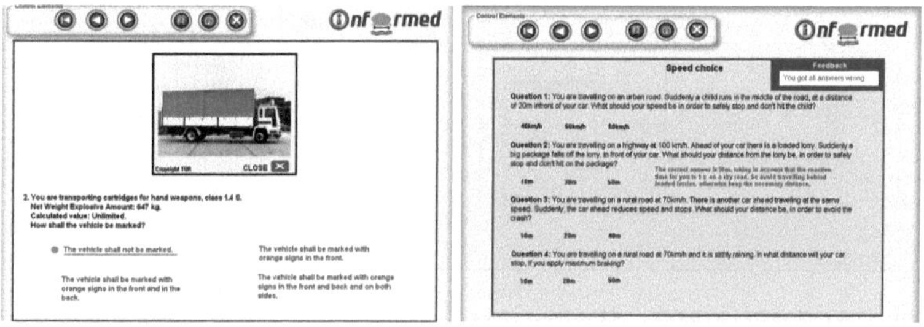

Fig. 9.17 INFORMED MMTs – Assessment in "Training Mode" (*left*: MMT for ADR training; *right*: MMT for advanced driving training)

User Assessment: The following elements are met in both MMTs:

- Text
- Pictures
- Multiple choice questions
- Videos
- Interactive Videos
- Simulations
- Interactive Simulations
- Multiple choice questions based on pictures
- Multiple choice questions

User performance tracking: In most cases the assessment (training or test mode) of the MMTs are realised by multiple choice questions. In some cases, the multiple choice questions are based on a picture, a video or a simulation. Within the assessment part of the "Training Mode" (multiple choice tests), the trainees are able to be aware of the correct answer to the question (Fig. 9.17) after they have answered it themselves and have ticked in the "Submit" button. In some cases relevant feedback messages are pushed. There are also cases when the assessment is

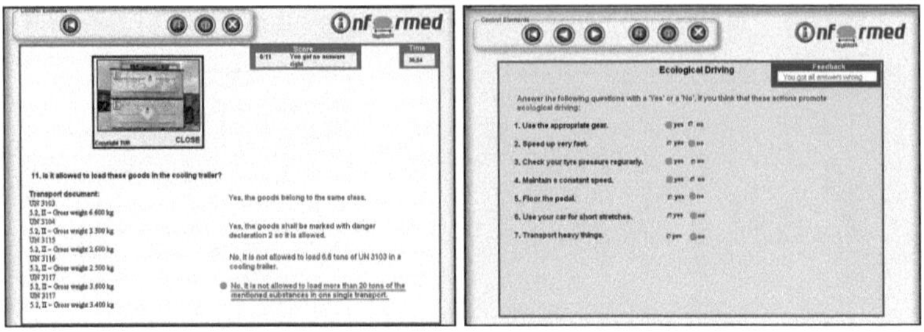

Fig. 9.18 INFORMED MMTs – Assessment in "Test Mode" (*left*: MMT for ADR training; *right*: MMT for advanced driving training)

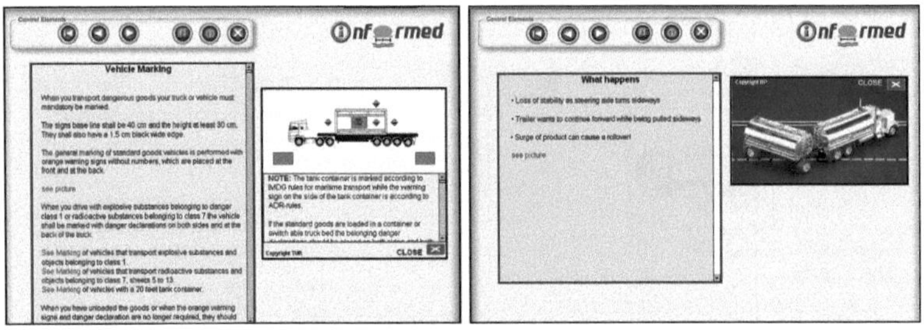

Fig. 9.19 INFORMED MMTs – Scrolling text accompanied by explanatory picture (*left*: MMT for ADR training; *right*: MMT for advanced driving training)

performed by means of interactive videos in which the trainees are requested to tick in certain spots of the video user interface (e.g. for risks identification). In addition to that, during the performance of the test mode's multiple choice questions, the interaction time of the user with each test scenario as well as his/her score are continuously visible on the MMT (Fig. 9.18). Thus, at the end of each test scenario, the overall score of the trainee at the specific scenario is visible.

Log files are created for both the exercise parts of the training scenarios and the test scenarios. Moreover, the end-users scores with the specific date of the corresponding end-user performance are being recorded in the created log files, as a performance percentage, as a performance ratio or as an indication of correct or wrong.

Pictures and text: A lot of material is included in the MMTs, as it is required for such tools, which has to cover effectively such a broad training material. In addition to the fact that attention has been paid so that the text presented in the user interface of each screenshot is not overloaded with text, the existing text is provided in scrolling format (Fig. 9.19), in order to ease the user training (training mode). Moreover, relevant pictures are included, considered to replace a great part of required text and at the same time helping the reader to understand the provided information more easily and effectively (Fig. 9.19). Besides, in some cases, pictures and other kind of figures are the only way to explain to the user certain didactical objectives.

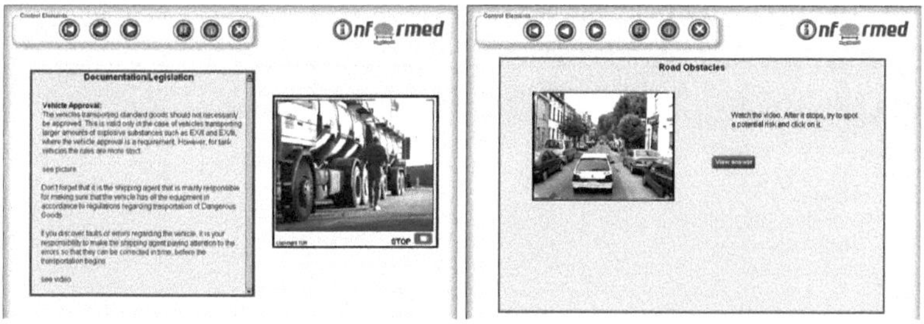

Fig. 9.20 INFORMED MMTs example videos (*left*: MMT for ADR training; *right*: MMT for advanced driving training)

Fig. 9.21 "Multimedia for advanced driving carrying dangerous goods"–"Training Mode" – "Gap Acceptance" scenario interactive simulation

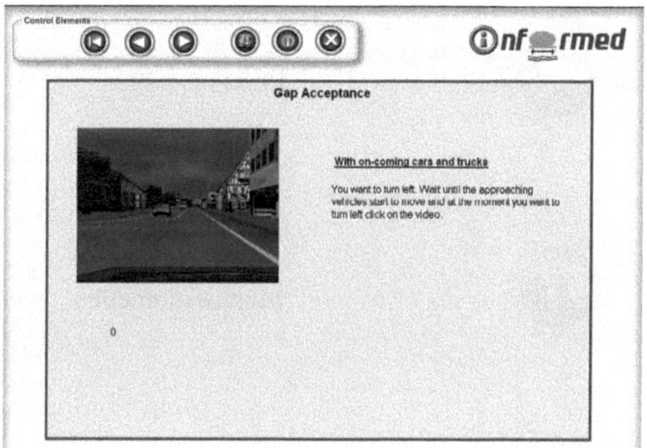

Videos: A series of videos are included in the MMTs (Fig. 9.20). These videos aim to depict dynamic situations, which cannot be efficiently illustrated in a different way.

Simulations: Simulations are movies that are not real, but consist of simulated components (job equipment, humans) and are used mainly as an alternative to real videos. The most important advantage of the animations is that precise information can be designed and given, which, in many cases, are not visible in real videos. Moreover, several risky situations can be simulated, which is not possible to be provided by means of a real video. In some cases, the simulations can be interactive (Fig. 9.21).

Control buttons and menus: The trainee is able to navigate through the pages, with the navigation buttons 'Next' and 'Previous'. In addition, command buttons exist to submit an exercise, to request for help, to exit the software, etc.

Table 9.1 summarises the most important command buttons that are embedded in the MMTs.

Help and support functions: As it is expected from a training program, support functions have been anticipated for the trainee, so that s/he will be able to autonomously operate it and overcome any difficulties occurred. Thus, by selecting the "Help Function" of the menu (Table 9.1), a window appears, containing guidance for the use of the MMT (Fig. 9.22).

Table 9.1 The MMTs command buttons

Button	Description
	First page
	Next page
	Previous page
	Menus/submenus
	Help function
	Exit the software. Before exiting, an intermediate message appears: 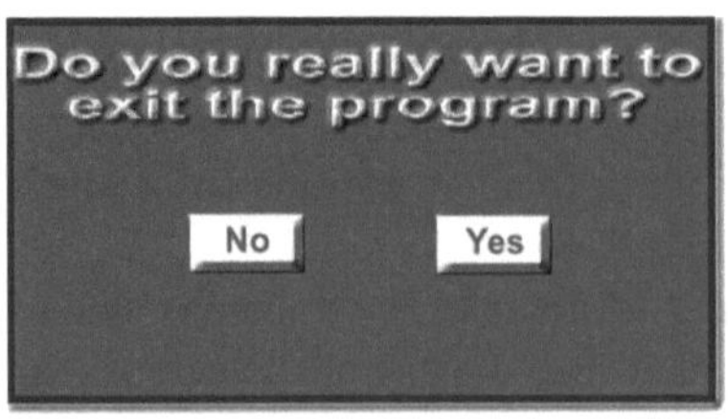
	Submit exercise/test
	Play the video

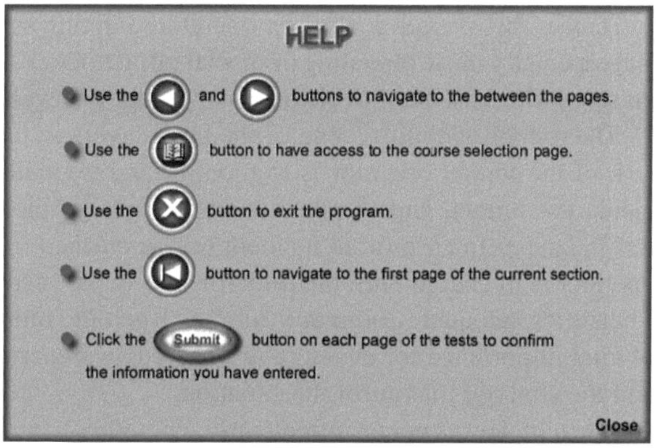

Fig. 9.22 Help functions of the MMTs

9.5.4 *Avoiding the Risks*

The multimedia software tools developed in the context of INFORMED project are both innovative and aim to support the training (in-class or self training) provided to professional drivers involved in the dangerous goods haulage sector. All guidelines, recommendations and regulations included are valid in all European countries and in full compliance with the ADR 2003 regulations.

However, on one hand, the INFORMED *Multimedia tool for ADR training* is much more enriched and covers in depth more training issues in comparison to the ones requested by the ADR 2003 regulations. On the other hand, the *Multimedia for advanced driving carrying dangerous goods* is an innovative software tool, which aims to train ADR and other drivers on the risks of driving, the reasons that lead to road accidents, ways to avoid and confront them, and the vehicles technical issues and fundamental physics concerning driving that each driver should be aware of.

A compendium of experts with great experience, some of them being also INFORMED Partners, have contributed to their development and are considered to constitute a good point of reference for end-users, MMT developers and experts belonging to the dangerous goods haulage sector.

9.6 Operators Training

9.6.1 *When to Train?*

The objective of a TMIC/TMC is to monitor and evaluate traffic across the major arterial roads and provide relevant, real time information about traffic conditions in the region. Traffic management centres also play a vital role in managing traffic in the event of traffic incidents. With cameras situated in strategic locations, console operators become a command centre for police and emergency services in co-ordinating all activities to effectively deal with the situation.

Hence, the overall goal of an operators training scheme is to provide operators and especially those operating in special infrastructures, i.e. tunnels, bridges, with the know-how to base their decisions on thoroughly developed and evaluated methods.

The tunnel operator is one of the key-players in tunnel incident situations. The task of the tunnel operator is to monitor and regulate the flow of traffic near and inside the tunnel, and to take actions in case an incident occurs (e.g. by closing traffic lanes). In case of an incident (e.g. a crashed truck catching fire), the tunnel operator is in charge until the principle fire-department officer arrives on the scene. Thereafter, adequate communication between the tunnel operator and the officer is of vital importance for developing a shared awareness on the nature of the incident and the strategy to control the situation.

Since incidents occur infrequently, operators have to train themselves regularly in order to maintain sufficient incident management skills. Especially TMI/TMC operators need internal expert training and training to handle computer based systems to run the control systems that are in use.

The term "operators" implies people that are involved in managing and operating infrastructure at different levels, such as road operators, tunnel operators, Traffic Management Centres (TMC) operators. Their role may vary from the everyday use and/or maintenance of the infrastructure and its equipment, to handling emergency situations and participating in decision making processes for the enhancement of the infrastructure under their responsibility and its equipment with the appropriate technological items and what this may impose to their work.

Thus, there is also the need for developing training schemes for the different categories of operators that would help them in recognising the potential of new technologies and their applicability in the infrastructure they are managing. Moreover, they may be motivated to install new equipment in the infrastructure, which would enhance its safety and/or efficiency level, as well as, once new equipment is installed, they would be aware of its technical and functional requirements, together with its maintenance needs for optimum performance. As far as in-vehicle equipment is concerned, this is also of interest to the operator, for its use and subsequently growing penetration rate, might modify the drivers' behaviour, and thus impose different needs from the infrastructure point of view for its safe and efficient use.

Recognising the above, relevant training schemes have been developed in different research initiatives. Within the scope of this chapter, the TMC operators' training schemes (manual and MMT), developed within IN-SAFETY STREP and the tunnel operators' training curricula, developed within GOODROUTE STREP are presented.

9.6.2 IN-SAFETY Operators' Training Manual and MMT

In the course of developing the IN-SAFETY operators' training schemes, a series of steps has been followed. Initially, research was performed in the currently existing schemes in Europe, their procedures, contents and training curricula. For this reason, a detailed questionnaire has been structured, which was used for interviewing TMC operators throughout Europe regarding the training schemes that are used in their company, their contents (especially regarding new technologies), the needs and gaps they find in the applied procedures and their ideas for an optimal training. On this basis, a content structure was elaborated, for the needs of the IN-SAFETY training schemes. Finally, relevant content has been collected and the Operators' Training Manual and Operators Training Multimedia Tool have been developed.

9.6.2.1 Review of Existing Training Schemes

The primary task was to collect the state of the art of training schemes for operators in Traffic Management Information Centres (TMI) and Traffic Management Centres (TMC) and to find definitions for an optimal operators' training scheme, as well as further needs in terms of training. Therefore, a questionnaire was developed and provided to various organisations dealing with road telematics. Responds were received by representative entities from several countries.

The operators' training is usually divided in two parts. Firstly, there is the training for everyday work with the existing equipment of a TMI/TMC. The operator is responsible for this training and this part of the training was not within the scope of IN-SAFETY. Secondly is the training on new equipment use and their functionalities.

Within this context, the aim of IN-SAFETY was to develop an additional training scheme for operators to use telematic applications. The staff of a TMI/TMC centre should obtain knowledge about providing target-oriented information to road users. Thus, their training should contain general knowledge about telematics as well as the state of the art and future potentials of ITS applications. This second part of the training supports the integration of new technologies in TMI/TMC and further deployment of various ITS.

From the interviews results it has been recognized that, in most cases, the operators use internal expert training and the training is often on the job, while specific training courses are periodically held. The topics included in the existing operators training schemes are related to both normal use and emergency situations. According to the majority of the respondents, the list of the contents considered necessary or useful to be included in a training scheme is much longer than the actual list of contents of the existing training schemes. This indicates that many among the operators would be willing to improve their training and they have certain ideas regarding which topics should be covered by an improved training procedure. It has also been noted that there is not very much about ADAS/IVIS in the list of existing training contents, however the respondents expressed their interest to learn about various ADAS/IVIS applications (Winkelbauer et al. 2008).

9.6.2.2 IN-SAFETY Operators' Training Scheme on ITS

Operators' Training Manual

Based on the above and within the scope to provide a manual, focussing on new technologies and telematic applications from different points of view for the needs of operators' training, the contents of the IN-SAFETY operators' manual were primarily defined. The main aim was to include basic knowledge on traffic engineering, together with up-to-date information on telematic applications, in-vehicle and infrastructure-based electronic systems and other traffic related technologies. Additionally, issues related to these systems' impacts and standardisation are also included.

Structure

The Manual is divided into three parts. The first part, "Introduction", provides some general knowledge on training and telematic systems. It includes background information on operators' training, general information on ITS and some basic definitions (i.e. telematics, forgiving road environment, etc.). It also incorporates an overview of relevant telematic systems, of the functions they perform and the areas of use of those systems. Different types of classification are described, along with specific system features.

In the second part, a specific classification is being proposed, which is followed in each system's description in Part 3. The components of each system's description are the following: timeframe of their action, type of support, benefits, quality of service, limitations, available standards, operation and maintenance guidance, implementation examples, future prospects and relevant scientific research.

Then, in the third part, the systems themselves are included, following the above classification. The systems are grouped in six categories:

- ITS for private vehicles
- ITS for public transport
- ITS for commercial vehicles
- ITS for infrastructure
- ITS for vehicle control
- Cooperative ITS

For each one of them, the following features are provided (where relevant or available), according to the classification described in Part 2:

- General information
- Functionality
- Time frame of action
- Support type
- Benefits

- Quality of Service
- Limitations
- Standards
- Guidance
- Examples
- Future prospects

In total, more than 45 systems are described in detail, including – to the possible extend – up to date information. Relevant references to each topic are also provided for further consultation.

The third part of the Manual is dealing with operational background knowledge on several issues of interest of road and TMC operators. More specifically, the following issues are included:

- Providing an optimal training.
- Maintenance, explainability and interoperability.
- Cross TMC cooperation (urban, highway, rural).
- Cost effectiveness (socio-economic evaluation).
- Impact assessment of ITS applications.
- Basics of traffic engineering.
- Conflict management.
- Standards and standardization bodies.

Moreover, a network, consisting of the project partners is suggested, as contact points for any further relevant information.

Operators' Training Multimedia Tool

As indicated above, the MMT was developed based on the same content that is included in the Operators' training manual, mostly focussing on ITS applications.

The MMT has been developed using the Macromedia Flash 8. A user friendly layout has been developed, with which the user can find the requested information easily, just with a few mouse clicks.

Each of the topics is listed in a drop-down list on the right of the screen. Especially for the systems, each of them is presented separately, with their specific components listed on the left side of the screen, as shown in Fig. 9.23.

Moreover, "Next" and "Previous" page buttons exist at the bottom of the screen, as well as a "Close" button. Additionally, a "Glossary" button exists at the bottom, which opens an additional feature, developed for the MMT. This is a glossary of main terms related with telematics and road safety, as can be seen in Fig. 9.24. Last but not least, another button, called "Standards Reference" is included, containing a list of relevant ITS standards and corresponding links.

Thus, the operators' training manual is striving to provide operator trainees – but not only – with an easy to use tool, whose application is not only addressed to the training of staff; the MMT addresses also the needs of upper hierarchy employees

Fig. 9.23 Snapshot of the IN-SAFETY MMT (Lane Departure Warning)

Fig. 9.24 Snapshot of the IN-SAFETY MMT, showing the Glossary feature

(i.e. TMIC planners and managers), as a consulting tool for enhancing their decision making procedure. The MMT could be of use in several cases as a consulting toolbox for decision making, in terms of applying a new system in the network, deciding among different alternatives for solving a particular problem, related either to safety or traffic efficiency enhancement. Another application area could be this of assessing the effects of already applied measures, by comparing with similar applications and taking into account the different parameters of each system's functionalities.

9.6.2.3 GOODROUTE operator's Training Schemes on Safety Routing and Emergency Handling of Dangerous Goods

Apart from the operators themselves, the training scheme developed within GOOD ROUTE, building among others on the experience of IN-SAFETY project, addresses also the higher level personnel of TMIC/TMCs as well as road operating companies, to provide them with a consulting tool aiming to enhance their decision making process, while adapting new safety systems in their operations.

After conducting an extensive benchmarking, the training curriculum for the infrastructure operators has been developed on the basis of the Guidelines for TMC Transportation Management Operations Technician Staff Development, also referred as the TMOT Guidelines (Bekiaris and Gemou 2009). New research into TMC staffing requirements have been combined with existing practices in order to create comprehensive guidelines for developing the operations personnel position descriptions needed to properly staff a TMC.

In Fig. 9.25, the selected functions are listed and the number of discrete tasks identified, for a total of 1,060 discrete tasks. In intermediate levels, the GOOD ROUTE dedicated sessions have been added, as shown also in Fig. 9.25. These are namely the following:

- Ontologies and interfaces to other modes of transport
- Traffic safety data collection and management
- Passport, routing and re-routing
- Emergency functionalities
- TMC and allied services monitoring and access levels
- Enforcement functionalities
- Control Centre operation and maintenance
- On the job training

The overall duration of the curriculum is 29 training hours. The training manuals, related to the installation, operation and maintenance of the enforcement system, the on-board unit and the operation of the Control Centre by the infrastructure operators and the Logistic Company (or driver, if own contractor) operators have been developed in order to support the training sessions.

What is very significant in this case is the on-the-job training part, which could last from 1 week to 6 months, and should focus on realisation of GOOD ROUTE scenarios, as done in the GOOD ROUTE Pilots (Bekiaris et al. 2009). Another alternative for the on-the-job training or pre-phase would be the use of training simulation tools which seem to be preferred more and more in the field (Bekiaris et al. 2009). The GOOD ROUTE Training Curriculum for Infrastructure Operators is shown in Fig. 9.25.

TMC Task Integral Function	Time Duration
F1. Provide Travel Information	2 teaching hours[6]
GOOD ROUTE: Ontologies and interfaces to other modes of transport	1 teaching hour
F2. Records Management	1 teaching hour
GOOD ROUTE: Traffic safety data collection and management	1 teaching hour
F3. Congestion Management	1 teaching hour
F4. Failure Management	1 teaching hour
F5. Incident Management	1 teaching hour
F6. Special Event Management	1 teaching hour
GOOD ROUTE: Passport, routing and re-routing	2 teaching hours
F7. Traffic Flow Monitoring	1 teaching hour
F8. Emergency Management	1 teaching hour
GOOD ROUTE: Emergency functionalities	2 teaching hours
F9. Provide / Coordinate Service Patrols	1 teaching hour
F10. Reversible & HOV Lane	1 teaching hour
F11. Traffic Signal System Management	1 teaching hour
F12. Transit Vehicle Monitoring	1 teaching hour
GOOD ROUTE: TMC and allied services monitoring and access levels	2 teaching hours
F13. APTS System Management	1 teaching hour
F14. Environmental & RWIS Monitoring	1 teaching hour
F15. Overheight Vehicle Management	1 teaching hour
GOOD ROUTE: Enforcement functionalities	1 teaching hour
F16. Rail Crossing Management	1 teaching hour
GOOD ROUTE: Control Centre operation and maintenance	3 teaching hours
GOOD ROUTE: On the job training	1 week to 6 months

Fig. 9.25 GOOD ROUTE training curriculum for Infrastructure Operators

[6]One training hour corresponds to 45 minutes

9.7 Least We Forget…

From all the above it is made evident that the existence of new technologies in the everyday life of both drivers and infrastructure and TMI/TMC operators is nowadays an unquestionable reality. In-vehicle (ADAS/IVIS) and infrastructure-based (VMS, VDS, etc.) equipment is currently more or less part of the everyday driving routine throughout Europe. The use of such systems is targeting the road safety and efficiency enhancement by assisting the drivers in recognising and preventing hazardous situations, guiding them safely and quickly to their destination, allowing the traffic managers to predict and handle traffic congestion, emergency situations, etc.

So it becomes of major importance for the users of these systems to be adequately trained, so as to make the most out of the use of new technologies, avoiding at the same time their misuse, which might lead to less benefits as well as potential negative consequences, such as the driver's distraction from the driving task or over-reliance to the system.

Towards this target, a series of training schemes and tools have been developed, which have been described above. These schemes are targeted to train either different drivers (novice, experienced, professional, etc.) or infrastructure and TMI/TMC operators in using ITS for their needs. At the same time, they can be used as consulting documents and decision making tools, either for drivers to buy such a system for their vehicle or for operators to choose the most appropriate one for the needs of the piece of infrastructure under their responsibility.

In the course of developing these training schemes, relevant research has been performed in order to investigate what similar schemes currently exist and their characteristics. Significant gaps have been identified in the training of both drivers and operators and, for the specific case of ITS there were almost no training tools in most of the cases. However, the drivers as well as the operators declared their interest to have such training included in their existing training curricula.

The training schemes that are presented in this chapter are using multiple means to address the trainees. Training manuals, multimedia tools, simulators and specifically designed curricula are proposed for the needs of the various user groups. The variety of knowledge channels addressed constitutes the training more efficient for the trainers and more interesting for the trainees.

Of course the list of schemes presented in the present chapter is not exhaustive, as this was not its scope. What should be depicted is the utmost importance of training the road users and managers and accustoming them on the use of ITS for their own benefit as well as for the benefit of the society. And here it is shown that there exist today the means and the methodology to perform it. The step forward is to implement them; and this usually requires a political decision.

References

E. Bekiaris, M. Gemou, S. Tagianoglou, E. Vandenberger, Deliverable 1.1, Training needs of drivers, trainers and companies involved in the dangerous goods haulage sector and accidents, INFORMED Project, EL/PP/2002-114010, Oct 2003

E. Bekiaris et al., LATERAL SAFE SP of PREVENT IP, D3.2.9, Evaluation and assessment, Jan 2007

E. Bekiaris, M. Gemou, Deliverable 8.2, Training schemes for DG drivers and traffic control operators, GOOD ROUTE project, IST-4-027873-STREP, Jan 2009

E. Bekiaris, M. Gemou, E. Chalkia, J. Kiuru, Deliverable 7.2, Pilot results consolidation, GOOD ROUTE project, IST-4-027873-STREP, Jan 2009

H. Christiansen, E. Bekiaris, M. Gemou, Deliverable 2.3, Towards a common European strategy for training trainers and drivers in the dangerous goods haulage sector, INFORMED Project, EL2002/B/F/PP-114010, Dec 2003

S.E. Fahey, W.W. Wierwille, Advisory and alarm stimuli optimisation for a drowsy driver detection system. NHTSA, DOT HS 808 299 Semi-Annual Report, Research on Vehicle-Based Driver Status/Performance Monitoring: Seventh Semi-Annual Research Report, National Technical Information Service, Springfield, 1995

P. Gaumet, E. Bekiaris, M. Gemou, E. Vandenbergen, Deliverable 3.3, State of the art on materials and tools in the market and proposed advanced on-the-road training schemes, INFORMED Project, EL2002/B/F/PP-114010, Aug 2004

M. Green, How long does it take to stop? Methodological analysis of driver perception – brake times. Transport. Hum. Factors **2**(3), pp. 195–216, 2000

W. Huiskamp, R. Jansen (TNO), R. Holzer (UP), I. Mandsouris (ICCS), Ph. Vanhulle (Thales), TRAIN-ALL Del. 2.1, Common system architecture for driving simulators based on interoperable federates, 2008

ISO standard ISO/DIS 26022, Road vehicles – ergonomic aspects of transport information and control systems – simulated lane change test to assess in-vehicle secondary task demand (under development; to be available on Sept 2011)

S.E. Lee, R.R. Knipling, M.C. DeHart, M.A. Perez, G.T. Holbrook, S.B. Brown, S.R. Stone, R.L. Olson, Vehicle-based countermeasures for signal and stop sign violation, NHTSA, DOT HS 809 716, 2004, pp. 98–111

A. Mousadakou, M. Gemou, Deliverable 3.1, Multimedia tool for ADR training, INFORMED Project, EL2002/B/F/PP-114010, June 2004a

A. Mousadakou, M. Gemou, Deliverable 3.2, Multimedia tool for the advanced driving of vehicles carrying dangerous goods, INFORMED Project, EL2002/B/F/PP-114010, June 2004b

M. Panou, *Advanced Personalized Travellers' Warning and Information System*, PhD Dissertation, Aristotle University of Thessaloniki, Thessaloniki, 2008

A. Papamathaiakis, S. Nikolaou, HUMANIST project, Deliverable F4, Multimedia software tool for driver's training in the use of new ITS, Apr 2004

S. Poschadel (editor, IFADO), J. Kapplusch, P.M. Lindberg Kristiansen, (IFADO), Ph. Vanhulle (THALES), J. Pfaffenzeller, Ch. Kainz (BPP), S. Espié (INRETS), B. Peters (VTI), E. Bekiaris, M. Panou (HIT), TRAIN-ALL Del.1.2, Training needs, scenario and curricula definition and specification of tools and curricula, 2008

A. Simoes et al., HUMANIST project, Deliverable F.2, Inventory of drivers' needs for training regarding ITS according to driving tasks affected, Dec 2006

H. Summala, Ergonomics of road transport. IATSS Res. **21**(2), 49–57 (1997)

D. Twisk, S. Nikolaou, Deliverable F.1, Inventory of ITS functionalities according to driving task models, Task Force F, HUMANIST NoE, Sept 2004

L. Warshawsky-Livne, D. Shinar, Effects of uncertainty, transmission type, driver age and gender on brake reaction and movement time. J. Safety Res. **33**, 117–128 (2002)

M. Winkelbauer, C. Nussbaumer (KfV), E. Gaitanidou, E. Bekiaris (CERTH/HIT), IN-SAFETY D3.3, Road operators training schemes and tools, Mar 2008

K.L. Young, M.A. Regan, M. Hammer, Driver distraction: a review of the literature. Report No. 206, Monash University, Accident Research Centre, 2002, ISBN 0 7326 1715 4

Part III
Forgiving Road Environments

Chapter 10
The Impact of Lateral ADAS in Traffic Safety

Tom Alkim

10.1 ADAS Considered

Many ADAS have come to the market the past decade, or are near market introduction. Electronic stability control (ESC) for instance is currently available on almost every new car. In the longitudinal support category Advanced Cruise Control (ACC) and Speed Alert or Intelligent Speed Adaptation (ISA) are well known. In the lateral support category Lane Departure Warning Systems (LDWS)/Lane Keeping Systems (LKS) are the most known. They either warn the driver or intervene when he or she is about to drift unintentionally from their lane. Also other systems (i.e. Blind Spot Detection and Overtaking Assistant) fall in the category of lateral support, and more specifically, in the (lateral) Collision Avoidance Systems cluster. For a more detailed overview on them please refer to Chap. 4.

In various field operational tests (FOTs) in The Netherlands LDWS/LKS, as well as other ADAS, have been evaluated to explore the effects on driving behaviour and traffic flow, in terms of safety, throughput and environment.

In this chapter, you will find a functional description of the tested LDWS/LKS. Several FOTs have been conducted with such systems in The Netherlands and their results are included in this chapter. They refer to the applications of "Lane Departure Warning in Trucks" (2002/2003), "The Assisted Driver" (2005/2006), "Accident Prevention Systems in Lorries" (2008/2009) projects.

10.2 Lateral Support Systems that Have Been Evaluated

Two important tasks for a driver are to maintain a certain velocity and keep a designated course. The driver needs both his/her hands and feet to do so. And let's not forget the head, because driving a vehicle involves a lot more than just

T. Alkim
Ministry of Infrastructure and the Environment, Directorate-General for Public Works and Water Management (Rijkswaterstaat), Centre for Transport and Navigation, Rijkswaterstaat, The Netherlands
e-mail: tom.alkim@rws.nl

E. Bekiaris et al. (eds.), *Infrastructure and Safety in a Collaborative World*,
DOI 10.1007/978-3-642-18372-0_10, © Springer-Verlag Berlin Heidelberg 2011

operating the vehicle. Many situations are encountered and many decisions have to be taken on the operational, tactical and strategical level, which together define the driving task (Michon 1971, 1985; Janssen 1979). Given the fact that approximately 90% of all traffic accidents are caused (sole cause or contributing factor) by human errors (Treat et al. 1977), assistance in the form of ADAS is welcomed, to reduce the number of accidents or the impact of accidents and thereby increasing traffic safety. They can support the driver in situations where the workload is either (too) high or (too) low, by taking over part of the driving task on the operational level. This can provide the driver with extra capacity to be used on the tactical and strategic level in situations where the workload is high. In situations where the workload is low (hypo vigilance), the driving performance on the operational level can be improved as well.

Assistance in both lateral and longitudinal tasks may be beneficial for traffic safety. In general this is due to the fact that ADAS use sensors to monitor the direct surroundings of the vehicle and therefore do not lose valuable reaction time in assessing potentially critical situations. The human driver typically needs around 1 s to react to such situations. Sensors, such as a laser system, radar systems, video camera, etc., typically need just milliseconds to recognize such a situation and therefore can yield a valuable time saving of approximately 1 s. Depending on the type of ADAS this time saving is used to either alert the driver earlier (passive or warning ADAS) or even to intervene automatically (active or intervening ADAS). In this chapter both passive (LDWS) and active (LKS) lateral support systems are described and discussed, as an introduction to the relevant evaluations performed with them in The Netherlands.

10.2.1 Lane Departure Warning Systems

LDWS are designed to warn the driver when s/he is about to drift unintentionally from his/her lane. LDWS is a passive ADAS and only issues a warning leaving the driver in full control of the vehicle. The driver decides whether s/he will issue a corrective steering manoeuvre or not.

Usually a LDWS uses a camera to detect the lane markings on the road, sometimes infrared is used. This implies that LDWS is dependent on a certain minimum level of quality of the lane markings. If this minimum level is not met, the system usually gives feedback to driver informing him/her that the system is not functioning under the current conditions. In order to decide whether a warning should be issued to the driver the Time to Lane Crossing (TLC) is monitored; if it exceeds a certain threshold (e.g. 0.5 s) a warning is given. To decide whether the driver is unintentionally leaving the lane or not, the status of the indicators is used. If they are switched on, the driver is intentionally leaving his/her lane. If they are not switched on, the driver is unintentionally leaving his/her lane or forgot to use the indicators; in both cases a warning is justified.

Warnings issued by LDWS have different forms; it can be audible, visual, tactile, haptic or a combination. The most common warning used is a combination of audible and visual; a beep or a sound resembling a car driving on a rumble strip is issued by the system while visual feedback is given on the system's display. In general the warning sound is given on the side of the car that is about to leave the lane, so that the driver intuitively knows where s/he's going and how s/he should correct. Other ways of warning the driver is by a vibrating seat, belt or steering wheel. These forms of warning are generally considered less intrusive by the driver and therefore more acceptable.

Due to the fact that LDWS is only designed to warn the driver and not to take over part of the driving task, there is no interference with the steering wheel, brakes or other parts of the vehicle. This makes it relatively easy to install LDWS on a vehicle allowing an aftermarket approach in addition to factory fitted systems. At least four suppliers offer such an aftermarket solution in Europe; Mobileye, Valeo, Iteris and Albrecht. The systems provided by these companies can be installed on both person vehicles and trucks. Factory fitted systems are also available on both person vehicles and trucks, these LDWS are completely integrated in the vehicles.

10.2.2 Lane Keeping Systems

LKS are designed to support the driver to keep his/her lane and to prevent drifting from the lane unintentionally. LKS is an active ADAS and supports the driver by taking over part of the driving task. The driver remains completely responsible for operating the vehicle and is kept "in the loop". The driver can however decide to overrule the support, for instance if an emergency manoeuvre has to be performed.

Usually an LKS uses a camera to detect the lane markings on the road, sometimes infrared is used. This implies that LKS is also dependent on a certain minimum level of quality of the lane markings. If this minimum level is not met, the system gives feedback to the driver informing him that the system is not functioning under the current conditions.

There are basically two approaches for LKS, one is to actively keep the vehicle in the middle of the lane by continuously giving torque feedback and the other one is to give only torque feedback when the vehicle approaches the lane marking closer. The latter situation feels for the driver as if the sides of the road curve upwards whereas in the middle of the lane driving conditions are normal. Therefore the driver is only assisted when he/she is getting close to the lane markings and is about to leave the lane (see Fig. 10.1). With the continuous torque feedback approach it is basically more difficult to steer away from the middle of the lane, the further away from the middle the more resistance is encountered, see Fig. 10.2.

Due to the fact that LKS is designed to take over part of the driver's task, and, thus, intervenes the steering action, no aftermarket approach is possible. Only factory fitted systems in cars that have electronic power steering are offered on the market.

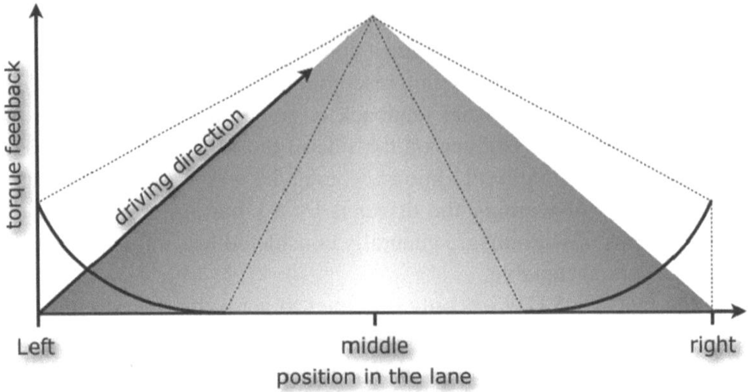

Fig. 10.1 Critical areas-based Lane Keeping Assistant torque feedback

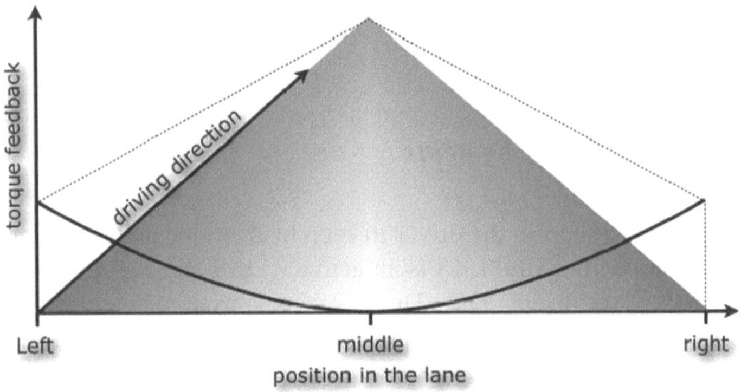

Fig. 10.2 Continuous Lane Keeping Assistant torque feedback

To prevent drivers from misusing the system and use their hands for other tasks than steering, LKS has an algorithm to assess whether the driver is paying attention or not. This algorithm uses as input either the minute steering corrections that are continuously provided by a driver to the steering wheel or it actually measures whether the driver's hands are touching the steering wheel. If, for an extended period (usually 5–10 s), no steering corrections are detected or no hands on the steering wheel are detected, than the systems gives an audible warning and switches off, so the driver has to take over again.

10.2.3 Market Penetration

LDWSs and LKS entered the market first in Japan (Nissan Cima 2001). The United States and Europe followed several years later (Bishop 2005). Currently there are many models that offer LDWS or LKS as an option in Europe. Only some models have LDWS standard (e.g. Citroen C5 and Lancia Delta).

In addition to factory fitted systems there are at least four companies (Mobileye, Valeo, Iteris and Albrecht) that offer an aftermarket solution, which can be installed in practically any vehicle. The aftermarket solution is always a warning system (LDWS) and not an active support system (LKS). It's relatively easy to install an aftermarket LDWS in your vehicle, it consists of basically three components: a camera to recognise the lane markings, a display and chip with computing power for the lane departure algorithm and a set of speakers to issue such a warning.

On trucks and lorries, LKS is not available yet because the vehicle dynamics of these vehicles are too powerful and complex to be dealt with by current available technology. Of course LDWS is available as either factory fitted or after market system.

10.3 FOTs in the Netherlands

In the past decade The Netherlands have organised several FOTs and demonstrations in the field of intelligent vehicles and cooperative systems. It all started with the very successful Demo 1998, where the world was shown what the possibilities are, from a technological point of view, regarding intelligent vehicles and automated highways. This event also created a lot of awareness amongst stakeholders and the public.

Due to the enormous investments necessary to create automated highways and autonomous vehicles, the following years the focus was more on ADAS and it's impact on traffic flow in terms of safety, throughput and environment. Specifically, the human behaviour aspect played an important role in these FOTs: will drivers accept these systems, will they use it as intended by the designers and are there any unexpected effects in driving behaviour? Much knowledge and experience regarding the aforementioned issues was acquired in The Netherlands through the following FOTs:

- Intelligent Speed Adaptation (1999–2000)
- Lane Departure Warning Assistant in Trucks (2002–2003)
- Roadwise (2005)
- The Belonitor (2005)
- The Assisted Driver (2005–2006)
- Accident Prevention Systems for Lorries (2008–2009)

The focus of this chapter is lateral support and therefore results from the following three FOTs are presented: "Lane Departure Warning Assistant in Trucks", "The Assisted Driver" and "Accident Prevention Systems for Lorries".

10.4 Lane Departure Warning in Trucks

In order to understand the impact of LDWS, The Ministry of Transport, Public Works and Water Management issued a FOT on lateral support systems in 2002. This FOT was focused on heavy goods vehicles driving with LDWS. The objective was to find out if, and if so, how much LDWS might improve traffic flow in terms of safety and throughput (Alkim et al. 2003).

10.4.1 The FOT

The overall objective of the FOT was 'to evaluate the selected lateral support systems from the perspective of users and stakeholders based on actual driving experience with these systems'. The FOT aimed to provide insight into the effects of driving with LDWS on traffic flow in terms of safety and throughput, acceptance and possible consequences for the infrastructure.

In order to obtain this information, the former Transport Research Centre (AVV) of the Dutch Ministry of Transport, Public Works and Water Management funded a FOT with 35 LDWS equipped heavy vehicles and one LDWS equipped bus. AVV was responsible for the FOT and commissioned a Consortium consisting of TNO (Netherlands Organization for Applied Science), ITS (Institute for Applied Social Sciences, University of Nijmegen), Arcadis, PATH (Partners for Advanced Transit and Highways, University of Berkeley) and the University of Minnesota to conduct the research program.

10.4.2 Research Framework

Specific behavioural changes over time as a result of driving with LDWS were monitored with data logging devices. This enabled the description of possible traffic safety effects resulting from driving LDWS. Within the group of 35 heavy-goods vehicles, a sub-group of six trucks was equipped with a data-logging device in order to monitor their behaviour more closely.

LDWS particularly relates to the driving task at the operational level. Manoeuvring of the vehicle within the lanes is an automatic task and involves the precise steering wheel movements executed unconsciously yet continuously. Changing the behaviour at that level is difficult, precisely because the behaviour has become so automatic. With respect to lane position it is known that drivers of heavy vehicles usually do not keep the middle position in a lane. This is the result of balancing the demands of the driving task, i.e. workload, and required safety level. However, by giving auditory feedback on possible unintended lane departures and as such informing drivers when they are about to drift from their lane, the system intends to evoke a conscious realisation of the driver's swerving behaviour. In order to avoid the warnings and the realisation of the swerving behaviour it is expected that drivers will either choose a more stable and central lane position, or turn off the system altogether when the warnings become too irritating. Adding a device like LDWS may have a positive impact on driving behaviour, but may also affect behaviour in a negative sense, i.e. compensating behaviour (increase in speed or a decrease of TLC values), resulting perhaps in an overall negative effect on traffic safety. Therefore behavioural data from the driver and from the system was collected to investigate possible positive and negative effects of driving with LDWS.

Research questions to be answered were:

1. How often do drivers use the system?
2. How many warnings for an unintentional lane departure do drivers receive on average per hour?
3. Do drivers change their position on the road after using an LDWS system over time?
4. Is there a difference between motorways and secondary roads and the amount of warnings drivers receive for unintentional line crossings?
5. Does the number of warnings change over time?
6. Do drivers show compensatory behaviour?
7. Is there a difference in length of unintentional line crossings and their amplitude over time?

To answer these questions three different systems were installed in various trucks: The SafeTrac system in DAF trucks, the Spurassistent in Daimler trucks and the Lane Guard System in MAN trucks. All three systems are vision based and get their information on road position from a camera mounted on the front window of the truck. Warnings were given in two different fashions: the Spurassistent and the Lane Guard System both issued their warnings through the radio speakers on either side of the driver corresponding with the side the truck crosses a road boundary. The SafeTrac system gave a general non-directional warning when a lane-boundary is crossed. All three systems did not give a warning when the indicators were used. All participating drivers had a minimum driving experience of 5 years and were between 28 and 55 years old. The duration of the FOT was 4 months.

The behavioural data gathered by the data-logging systems were:

- Line crossings (amount, amplitude and duration in), related to type of road.
- Number of alarms of issued by the LDWS (overall, left, right and separated for highways and secondary roads).
- Lateral position (with regard to lane markings).
- Lateral velocity and acceleration.
- TLC (derived from position, speed and acceleration).
- Steering wheel movements (amplitude, frequency).
- Speed.
- Acceleration/deceleration.
- LDWS on/off.
- Indicator on/off.
- Windshield wiper on/off.

10.4.3 Results

The answers to the aforementioned research questions, based on the analysed data were as follows.

1. *How often do drivers use the system?*
   Drivers switched the system on between 26 and 66% of the total time they participated in the FOT. The time where the driver drove slower than 60 km per hour or the data contained errors was not included in any of the analyses and thus also not included in this percentage.
2. *How many warnings for an unintentional lane-departure do drivers receive on average per hour?*
   Drivers received between 11 and 62 warnings per hour with an average amount of 29 warnings per hour, and between 0.1 and 0.7 warnings per kilometre for all roads and routes driven during the FOT.
3. *Do drivers change their position on the road after using LDWS over time?*
   Drivers changed their lateral position on the road over time but they did so very differently. In the end, however, they returned to their original position.
4. *Is there a difference between motorways and secondary roads and the amount of warnings drivers receive for unintentional line crossings?*
   The results showed a difference in amount of warnings received on motorways and secondary roads per hour. However, the calculated number of warnings per hour on secondary roads cannot always be considered as valid, since drivers did not drive on secondary roads very often and when they did, they turned off the system frequently.
5. *Does the number of warnings change over time?*
   The number of warnings changed over time: they first decreased but increased later over time to return to almost base-line levels.
6. *Do drivers show compensatory behaviour?*
   Drivers did not show any sign of compensatory behaviour based on the variables measured and analysed here. It could have been the case however, that drivers did perform different tasks in the cockpit we don't know of.
7. *Is there a difference in length of unintentional line crossings and their amplitude over time?*
   There was no clear evidence for a change in the nature of the line crossings themselves over time, neither in their maximum amplitude nor in their duration.

10.4.4 Conclusion

For the FOT it could not be concluded that influencing the driving task at the operational level in the long run was possible with the type of LDWS used. However, no negative influence was noticed either. The impact on traffic safety was not clear since over time drivers did not receive less warnings and did not decrease the amplitude and duration of line crossings, nor did their lateral position change over time. As such, one cannot infer a positive result for traffic safety.

In conclusion, the influence of the LDWS appeared to have neither positive nor negative influence on driving behaviour at the operational level in the long run. At first, drivers seemed to adjust their driving mostly by adjusting their lateral position to avoid warnings. Later, they stopped doing so and the influence of LDWS on the operational level disappeared. Possibly, because of the high effort required. At the strategic level however, there seemed to be an indication of influence on using LDWS over a longer period of time. At this strategic level, drivers seemed to use the system as an indicator for their lane position. Drivers initially changed their driving task at the operational level, but in the long run found this to be too strenuous. Interestingly so, they did not all choose to turn off the system either and accepted the high number of warnings without trying to prevent them. As such they changed their monitoring strategy, a change at the strategic level of driving.

10.5 The Assisted Driver

In 2004 the innovation program of the Dutch Directorate-General for Public Works and Water Management (Rijkswaterstaat) "Roads to the Future" decided that it would be time for a pilot project where Dutch car drivers would be given the opportunity to drive ADAS. The project was named "The Assisted Driver" and the goal was to find out if these systems, offering longitudinal and lateral support, would actually improve traffic flow in terms of safety, throughput and the environment when tested in real life, with real people on real roads. Also increasing awareness of ADAS was an objective of the project.

On highways in the Netherlands approximately 45% of the accidents are head to tail collisions, about 20% are single accidents and about 12% are aside accidents. This was the motivation to choose a combination of a lateral and longitudinal ADA system to be tested. The combination of ACC and LDWS was the most realistic. In addition there was even a test with a combination of ACC and LKS, in a 3-day clinic (Alkim et al. 2007a, b, c).

10.5.1 The FOT

In this section three elements of the FOT "The Assisted Driver" are discussed. These are named "VANpool", "Full Traffic" and "Clinic". In all these separate tests, lateral assistance was used, sometimes in combination with longitudinal assistance:

- "VANpool" = LDWS and Headway Monitoring and Warning.
- "Full Traffic" = LDWS and Headway Monitoring and Warning.
- "Clinic" = LDWS and LKS.

10.5.1.1 VANpool

The VANpool project was running in and around the city of Amsterdam. Employees were given the opportunity to carpool to work in a Volkswagen Sharan (a family car that seats a maximum of six people). These "vans" were allowed to travel in the bus lane alongside the usual traffic jam from Flevoland to various business areas in Amsterdam. VANpool was an initiative of Bureau Verkeer. advies commissioned by Stichting Amsterdam Zuidoost Bereikbaar. The Dutch Directorate-General for Public Works and Water Management (Rijkswaterstaat) joined this project by installing ADAS in several vehicles involved in the project. A pilot was held from September 2005 until January 2006 and enabled drivers and passengers to experience these technologies. The aim of the pilot was to acquire an insight into the behaviour and acceptance of these systems among drivers as well as passengers.

The participating vans (20 in total) were equipped with two types of ADAS, both of which form part of Mobileye's Advanced Warning System (AWS). The LDWS warns the driver when s/he is about to drift from his/her lane unintentionally. The Headway Monitoring & Warning system (HMW) indicates the distance in seconds until the vehicle in front on a dashboard display. When the car gets too close to the vehicle ahead, the system emits an audio warning. Both systems only provide information (warnings) and do not actively intervene.

10.5.1.2 Full Traffic

During a 5-month trial (February 2006 until June 2006), nineteen people – living in various places throughout the Netherlands – drove around in a Volkswagen Passat equipped with driving assistance technology. During the first month, no ADAS were used in order to chart the standard driving behaviour of the driver in question (reference situation).

Data loggers are used to compile objective information. An analysis of this data then reveals what effect driving assistance systems have on individual driving behaviour and the consequences thereof for traffic flow in terms of safety, throughput and environment.

Two types of ADA systems were installed in the participating cars: LDWS and ACC. For this chapter only LDWS is regarded.

The participants in the trial were selected by Pon's Autolease. Customers who wanted to drive a Passat were asked to participate in the project. In exchange, ACC and LDWS were installed in the cars for free. The group that participated in the full-traffic trial (16 men and three women) consisted of lease drivers. Every year, they drove between 25,000 and 75,000 km and are mostly very interested in technical devices for cars. We can therefore assume that they use ADAS more intensively than average drivers, which means they become more quickly accustomed to this type of technology.

The full-traffic trial had two objectives. The subjective component of the full-traffic trial answers the question whether participants personally experience a change in their driving behaviour by using these systems and whether they appreciate and accept ADAS. This part of the trial shall not be discussed in this chapter (for more information, see Alkim et al. 2007a, b, c). Objective information has also been compiled with the help of data-loggers. An analysis of this data revealed what effect ADAS have on individual driving behaviour and the impact on traffic flow in terms of safety, throughput and the environment.

10.5.1.3 Clinic

Over three consecutive days, several people test-drove a Lexus equipped with ACC, LDWS and LKA (Lane Keeping Assist from Toyota). Participants in the full-traffic trial, who drove a Volkswagen Passat equipped with ACC and LDWS, were also involved in the clinic. The aim of the clinic was to acquire an insight into the individual driver's experience with ADAS with regard to driving comfort and safety. The key question was whether the combination of two active systems is not too much for the driver. Experiences with LDWS (Passat) were also compared with LKA and LDWS (Lexus).

LDWS in the Lexus emits an audio warning and corrects the steering briefly as soon as the car is about to drift from its lane. The system in the Passat only emits an audio. LKA provides active assistance. The system corrects the steering automatically when the car drifts away from the middle of the lane.

10.5.2 Research Framework

10.5.2.1 VANpool

All participants were trained before the pilot started in September 2005. Part of the training focused on driving the car in the bus lane and on the hard shoulder. Participants were notified about applicable rules that they had to observe. Drivers were also told how LDWS and HMW should be used in practice.

Three measurements were carried out shortly before and during the pilot (September 2005–January 2006), during which both the van drivers and passengers were surveyed. The baseline measurement was performed in September 2005. During this first month of the pilot, the systems were not yet activated. This measurement involved 16 participants (9 drivers and 7 passengers).

The first measurement was conducted in November 2005, once the participants had acquired some experience. Nineteen participants completed this survey: 14 drivers and 5 passengers. The second measurement was conducted in January 2006, by which time the participants had acquired a reasonable amount of experience.

A total of 13 participants (9 drivers, 4 passengers) participated in the second measurement. The surveys were carried out using paper questionnaires.

In addition to the surveys, two meetings were held in November 2005 involving a selection of participants. The aim of these focus groups was to acquire a deeper insight into experiences of driving with LDWS and HMW. Nine people in total participated in these two group discussions, which comprised an afternoon session with five participants and an evening session with four participants.

10.5.2.2 Full Traffic

The first phase of the study entailed the compilation of vehicle data with the help of data-loggers. All vehicles were equipped with hardware and software, so that data could be saved on a vehicle PC and sent to a central server. Special software was written in order to process this crude data.

The manner in which ADAS are used and how they influence the driving behaviour of the motorist were also examined. The impact study investigated what effect changes in driving behaviour would have on road safety, throughput and the environment.

The aim of the study was formulated as follows:

> Carry out an objective analysis of the effect of driving with ACC and LDWS on individual driving behaviour and the consequences thereof for traffic flow as a whole, including the compilation of data on driving behaviour required to this end.

The following related study questions have derived:

1. How do drivers use the ADAS?
2. What effect does the use of these systems have on individual driving behaviour?
3. How do changes in driving behaviour influence traffic flow throughput?
4. How does driving with these ADAS influence road safety?
5. To what extent do fuel consumption and emissions change because of driving with the ADAS?

The primary focus of the study was on the immediate, quantifiable effects in relation to the use of ADAS, the effect on driving behaviour, the effect on traffic flow throughput and the effect on road safety and the environment.

In order to predict the effects of driving with LDWS and ACC, it was essential to obtain an insight into the extent and manner in which drivers used the various systems. After all, if the driver hardly uses the systems, no major effects could be expected. To acquire this insight, the following sub-questions were formulated for LDWS:

- On what type of road is LDWS used (motorway, provincial road, city road)?
- Under which traffic conditions is LDWS used?
- Do drivers turn off LDWS?

Sub-questions regarding the effects on driving behaviour:

- How do weaving and the standard deviation of lateral position (SDLP) change as a result of the available ADA systems?
- How does lane changing behaviour alter?

Sub-questions regarding the effects on traffic flow throughput:

- To what extent does the distribution of traffic across lanes change?
- What is the anticipated effect thereof on congestion and travelling times?
- What are the anticipated indirect effects on traffic flow throughput as a result of the change in road safety?

Sub-questions regarding the effects on road safety:

- To what extent is the risk of an accident involving one vehicle/party influenced?
- Will the likelihood of a sideways collision change due to LDWS?

Part of the method that was used involved determining the anticipated impact beforehand. This was done partly on the basis of the results from an expert workshop and partly on the basis of the personal expertise of the project group. This pre-assessment resulted in the formulation of several hypotheses, which are outlined below (only regarding LDWS):

LDWS is a passive system that does not intervene directly while driving. Under certain conditions, however, the system can be considered as disruptive. It is therefore expected that LDWS will be deactivated on narrow roads. Thus:

- LDWS is deactivated on narrow roads (access roads).

LDWS issues a warning every time the vehicle crosses a line (desired or undesired), which is not accompanied with the use of the direction indicators. LDWS therefore encourages the use of indicators. Thus:

- Direction indicators are used more often and more effectively.

In addition to more effective use of direction indicators, it is likely that drivers will maintain a greater distance from line markings due to LDWS. This is reflected in both SDLP and the average position of the vehicle in relation to the middle of the lane. Thus:

- SDLP decreases as a result of LDWS.
- The average distance to the middle of the lane decreases due to LDWS.
- LDWS issues more warnings on provincial roads and access roads.

The expected changes in the behaviour of motorists will also have an effect on road safety, just like on throughput. A positive effect is expected due to the decrease of unintentional line crossings, and an increase in the use of direction indicators.

On motorways, 4% of accidents are caused by the lateral position of the vehicle in relation to the road (too far to the left or to the right). On provincial and urban roads, this is significantly higher, 13 and 10%, respectively.

The number of side-impact collisions will decrease, because:

- There is an increase in the use of direction indicators, which ensures that lane changes are indicated more effectively.
- There is less SDLP;
- There are fewer lane changes.

The number of collisions involving one vehicle/party will change, because:

- There is an increase in speed.
- There is a decrease in the number of unintentional line crossings.

10.5.2.3 Clinic

The clinic was held on May 30th, May 31st and June 1st, 2006 on the premises of the distributor for Lexus and Toyota in the Netherlands, Louwman & Parqui in Raamsdonkveer. On each day, participants in the clinic took trips lasting approximately 1 h. The seven participants involved in the study were driving a Volkswagen Passat and had experienced driving with ACC and LDWS. They drove between 25,000 and 60,000 km annually.

Before the test-drive, an employee from Toyota Motor Europe (the manufacturer of Lexus) explained how the systems function. During the first part of the trip, the test subject watched from the passenger's seat while the functioning of the systems was explained in greater detail. During the clinic the participants were seated behind the steering wheel. After every trip, the participants were interviewed.

10.5.3 Results

10.5.3.1 VANpool

It is worth mentioning beforehand that the following results are based on the opinions and experiences of participants (drivers and passengers). In other words, these are not objective facts. Moreover, it's the verdict of a relatively small group.

Participants in the VANpool pilot were positive about the AWS due to its functional values in particular (increased alertness, practical and helpful). Drivers and passengers thought the system enhances safety and that LDWS as well as HMW reduce the likelihood of accidents.

Participants thought that the warning signals of both systems are clear. Drivers found the signals provided by HMW more superfluous than those of LDWS. During the second measurement, drivers and passengers considered the warning signals less irritating and less often superfluous than during the first measurement. They also considered LDWS to be more useful when driving in narrow lanes compared to the first measurement.

Drivers' experiences with LDWS and HMW varied from negative to positive. In general, they do not like being corrected, but do perceive the systems as an asset.

10.5.3.2 Full Traffic

Almost every driver used the LDWS during the post-measurement under all conditions, on highways and secondary roads and during free flow and congested conditions. In other words they did not turn off the system. This means that the hypothesis that LDWS would be switched off on narrow roads is not valid.

The presence of LDWS had an effect on the use of direction indicators. It was implicitly assumed that drivers do not always use direction indicators during the pre-period when this is strictly speaking necessary. Tables 10.1 and 10.2 show the percentage of all lane changes logged for which the indicators were used. From these tables, it could be concluded that there is a slight increase in the use of indicators in all situations. The increase appeared to be most substantial on provincial roads. The hypothesis that the use of indicators would increase while driving with LDWS is therefore confirmed.

To explore what effect driving with LDWS would have on the position within the lane the number of warnings in the pre- and post-period were compared. In the pre-period LDWS was switched of and only "virtual" warnings were logged, meaning that the data-logger was able to monitor the position of the vehicle and assess whether a warning should have been issued. There was a distinct reduction in the number of warnings issued by LDWS (see Fig. 10.3). Figure 10.3 indicates changes in the number of warnings per hour for various periods. Remarkably, the number of warnings on motorway was the largest on average and therefore contradictory to the expectations.

After an initial significant decrease in warnings (for motorways and secondary roads) there was a limited increase in the number of warnings in April. This result is similar to the findings in the FOT with LDWS in trucks 3 years earlier.

Table 10.1 Percentage of lane changes for which the indicator was used during the pre-period (average percentage for all drivers)

	City (%)	Provincial (%)	Motorway (%)
To the left	13.0	21.6	43.9
To the right	14.9	16.9	49.6

Table 10.2 Percentage of lane changes for which the indicator was used during the post-period (average percentage for all drivers)

	City (%)	Provincial (%)	Motorway (%)
To the left	15.0	26.5	44.9
To the right	17.1	33.2	52.0

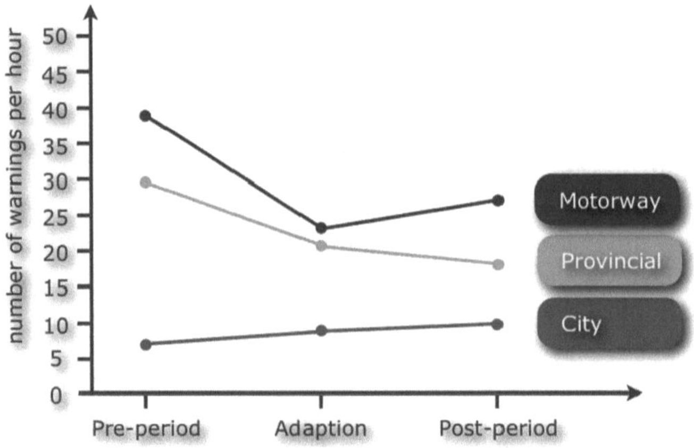

Fig. 10.3 Number of warnings per hour for various periods, for urban areas, provincial roads and motorways

This explicitly showed that there were significantly fewer unintentional line crossings with LDWS than without LDWS. This is in accordance with the anticipated effect of LDWS reducing the SDLP. Thus:

- The number of unintentional line crossings decreases thanks to LDWS (for provincial roads by 35% and for motorways by 30%).
- LDWS issues more warnings on motorways than in urban areas and on provincial roads.
- Drivers reduced the SDLP of their vehicles to avoid warnings. This has direct consequences for the driving task load of drivers (they have to concentrate better).
- Direction indicators are used more often and more effectively.

There is no direct effect on throughput by LDWS, indirectly however congestion can be reduced by less accidents and their associated traffic jams.

The safety potential of LDWS was also considered. The focus was particularly on the number of line crossings and the use of direction indicators. It may be concluded that a reduction in SDLP and the number of unintentional line crossings can increase safety. After all, roughly 6% of accidents on road sections occur because drivers keep to the right too much or too little (depending on the road type).

On the basis of the results referred to earlier relating to the distribution in lateral position and a reduction in the number of unintentional line crossings, it can be concluded that SDLP does indeed decrease while the average distance to line markings increases. If only the decrease in the number of unintentional line crossings is considered, an average reduction of roughly 20% applies. There is also a small increase in the (correct) use of direction indicators (about 20%). This has a small positive effect on road safety. After all, the primary circumstance for roughly 0.5% of the total number of accidents (on road sections) is the failure to indicate.

On the basis of changes in the use of direction indicators (increase of roughly 20%, depending on the road type) and the number of unintentional line crossings (also 20% approximately), an effectiveness of 20% for LDWS was concluded.

The overall percentage of accidents caused by the following circumstances was used to estimate the safety effects of LDWS. Relevant accidents include the ones caused by the driver:

- Not indicating.
- Indicating incorrectly.
- Not driving enough on the right.
- Driving too much on the right.

An estimated reduction in the number of accidents of 0.9, 2.7 and 2.3% for motorways, secondary roads and urban roads respectively was estimated.

On the basis of the analysis above, it can be concluded that LDWS has a positive effect on road safety. This positive effect can be attributed to changes in the following safety indicators:

- Decrease in SDLP and line crossings.
- Increase in the use of direction indicators.

10.5.3.3 Clinic

All participants indicated that the test drive was a pleasant and positive experience. The majority found that they became accustomed to the systems quickly and thought the systems were configured nicely. All of them expected that these types of systems would have a future. The main reasons for this, in their opinion, are comfort and safety. According to them, ACC, LDWS as well as LKA contribute to road safety (with LKA contributing the most and LDW the least). All participants preferred LKA to LDWS. The active steering corrections made by LKA appeared to appeal to the majority. Incidentally, it must be emphasised that this was the opinion of a small group. The small-scale clinic was not intended as a representative study.

LDWS met the expectations of practically all of the participants. The reason for this may have been that at the time of the study they were driving a Passat also equipped with LDWS. The participants mentioned the following positive points about LDW: it increases the use of indicators (3×), you become accustomed to it (1×), the system warns you if you unintentionally leave the lane (2×), and the alarm and corrective steering function well (1×). They also pointed out various negative points about the system: the alarm can be irritating at times (3×), the LDW system in Passat responded too quickly (2×), lines were not always recognised properly, in the Passat lines on bicycle paths were also recognised in situations where the driver had to cross the lines (2×) and the audio warning in the Passat is insufficient (1×).

Five of the seven participants indicated that LKA performed better than expected. The other two said that the system met their expectations. The participants mentioned the following positive points about LKA: the steering correction occurs smoothly with the correct amount of power and is not disruptive (1×), the system keeps the vehicle within the lines via steering corrections (3×) and it enhances safety and comfort levels (2×). Negative points were as follows: it takes some time to get used to the system (2×), you do not feel that you are driving in the middle of the lane (2×), the system creates a sensation of crosswinds (2×), the car tends to drift slightly when you counter-steer too late (1×), the car tends to drift slightly due to prolonged correction if you release your hands (1×), and there is a risk that you will drive with less concentration (1×).

A positive aspect of the LKA system, according to the participants, was that it really does provide assistance and comfort. LDWS does not offer a similar level of comfort. The participants also said that feedback from LDWS to the steering wheel was preferred over an audio warning only.

Six of the seven participants in the clinic indicated a tendency to perform other activities in a car equipped with ADA systems. Examples of other activities include using a mobile phone more often, talking to passengers more, consulting one's diary, spending longer searching for CDs and looking at the navigation system.

10.5.4 Conclusions

The conclusions of the three described elements of "The Assisted Driver" as well as the overall conclusions of the whole project are described below.

10.5.4.1 VANpool

The majority of participants in the pilot were satisfied with AWS. They found the system easy to use and believed that driving with both LDWS and HMW is conducive to road safety. This is due to the fact that drivers adapt their driving behaviour (in a positive sense) in order to minimise the number of warnings. Participants used their indicators more often and maintained their direction on the road more effectively. They also thought that the driving task was less demanding overall. To summarise, the system was considered irritating but effective.

10.5.4.2 Full Traffic

LDWS was used more or less throughout the trial and the number of unintentional line crossings decreases thanks to LDWS (for provincial roads by 35% and for motorways by 30%). Drivers reduce the SDLP of their vehicles to avoid warnings. This has direct consequences for the driving task load of drivers (they have to concentrate better). Direction indicators are used more often and more effectively. The changes between the pre- and post-measurement are minimal, however. It cannot be concluded that there are fewer lane changes due to driving with ACC and LDWS. But drivers do continue driving in the left lane and particularly in the middle lane for longer.

10.5.4.3 Clinic

Overall, the participants responded positively to the tested systems. They all indicated that the systems help increase safety and comfort levels. With regard to the level of safety that was experienced, LKA and ACC were rated better than LDWS. With regard to comfort, ACC and LKA are significantly better than LDWS. This is primarily because ACC and LKA are systems that provide active assistance instead of only a warning. For them driving with LKA was more pleasing than driving with LDWS and active corrective steering was considered an enjoyable aid.

The results of the clinic revealed a clear contrast. One of the expected advantages of the systems is that drivers will focus more effectively on traffic due to a lower workload. The answers provided by the participants indicated that six out of seven drivers used the available capacity for other activities (e.g. making phone calls). This would not be positive to road safety. Nevertheless, the participants expected the systems to make a positive contribution towards safety.

10.5.4.4 Overall Conclusions of the "The Assisted Driver" FOT

The three aforementioned components of "The Assisted Driver" revealed that participants appreciated active assistance (intervention) more than warnings. This was unexpected given that people usually indicate beforehand that they would prefer to have an informative system in the car instead of a system that takes over part of the driving task. This stated preference is not based on experience but on expectation.

Two warning systems were used in the VANpool pilot, LDWS and HMW. Participants indicated a preference for LDWS because the system issues warnings less often than HMW on average, including situations in which driving behaviour had been adapted. In the full-traffic trial, a passive (LDWS) and active (ACC) system were used. Practically all participants preferred ACC and indicated that it is more pleasant and comfortable than driving with LDWS. The lack of warnings and active support are the main reasons for this preference. During the clinic, a number of participants from the full-traffic trial, i.e. people who had experience with ACC and LDWS, test-drove a vehicle equipped with ACC and LKA: two active assistance systems. All participants agreed that LKA, in relation to LDWS (same functionality, other approach: warning versus active assistance) was an improvement. Here too, a preference was indicated for intervention instead of warning. Driving with two active assistance systems was not considered "excessive".

A thorough analysis of the wealth of data compiled with the help of data-loggers during the full-traffic trial provided an insight into the effects of driving with ACC and LDWS on traffic flow. The effects on road safety, throughput and the environment were examined.

Road safety is expected to increase by the tested ADAS due to the following effects in particular:

- Less tailgating.
- Fewer undesired line crossings.
- More uniform speed.
- More uniform acceleration.
- Better use of the indicators.

The pilot results leaded to the estimation that accidents on motorways and secondary roads would decrease by approximately 8% if everyone in the Netherlands was to use both ACC and LDWS.

10.6 Accident Prevention Systems for Lorries

In 2008 and 2009 another FOT with ADAS took place in the Netherlands. This time the sheer magnitude of the FOT was unprecedented as well at the national and international level. Two thousand four hundred and two trucks and lorries belonging to 123 participating companies were involved. The FOT was commissioned by the Dutch Ministry of Transport, Public Works and Water Management and conducted by Connekt (ITS Netherlands) (Connekt 2009).

The reason for choosing a FOT with trucks and lorries was the perception that incidents with these heavy vehicles often result in heavy and unexpected congestion. It's not that trucks are more involved in accidents than passenger cars, the consequences are just larger.

10.6.1 The FOT

Even though the size of this FOT was much larger than the previous FOTs in The Netherlands with ADAS, the number of participating trucks is still much too low to measure a direct effect of driving with these systems on the traffic flow. The penetration rate is just too low. For reference, there are currently between 150,000 and 170,000 heavy goods vehicles in The Netherlands and the total number of vehicles is 8,881,800. This means that if 2,402 trucks are equipped with ADAS, this is just 1.5% of all trucks and 0.03% of all vehicles in The Netherlands. These numbers imply that direct effects are difficult to measure, but the effect of driving with ADAS on driving behaviour can be monitored and extrapolated to effects on the traffic flow in terms of safety and throughput.

The purpose of the FOT was to acquire a better understanding of how ADAS (in this FOT called Accident Prevention Systems) can contribute to improved traffic safety and better throughput. An additional goal was to see whether the government could serve as a launching customer for these systems. To achieve these goals an FOT with 2,402 trucks from 123 participating companies was carried out, 1,671 trucks were equipped with at least one ADAS and the remaining 731 vehicles served as a reference group. All vehicles were equipped with data loggers and monitored for a period of up to 8 months. This resulted in vast amount of data as a result of approximately 77 million kilometres travelled by the whole fleet.

The systems used in this FOT were both factory fitted and after market. Five different systems have been used:

- Advanced cruise control.
- LDWS.
- Forward Collision Warning/Headway Monitoring and Warning (FCW/HMW).
- Directional Control/Roll Over Control.
- Black Box feedback.

For this chapter only the results of driving with Lane Departure Warning are discussed.

10.6.2 Research Framework

There were four primary questions formulated, to be answered in this FOT.

1. How effective are the systems? Do they detect a (dangerous) situation in a correct manner? Is a warning issued in time and is it an appropriate warning? Or if the system is an active support system, is the support appropriate?
2. What is the effect on traffic safety if ADAS are used by a (large) part of the trucks on the Dutch road network?
3. What is the effect on throughput if ADAS are used by a (large) part of the trucks on the Dutch road network?
4. Can the government act as a stimulator of the use of ADAS?

For this chapter ADAS can be replaced by LDWS. The system used in the FOT is the Mobileye system from Clifford Electronics, the same as in the previous FOT "The Assisted Driver".

In order to answer the aforementioned research questions, three different sub projects (SP) were set up:

- SP1 – Retrofit project; consisting of trucks equipped with the after market systems LDWS or FCW/HMW.
- SP2 – Chauffeurs project; consisting of trucks with both LDWS and FCW/HMW (aftermarket).
- SP3 – OEM project; consisting of trucks with both factory fitted systems and after market systems.

The number of trucks in SP1 with LDWS was 439 with 411 unequipped trucks as reference. The number of trucks in SP2 with LDWS and FCW/HMW was 186 with

234 unequipped trucks as reference. The number of trucks in SP3 with LDWS (factory fitted) was 54, with LDWS (after market) 100 and with 86 unequipped trucks as reference.

The data gathered by the data loggers consisted of event based data, triggered by an event (such as TLC < 0.5 s) and standard data collected every 2 km of a trip. Parameters that were logged consisted of GPS location, headway, position in the lane, velocity, acceleration, deceleration, harsh accelerations, harsh brakes, TLC, number of warnings, etc.

To answer research question 1, two tests were performed. One on a dedicated test track to analyse the specific settings of LDWS and basically to see if it performs as it should. The second test was during the 8 months of driving with LDWS, by means of logging specific data.

The second research question, regarding the effect of driving with LDWS on traffic safety, was answered by performing four different research approaches. A desk research into available literature was carried out to research the relation between ADAS and traffic safety. An analysis of traffic accidents involving trucks in the Netherlands was done to find out what the potential of specific ADAS is regarding the type of incidents that can be avoided or mitigated. Also an extensive analysis of the vast amount of data from the data-loggers was carried out. And, finally, a conceptual and quantitative model was developed to predict the effect of driving with specific ADAS on the traffic flow in terms of safety, based on the empirical data gathered in this FOT.

The third research question, regarding the effect of driving with LDWS on throughput was answered by performing four different research approaches. A desk research into available literature was carried out to research the relation between ADAS and throughput. An analysis of throughput effects regarding ADAS and trucks in the Netherlands was done as well as an extensive analysis of the vast amount of data from the data-loggers was carried out. And finally a conceptual and quantitative model was developed to predict the effect of driving with specific ADAS on the traffic flow in terms of throughput, based on the empirical data gathered in this FOT.

The fourth and final primary research question, to see whether the government can act as a stimulator of the use of ADAS, was answered on the basis of enquiries and interviews with participants (drivers and companies).

10.6.3 Results

Test track results indicated that LDWS performs as it should. Driving with LDWS resulted in a decrease of warnings per hour compared to the reference groups and therefore the number of unintended line crossings decreased. Driving with after-market LDWS resulted in a decrease of 30% whereas driving with factory fitted LDWS led to a 62% decrease. Also the use of indicators when changing lanes increased. As a negative side effect, the percentage of short headways (<1 s) increased with 5.9%. The average speed when driving with LDWS also decreased with 0.4 km/h compared to the reference group.

The literature study revealed in general that driving with ADA systems has a positive effect on traffic flow in terms of safety and throughput. For LDWS specifically a potential reduction of accidents was described up to 13%. This effect can mainly be attributed to the fact that the number of (unintended) line crossings decreases and the use of the indicator increases.

The direct effect of driving with LDWS on throughput seems neutral to slightly positive. The indirect effect could be much bigger due to the avoided congestion as a result of lesser accidents (with trucks).

During the 8 months of the FOT only five accidents with trucks were recorded, all in the reference group. If this were extrapolated to the fleet equipped with ADAS, 11–13 accidents would be expected. However, none occurred.

10.7 Traffic Safety Impact of Lateral ADAS: Early Conclusions

In general driving with lateral support seems to improve traffic safety without any adverse (direct) effects on throughput. Fewer accidents due to the use of lateral support systems could even have a positive (indirect) effect on throughput because of less congestion caused by accidents. This conclusion can be drawn after a decade of experience with FOTs in the Netherlands, which has yielded a wealth of empirical data, objective and subjective, qualitative and quantitative. The main contributing factors to this increased traffic safety and possibly decreased congestion are the fact that the number of unintended lane changes decreases, the indicators are used better and the driver is able to maintain his/her course within the lane better and with more comfort. The magnitude of these effects seems to be larger when LKS is involved versus LDWS.

When asked if one would like to have a lateral support system that only warns the driver versus an active lateral support system that helps the driver to maintain his/her course within the lane, drivers usually indicate they would like to have the informative system. This stated preference changes however if the driver has experienced both types of systems. For LDWS and LKS this means that after driving with both systems, most of those drivers would prefer to have LKS because it actively supports the driving task.

The key word here is experience. The driver has to gain experience with an ADAS in general or a lateral support system in particular in order to assess its merits. This is also an important factor in deciding whether or not to purchase a (new) car equipped with an ADAS or an aftermarket system. Because most of these systems are only offered as an option, consumers have to be aware of the existence of ADAS in the first place. Training on ADAS functionalities and limitations, as suggested in Chap. 9, may thus play a decisive role in enhancing driver awareness and performance with such systems.

Increasing awareness amongst all stakeholders, including consumers, has been a complimentary goal of all recent FOTs. Publications, presentations, workshops, and other forms of dissemination have been used. Also events, such as demonstration days, where participants could actually experience driving with ADAS were organized, as well as being present at events (such as the ITS World Conference) with a demonstration vehicle (equipped with ACC and LDWS). All these activities, including the aforementioned FOTs to form evidence based opinions and policies regarding intelligent vehicles and intelligent infrastructure, as well as the active participation in several European projects and platforms has earned Rijkswaterstaat the third annual eSafety award in the category "Policy" (see www.esafety.org).

But in the end it's either the consumer or fleet owner who decides whether LDWS or LKS will be purchased and used, or not. Positive experiences, training and more evidence regarding the effects of driving with these systems are key in this process.

References

T.P. Alkim, M. Korse, S. de Ridder, Field operational test with lane departure warning assistant systems, behavioural effects, in *Proceedings of the 10th World Congress on ITS*, Madrid, 2003

T.P. Alkim, G. Bootsma, P. Looman, The Assisted Driver. Final Report, Rijkswaterstaat, The Netherlands, 2007a

T.P. Alkim, G. Bootsma, S. Hoogendoorn, Field operational test "The Assisted Driver", in *Proceedings of IEEE Intelligent Vehicles*, Istanbul, 2007b

T.P. Alkim, G. Bootsma, S. Hoogendoorn, Dutch field operational test experience with "The Assisted Driver", in *Proceedings of 14th World Congress on ITS*, Beijing, 2007c

R. Bishop, *Intelligent Vehicle Technology and Trends* (Artech House, Norwood, 2005)

Connekt, Accident Prevention Systems for Lorries. Final Report, Connekt/ITS Netherlands, 2009

W.H. Janssen, *Route planning en geleiding: een literatuurstudie*. Rapport IZF 1979 C-13. (Instituut voor Zintuigfysiologie TNO, Soesterberg, 1979) (In Dutch)

J.A. Michon, *Psychonomie onderweg*. Inaugural lecture. (University of Groningen, Groningen, Wolters Noordhof, 1971) (In Dutch)

J.A. Michon, A critical view of driver behavior models: what do we know, what should we do? in *Human Behavior and Traffic Safety*, ed. by L. Evans, R.C. Schwing (Plenum, New York, 1985)

J.R. Treat, N.S. Tumbas, S.T. McDonald, D. Shinar, R.D. Hume, R.E. Mayer, R.L. Stansifer, N.J. Castellan, Tri-level study of the causes of traffic accidents, Vol. I. Causal factor tabulations and assessments, Vol. II., in *Special analyses Final Report on U.S. Department of Transportation (Contract No. DOT-HS-034-3-535-77)* Washington, DC: Government Printing Office (Indiana University, Institute for Research in Public Safety, 1977)

Chapter 11
Easy Going. Multi-Level Assessment of ISA

Sven Vlassenroot, Jan-Willem van der Pas, Karel Brookhuis, Johan De Mol, Vincent Marchau, and Frank Witlox

11.1 ISA-The Concept

Speeding is a widespread problem and is a major source of road traffic externalities. Speeding affects road safety; it not only increases the risk of getting involved in an accident, but it also affects the outcome or severity of an accident. Moreover, higher vehicle speed also contributes to increased greenhouse gas emissions, fuel consumption and noise, and to adverse impacts on the quality of life (OECD 2006). Speed management can help achieve appropriate speed, taking into account mobility and economic needs, as well as safety and environmental requirements. Speed management implies a consistent policy that integrates information, education, road design, road signs, enforcement and vehicle technologies.

Intelligent Speed Adaptation (ISA) is an Advanced Driving Assistance System (ADAS) that may help the driver to cope with the (posted) speed limits. ISA can be

S. Vlassenroot (✉)
Department of Transport Policy and Logistics Organization, Delft University of Technology (TUDelft), Delft, The Netherlands
and
Ghent University, Institute for Sustainable Mobility, Ghent, Belgium
e-mail: sven.vlassenroot@ugent.be

J.-W. van der Pas and V. Marchau
Department of Transport Policy and Logistics Organization, Delft University of Technology (TUDelft), Delft, The Netherlands

K. Brookhuis
Department of Transport Policy and Logistics Organization, Delft University of Technology (TUDelft), Delft, The Netherlands
and
Department of Psychology, University of Groningen, Groningen, The Netherlands

J. De Mol
Ghent University, Institute for Sustainable Mobility, Ghent, Belgium

F. Witlox
Department of Geography, Ghent University, Ghent, Belgium

E. Bekiaris et al. (eds.), *Infrastructure and Safety in a Collaborative World*,
DOI 10.1007/978-3-642-18372-0_11,

Table 11.1 Overview of different types of ISA (Morsink et al. 2006)

Level of support	Type of feedback	Definition
Informing (open)	Visual	The speed limit is displayed and the driver is reminded of changes in the speed limit
Warning (open)	Visual/auditory	The system warns the driver when exceeding the posted speed limit at a given location. The driver decides whether to use or ignore the information or warning
Assisting (half-open)	Haptic throttle	The driver gets a force feedback through the gas pedal if he/she tries to exceed the speed limit. Overruling of the system is still possible
Restricting (closed)	Dead throttle	The speed of the vehicle is automatically limited and the driver can not overrule the system

described as a system that (1) "knows" the real time location of a car (with the aid of GPS), (2) "knows" the (posted) speed limit at that specific location (e.g. using in-vehicle speed database) (3) compares the speed with the (posted) speed limit and (4) if the speed is inappropriate intervenes with the driving task (Brookhuis and De Waard 1999).

ISA can use three types of limits (Carsten and Tate 2005); static speed limits (posted speed signs), variable speed limits (information about speed limits depending on the location) and dynamic speed limits (information based on actual road and traffic conditions).

Many terms are used to describe these kinds of ADAS, like Speed Alert or Warning system, external vehicle speed control, intelligent speed information, intelligent speed assistant, etc.

ISA can be categorized in different types, depending upon how intervening (or permissive) they are (Morsink et al. 2006) (Table 11.1).

Since the early 1980s the effects of ISA have increasingly been studied through different methodologies and data collection techniques varying, from traffic simulation, driving simulators, instrumented vehicles to field trials. Based on the outcome of the research, the relevant social, ecological, economical, political and technical aspects are described. This leads to an overview of barriers and issues that have to be resolved to enable large scale implementation.

11.2 Research on ISA

11.2.1 ISA Field Trials

One of the very first trials that were held was in France in the 1980s. Drivers tested a system related to a cruise control, which did not automatically set the correct speed (Saad and Malaterre 1982). In the 1990s, new small tests were conducted in Sweden and the Netherlands. The drivers mostly drove in an instrumented vehicle on a test-route (Wallén Warner 2006 De Waard and Brookhuis 1997).

In the late 1990s different trials within a larger setting and with more vehicles started around Europe and in Australia. The most recent trials are described below. Between 1999 and 2002 the Swedish National Road Administration (SRA) conducted a large-scale trial involving ISA in urban areas (Biding and Lind 2002). The aim of the trial, which was conducted jointly by the SRA and four Swedish municipalities, was to learn more about driver attitudes and how they use the systems, the impact on road safety and the environment, the integration of the systems in vehicles and the prospects for ITS on a large scale. The systems were tested in Borlänge, Linköping, Lund and Umeå, where the local authorities were responsible for running the trials in their respective municipalities. Different systems and technical solutions have been tested at the different trial sites. In Umeå a warning system was tested, where the driver received a warning signal (audio and visual) when the legal speed limit was exceeded. In Borlänge, a system was tested that used audio and visual warnings for violations of the speed limit and, in addition a display informed the driver about the existing speed limit on the road in question. In Lund, a system was tested that supported the driver's speed adaptation through an active accelerator, which implies that when the driver has reached the legal speed limit a counter pressure is applied to the accelerator. In Linköping, both informative and active accelerator systems were tested. The results showed that, in general, positive effects on speeding behavior were noted. The average speed on stretches of road has clearly fallen with ISA. The ISA vehicles ran more homogeneously and with less variation in speed, which probably increased safety even more. The acceptance of ISA in urban area was noted as rather high, compared with the acceptance of seat belts' use. Effects on speed differed very little between the systems. The driving speed fell on stretches by up to 3–4 kph for each of the systems. The difference between the systems for the entire road system at 30–50 kph, which is the main focus of the trial, amounted to only 0.3–0.4 kph.

Around the same period, in 1997, a national study started in the UK, using field trials and driving simulator studies during 3 years. In the field trials, test drivers tested two different intervening systems (Carsten and Fowkes 2001). One of the systems could be switched on and off at will, while another was on at all times. The test drivers were divided into three groups, with eight test drivers in each group. One group tested the system that could be switched off and another tested the system that could not. The third group was a control group. The systems were tested on a 67 km long test route including 30, 40, 60 and 70 mph speed limits. The results showed that the test drivers who were able to switch off the system tended to do so when the traffic conditions gave them an opportunity to violate the speed limit. In 2001, a new project started, called ISA-UK (Carsten et al. 2008). Four field trials were conducted in different parts of the UK. In these trials, 80 private and professional test drivers drove 20 vehicles that had a system installed for 6 months (during the first and last month the system was not activated). The system disabled the test drivers to exceed the speed limit without using kick-down or pressing an emergency button. The behavioral results from the car trials showed that the overridable ISA that was used by the participants reduced the amount of speeding among every category of user. It also affected driving on every road category, except for the 60 mph rural roads, where comparatively little speeding by the participants in the pre-ISA period was found.

Between October 1999 and October 2000 (AVV 2001), 20 private cars and one bus equipped with a death throttle system (closed ISA) drove in a suburb of the city of Tilburg, the Netherlands. The goal of the trial was to demonstrate the feasibility of ISA as a speed management measure. Public acceptance and support for ISA was measured and information was collected about the technical requirements and functionalities, and the effects on driving behavior. The trial consisted of 30, 50 and 80 kph speed limits. In total 120 drivers participated in the trial, each for 8 weeks. In the first 2 weeks, the system was switched off. After the first 2 weeks of each period, the system was activated, making it impossible for the test drivers to exceed the speed limits (unless the emergency button was pushed) whenever they drove within the test area roads. The speed limits could only be exceeded by use of the emergency button for deactivating the system. The results showed that the average speed, as well as the speed variation decreased.

In Denmark in 2001, 24 cars were equipped with an informative "sound and light" system and the test-drivers drove for 6 weeks (Lahrmann et al. 2001). The results of the trial showed a mean speed reduction of 5–6 kph in general but large variations between individual drivers were noted. The speed violations reduced from 9–13 kph without driving with ISA, to 4–7 kph during the test period (Nielsen and Lahrmann 2005).

In 2000, during a field trial in Finland (Paatalo et al. 2001), three different ISA-types, namely informative, compulsory and recording, were tested. The information system provided information regarding the current speed limit on a visual display and gave an audio warning. The compulsory system was a closed system and limited the maximum speed of the vehicle to the posted limit. The recording system displayed the percentage of speeding of the total driving time. The 24 participants drove the car along a test route on four separate occasions. The results indicated that drivers spent less time speeding when driving with one of the ISA systems operating and the reduction was the most for the compulsory system (6.7 km/h). Results from the workload data revealed that drivers found driving with the compulsory system most demanding with regard to required attention and concentration.

In France, a series of field trials with ISA started in 2001 and this time two large car manufacturers, Renault and PSA, were participating in the project (Ehrlich 2006). A pre-assessment phase was first carried out using two prototype vehicles. The study was then extended to 100 test drivers who drove an instrumented vehicle for 8 weeks. After the first 2 weeks, when no system was activated, each test driver tested three different systems for 2 weeks each. The first system tested informed the test drivers of the current speed limit and warned them if this limit was exceeded. The second system made it impossible for the test drivers to exceed the speed limit without using kick-down. The third system also made it impossible for the test drivers to exceed the speed limit, but this system did not have any kick-down function. First results indicated that the informative system is less effective than the other systems. Speeding decreased with every system.

In October 2002, an ISA-trial in Belgium was started in Ghent (ISAweb.eu 2004; Vlassenroot et al. 2007). Thirty-four cars and three buses were equipped with the "active accelerator pedal". In this system a resistance in the accelerator is activated when the driver attempts to exceed the speed limit. A total of 90 drivers participated in the field trial. The test area covered roads with speed limits of 30, 50, 70 and 90 kph. Data analysis showed a reduction in the amount of speeding due to the ISA-system. There was, however, still a large remaining percentage of speeding offences, especially in low speed zones. Differences between drivers were large. For some drivers speeding even increased despite activation of the system. For less frequent speeders average driving speed almost always increased and for more frequent speeders average speed tended to decrease. With the system, less frequent speeders tended to accelerate faster towards the speed limit and drove exactly at the speed limit, which caused average speeds to go up.

A cross-cultural study with ISA was held in 2003 and 2004 in Hungary and Spain (Varhelyi et al. 2005). In this study 20 Hungarian and 19 Spanish test drivers had two different systems installed in their vehicles for two months each (1 month with the system activated and 1 month as a control period). Both systems informed the test drivers of the current speed limit. The advisory system used a sound and light system whereas the intervening system was an active accelerator pedal. The results showed that both mean speed and speed variation were reduced when driving with any of the two systems. The results also showed that the intervening system tended to be more effective than the advisory while, at the same time it was less accepted by the test drivers.

From February 2003 to March 2005 a trial was organized in Australia (Regan et al. 2006). In this trial 15 test vehicles were equipped with a warning ISA (visual and auditory signals), turning into an intervening ISA (upward accelerator pressure), if warning signals were ignored for more than 2 s. The vehicles were also equipped with a Following Distance Warning (FDW) system (aimed at preventing tailgating), a seatbelt reminder, a Reverse Collision Warning system (aimed to prevent collisions while driving backwards), and daytime running lights. A control group of 8 drivers was used. The control vehicles were not equipped with ISA or FDW. All 23 drivers drove at least 16,500 km annually. The results showed a reduction in speed and speeding with ISA. The combination of ISA with FDW tended to have a better result than ISA alone.

In 2004, a new Danish trial started in North-Jutland (Lahrmann et al. 2007). The project is based on a "Pay As You Drive" principle, which means that the ISA equipment not only gives a warning when the driver is speeding, but also gives penalty points which reduce a promised bonus of 30% on the insurance rate. The project proceeded in a 3 year test period with the goal to involve 300 car drivers as participants in the project. Results from 90 test-drivers showed that the percentage speeding more than 5 kph on 80 km roads is reduced from 28 to 2%.

In December 2004 around 20 vehicles in the city of Stockholm in Sweden were equipped with two types of ISA: an active accelerator pedal and a vibrating accelerator pedal, which vibrated when the speed limit was exceed (Myhrberg 2006). The purpose of the trial was to bring ISA knowledge and acceptance to

Stockholm, necessary for the development of future ISA implementation. On average, ISA reduced speeding by 30%. The effect was noted better at higher speed limits.

In August 2006, a trial started in Karmøy, situated on an island in Norway, with young drivers between the age of 18 and 25 years (Berg et al. 2008). Insured customers of a certain company were invited to participate in the project. The participants received a 30% discount on their car insurance premium during the 17 months test period. The ISA equipment was a warning system with sound and light. The participants were divided into three groups: participants motivated by traffic safety, participants motivated by the 30% discount on the car insurance, and participants motivated by both. The analysis shows that safety motivated drivers drove more carefully in terms of the amount of speeding than economically motivated drivers. Attitude data and participants' expectations with respect to the technology showed the same distinction.

Not much is known concerning the long term effects of ISA use on the drivers' speed choice behavior (long term being considered as more than a year's period). A study in Sweden (Börlange) showed that for the Assisting (half open) ISA the effect of ISA decreased year after year (Wallén Warner and Aberg 2008), between 1999 and 2003.

11.2.2 ISA Main Results

Generally, ISA seems to have positive effects on driving speed and speed violations. The effects depend on how intervening the systems are set. A restrictive ISA seems more effective in reducing speed and speeding than an advisory ISA. Below a number of aspects concerning ISA are elaborated.

11.2.2.1 Safety Effects

The most detailed prediction of overall network savings with ISA is provided by Carsten and Tate (2005). Table 11.2 shows the estimates of the overall system-wide collision savings for Great Britain, at various levels of collision severity, for various permutations of ISA. The scenario envisaged is that 100% of vehicles are equipped with ISA overnight.

ISA systems are divided into the broad classes of Advisory, Driver Select, and Mandatory systems. Advisory ISA displays the speed limit and reminds the driver of changes in the speed limit. Voluntary ISA is linked to the vehicle controls but allows the driver to enable and disable control of the vehicle's maximum speed. With Mandatory ISA, the vehicle is limited at all times. Each broad class of ISA can have speed limits in fixed, variable or dynamic forms (where dynamic also includes variable capability). With "Fixed" speed limit data, the vehicle is informed of the

Table 11.2 Best estimates of crash savings by ISA type and crash severity, assuming a penetration rate of (nearly) 100% (Carsten and Tate 2005)

System type	Speed limit type	Best estimate of injury crash reduction (%)	Best estimate of fatal and serious crash reduction (%)	Best estimate of fatal crash reduction (%)
Informing	Static	10	14	18
	Variable	10	14	19
	Dynamic	13	18	24
Voluntary	Static	10	15	19
automatic	Variable	11	16	20
control	Dynamic	18	26	32
Mandatory	Static	20	29	37
automatic	Variable	22	31	39
control	Dynamic	6	48	59

posted speed limits. With "Variable" data, the vehicle is additionally informed of certain locations in the network where a lower speed limit is implemented. With "Dynamic" data, additional lower speed limits are implemented because of network or weather conditions, to slow traffic in fog, on slippery roads, around major incidents, etc. Thus, with a Dynamic system, speed limits are current in terms of time.

11.2.2.2 Effects on the Environment

It is expected that ISA will have a positive effect on fuel consumption, emissions, dust and noise. Not many research initiatives focused on the effects of ISA on the environment, although some results can be mentioned from field trials. In Sweden, Varhelyi et al. (2004) noted a reduction of CO by 11%, NO_x by 7% and HC by 8% in a 50 kph-area. In the Australian trial, Regan et al. (2006) noted a 4% reduction of fuel consumption and a 4% reduction of CO_2 emissions, when ISA is used in combination with FDW on 80 kph zones.

It is also noted that the effect of speed on emissions is complex. The optimum speed, the speed at which emissions are minimized, varies according to the type of emission and type of vehicle. Typically, pollutant emissions are optimized for constant speed of 40–90 kph. It should also be noted that, in steady driving conditions, CO and CO_2 emissions, in terms of g/km travelled, are highest at very low travel speed (15 kph or less) (OECD 2006).

Liu et al. (1999) studied the ISA effects on network efficiency, fuel consumption and emissions through detailed micro-simulations. The ISA effects were modeled for the urban network in the morning peak and in the off-peak, rural two-lane road and motorway. Predicted fuel savings were 8% for urban peak, 8% for urban off-peak, 3% for rural road and 1% for the motorway at an ISA penetration level of 100%. Furthermore, they found that the emissions of CO, NOx and HCs varied by only ±2% for all ISA penetration rates.

11.2.2.3 Effects on Traffic Efficiency

Research concerning the effect of ISA on traffic efficiency is limited but the overall perspective seems good. Biding and Lind (2002) did not find any effects of ISA on travel times. It is assumed (Hogema et al. 2000) that a higher capacity and a more homogeneous traffic flow would be achieved due to ISA adjustments of the speed. Swedish trials also showed that drivers of vehicles equipped with ISA approached roundabouts, intersections and curves smoother in terms of deceleration (Vàrhelyi and Mäkinen 2001).

11.2.2.4 Side Effects

Different studies indicated that the vehicle following gap is reduced (Persson et al. 1993; Comte 2000), this leads to closer car following behavior. Várhelyi et al. (1998) conclude that safer car following behavior (bigger vehicle following gap) occurred on urban roads (30–50 kph). However, on 70–90 kph roads the tendency was the opposite and driver vehicle gaps decreased (meaning riskier car following behavior). Varhelyi et al. (2004) found no evidence that the behavior of ISA drivers towards other road users improved. The assumed effect of ISA on give-way behavior varies. Early research by Persson et al. (1993) indicated a slight increase in incorrect give-way behavior at intersections. Others found no negative effects (Várhelyi et al. 1998, 2004) or even a slightly positive effect (Almqvist and Nygard 1997; Várhelyi et al. 1998). It is concluded that overtaking behavior did not change (Comte 2000; Vàrhelyi and Mäkinen 2001), also no loss in vigilance was found (Comte 2000).

Different trials indicated an increase in travel time. In 1998, Varhelyi et al. conclude that the travel time increase due to ISA was 2.5–2.8%, depending on the country (Netherlands, Spain or Sweden). Other research also reports an increase in travel time due to ISA (Varhelyi and Makinen 2001; Liu and Tate 2004). A small relevant effect was found by Broekx and Panis (2004) Despite the increase in travel time, micro simulation showed that ISA does not lead to increased traffic jams (Liu and Tate 2004).

Most studies indicate that ISA results in a reduced driver comfort. Vàrhelyi and Mäkinen (2001) mention that drivers report to feel an increased frustration. Trials in the Brookhuis and De Waard studies (1997, 1999) indicate a slight increase in mental workload. Rook and Hoogema (2005) looked at the effects of ISA feedback force (for haptic throttle) on frustration level and workload. They found, amongst others, that high force ISA leads to more workload and frustration than low-force ISA. Comte and Jamson (2000) found no increase in workload.

11.2.2.5 Acceptance of ISA

Acceptance of ISA is one of the key elements for the (potential) success and effectiveness of the system. We can distinguish the users' acceptance, which gives an indication on how users (test-drivers) cope with the system and their acceptability or support, which indicates in turn how potential users will react when ISA is implemented (Vlassenroot et al. 2008a, b).

Morsink et al. (2006) describe an "acceptance versus effectiveness" paradox, the more effective ISA is on road safety (e.g. restricting ISA), the less accepted it is by the users. Brookhuis and De Waard (1999) showed that the acceptance of the system strongly depends on the mode of the used feedback. In the field trials in Hungary and Spain, a comparative study was made between an auditory warning system and active accelerator pedal (assisting system). In Hungary, most drivers preferred auditory and visual feedback to the haptic feedback pedal (Falk et al. 2004). However, it must be noted that comparison between the different systems is not that much researched (Carsten 2002; Morsink et al. 2006).

Also drivers' characteristics are important for the acceptance of ISA. Jamson (2006) noted that frequent speeders were less likely to support an ISA system. Hjälmdahl (2004) found that drivers, who were willing to use ISA, already drove at a speed close to the speed limit, while those who drive fast wanted to abort the trial after using the system.

In most trials the acceptance of ISA increased after using the system in the trial, compared to the opinions they gave before they used ISA (Biding and Lind 2002; Vlassenroot et al. 2007; Harms et al. 2007; Young et al. 2003). This indicates that trying the system and having experience will influence the user acceptance of ISA.

It has to be noted that, in general, the research on user acceptance varied a lot between the different trials (Vlassenroot et al. 2008a, b) and no coherent acceptance indications were described. Carsten (2002) noted that the attitudinal research on acceptance of ISA could be criticized for not being sufficiently rigorous.

Over the past years, some studies were done to determine the willingness to pay for ISA. Interesting are the studies performed after trials, questioning people regarding their willingness to pay before and after they used ISA. After the Swedish pilots, people who had the experience with driving with ISA were asked whether they wanted to keep the system after the trial. Only 28.4% indicated to be willing to keep the system. Drivers indicated to be willing to pay an average of 90 Euros' to keep the system. The market value was estimated to be 180 Euros on new cars, 155 Euros in case of retrofit. Over the past decade, other studies looked into the market price of ISA as well (Marchau 2000; Argiolu and Van der Pas 2006).

11.2.2.6 Public Support or Acceptability of ISA

Not much research was conducted during the trials on the acceptability of ISA by non-ISA users. De Mol et al. (2001) did a large-scale questionnaire in Belgium about the public support for speed measures, including ISA. Most of the respondents did recognize that speed and excessive speed is a problem. The acceptability of ISA was quite large; the mandatory ISA-system was not accepted by 30%, advisory ISA was accepted by 82%. Outside built-up areas 47% were not in favor of a mandatory ISA and on motorways 60% did not accept mandatory ISA. In built-up areas almost 70% accepted mandatory ISA.

In the SARTRE project (SARTRE 2004) over 24,000 drivers in 24 European countries were interrogated about road safety issues. One of the questions concerned the perceived usefulness of a system that prevents exceeding the speed limit. Less than 50% in Northern Europe, about 55% in Western and Eastern Europe and about 65% in Southern Europe would find such a system very or fairly useful. Piao et al. (2005) report results from a survey on ISA in three European cities. In all three cities there was a strong support for an informative ISA but very little support for a haptic throttle or restricting ISA. Up to 70% of the drivers said they would like to use ISA systems in residential areas.

11.2.2.7 Legal Aspects

Legal aspects are often mentioned as a barrier for ISA implementation (Marchau et al. 2005). In general, research shows that most ISA system do not intervene more with the driving task than other, already available, systems on the market. Based upon this some authors argue that the clarification of the product liability will not be a problem (Goodwin et al. 2006). Jamson (2006) mention that there are regulations that label it an offence to modify braking systems. This makes it complicated to implement a system that more strongly intervenes with the driving task by braking the vehicle. Furthermore, systems that draw power from the vehicle need to be approved and tested by an approved test organization before they can be implemented. Van Wees (2004) did a very elaborate study into this subject for the Netherlands and clearly pinpoints which additional legislation is desirable. Van Wees argues that there are some complex legislative problems before the ISA can be implemented in the EU. Furthermore, Van Wees argues that in case of ISA malfunctioning the user can give reason for an imputability defense, which has the likelihood of succeeding. In order to implement ISA Van Wees (2004) advised to implement explicit legal regulations; either risk and liability regulations or traffic insurance regulations. To which extent the absence of this legislation is a barrier for ISA implementation remains unclear.

In the PROSPER-project (Project for Research On Speed adaptation Policies on European Roads), SWECO (2005) did a study on legal matters concerning ISA, based on expert opinions. They concluded that for systems that would be introduced on a voluntary basis, no major legal risks would appear, since the actors concerned will mainly have the same responsibilities/liabilities as of today. Common for all the respondents is that the driver is always responsible for her/his driving. However, if intervening ISA systems would be put on the market in combination with a mandatory introduction, the legal situation would change. They noted that then the driver wouldn't be in complete control of the vehicle at all times while driving. SWECO (2005) also noted that the industry is more in favor of an informative ISA. The authorities responsible for road safety are more supportive to the principle of ISA system controlling the vehicle speed. A main conclusion in PROSPER was that

ISA implementation on the European road network is more connected with organizational difficulties and challenges than with legal risks and constraints.

11.3 Ongoing Issues Towards Implementation

The potential of ISA has been recognized, trials indicated that the ISA technology works, and that ISA has a considerable potential to contribute to traffic safety. Furthermore, it is generally considered that effects of ISA on road safety, the environment and the quality of life are beneficial and, as indicated above, policy-makers are shifting more and more from technological and behavioral research towards the implementation aspects of ISA. Traffic safety problems are huge and, moreover, traffic safety goals are not met, making the question "Why does ISA implementation go so slow?" a relevant and unavoidable one.

To reach the stage of a ready-for-implementation ISA, a lot of research was conducted during the past 15 years. However, considering the research setup of the trials it is noted that every research and every trial had its own method and approach. This makes comparison between results of different trials very difficult. Carsten (2002) noted missed chances within the ISA trials. Until today no systematic investigation of the impact of the different levels of ISA intervention has been made. Long-term effects of ISA on driving behavior are poorly investigated. The acceptability of ISA or what kind of system would be preferred by potential users has only been investigated on a small scale and no in-depth analysis has been made.

Discussions about implementation of ISA have been carried on since 2002 (Carsten 2002), but there are still issues that have to be resolved, e.g. regarding the technical architecture and speed limit databases implementation and maintenance.

11.3.1 Technology Developments and Speed Information Databases

Although there is no sign of ISA implementation in the road transport system yet, policymakers have not sat still the last years. They recognized the potential of ISA and stimulated research on different levels.

Many research activities funded by the European Union have constructed a framework which is of great use in the development of a speed limit database: SpeedAlert (2005) investigated and developed a framework to harmonize the in-vehicle speed alert concept definition and to investigate the first priority issues to be addressed at the European level, such as the collection, maintenance and certification of speed limit information. In the research of ActMap (Flament 2005), mechanisms for online incremental updates of digital map databases in the vehicle were created and investigated. In the MAPS&ADAS subproject of PREVENT, the use of digital maps as primary and/or secondary sensors for Advanced Driver Assisting Systems (ADAS) was investigated.

Besides these European projects, many national initiatives were undertaken, e.g. in Sweden (NVDB 2000), and Finland (DIGIROAD project 2006, Finish Road Administration (2006)), where the speed limit database is seen as a part of the national road database, which contains different kind of road information. In Denmark the registration is based on all speed signposts in the county of North Jutland, including approximately 22,000 km of roads. A GPS logger, with a special designed keyboard, has been used for this purpose. This special keyboard made it possible to gain this information in only about 4 weeks. In the Netherlands, a speed limit database has been made available on the Internet, which should become 98% accurate in 2 years time. The information could be filled-in online. In Belgium (De Mol and Vlassenroot 2006), the Flemish Government started to make a digital inventory of every vertical road sign, including speed limits on all types of roads.

It can generally be concluded that, at European level, the major technical guidelines and protocols have been developed. Within the national initiatives the focus was more on an operational level, concluding in legislations, national protocols, basic tools and field practices.

It must be noted that still most of these activities are not fully known by policymakers. If it can be said that today, the focus on ISA research has shifted more and more towards developing implementation strategies for ISA, a central notion is that policymakers do not have a clear picture of the ITS conditions, goals and concepts contributing to road safety or mobility. A certain risk-avoiding attitude towards ISA can be noted among policymakers, who still are the key-figures in conducting implementation of ITS.

11.3.2 Implementation Barriers

Over the past decade research has been carried out regarding barriers for implementation. When it comes to ISA implementation several barriers can be derived from the literature (ETSC 2006; Marchau 2000). In general it can be said that legislation, technical reliability, and the benefit to the user were important barriers for implementation:

- *Liability aspects*: Both in the PROSER as in the FADAS research, experts indicated that liability issues and legislation were the most important barriers for implementation of ISA. Most investigators (legal experts), however, point out that the reliability issues for the informing and warning types of ISA are by no means different than that of other driver support technologies that are currently implemented on a large scale (Albrecht 2005, cited in Goodwin et al. 2006). For the more intervening types of ISA (half-open and closed ISA), this might be more difficult, especially when introduction takes place of a mandatory system (Sweco 2005).

- *Reliability issues*: Trials in many countries have indicated that ISA is a proven technology. There is still room for improvement but there is no reason for extending implementation for reliability issues. Technology will keep on being improved in the meantime. Important issues are indicated to be related to the HMI (Human–Machine Interaction) interface (FADAS, PROSPER, ETSC). When it comes to reliability of the speed limit database, research shows that only few countries have a speed limit database that is accurate enough to use. However, as mentioned above, it is possible to create such a database within a relatively short term.
- *The perceived benefit by the users*: Although experts indicate that a major barrier for ISA implementation is the fact that users do not see the benefit of the system, this is contradictory to the results of different studies. The SARTRE 3 research (2003) interrogated drivers across the EU (23 countries). Overall, 60% of the drivers indicated to support more severe penalties for speeding (varying levels between 19 and 80%), contradicting the idea that people do not see the benefit of ISA. Furthermore, the SARTRE 3 survey also demonstrated that across Europe, about 55% of drivers would find a system preventing them to exceed the speed limit, "useful" or "very useful". Research performed in the UK shows similar results (MORI 2002). Furthermore, research shows that people who tried ISA are willing to use ISA on a voluntary basis. In Europe, between 60 and 75% of drivers who have tried ISA technologies, said they would like to have the system in their own cars (Peltola and Tapio 2004). Research performed as part of the pilot in Sweden showed that ISA drivers indicate that they are willing to keep ISA on a voluntary basis and are even willing to pay for it (Adell 2008).

So, on the one hand experts indicate that there are major barriers for implementing ISA; on the other hand experts prove that none of these barriers are really a barrier for implementation. To cope with these barriers over the past decade, researchers started gradually researching implementation strategies.

11.3.3 Implementation Policies

Tate and Carsten (2008) made a study based on their field trials in the UK to predict the safety-impacts of ISA. It also examined hypothetical scenarios for ISA implementation and investigates how those scenarios might affect overall safety gains with ISA. Two alternative scenarios were examined, a market driven scenario in which drivers choose to adopt ISA and an authority driven scenario with more encouragement of ISA adoption. The analysis indicated that over a 60-year period from 2010 to 2070, the market driven scenario is expected to reduce fatal accidents by 10%, serious injury accidents by 6%, and slight injury accidents by 3%. The authority driven implementation scenario is expected to reduce fatal accidents by 26%; serious injury accidents by 21%; and slight injury accidents by 12%. The economic benefit associated with the predicted crash reductions under both the implementation scenarios outweighed the costs, thus justifying the deployment of ISA. The market driven implementation scenario resulted in benefit-to cost ratios in the range of range 1.8–3.0. The authority driven implementation of ISA produced benefit-to-cost ratios in the range 2.8–4.8.

Different investigators looked at new policymaking approaches to deal with the uncertainties surrounding ISA implementation (Agusdinata et al. 2005; Marchau et al. 2009; Van der Pas et al. 2007). They suggest new adaptive approaches to implement ISA. This involves strategies where you start implementing ISA on a small scale (e.g. only for young drivers in an area), learn regarding the uncertainties over time, and adopt the policy over time. To support building such adaptive policies new modeling approaches are used (exploratory modeling), this allows decision making regarding ISA implementation despite large uncertainty.

At the EC level (European Commission 2008a, b) a proposal for a Directive is made called "laying down the framework for the deployment of Intelligent Transport Systems in the field of road transport and for interfaces with other transport modes". In this Directive, together with the action plan for the deployment of ITS in Europe (European Commission 2008a, b), optimization of the collection and provision of road data, which also will include speed limit data, is one of the action points. Some countries, as described before, already have these data but it seems that large-scale implementation actions are not made. In the mean time, it seems that some local governments are not waiting for the initiatives made on national or regional level. It is noted that local governments are directly confronted with the problems that speed causes and therefore they want to take the initiative for themselves (De Mol and Vlassenroot 2006). For example in the city of London (Keith 2008), Transport for London (TfL) has produced, and will continue to maintain a digital map of all London's speed limits. This map has been made available free of charge to anyone who wishes to use the map for personal use in their own navigation systems, or to create commercial applications. Also TfL works together with companies and fleet owners to promote the use of ISA.

11.4 Time for ISA

The last decades a lot of research concerning ISA has been performed. Several trials with different types of ISA have shown that ISA can be an efficient and effective way to reduce speed and speeding and, as such, have a positive effect on traffic safety. It is also expected that ISA will have a positive effect on fuel consumption, emissions, dust and noise. There is no doubt about the fact that ISA is among the most tested and investigated ITS. However, large-scale implementation of the most effective ISA (assisting and restricting ISA) seems far away. A point of criticism across all the researches is that the research on user acceptance varied a lot between the different trials and no coherent acceptance indications were described. It can be noted that the attitudinal research on ISA was not sufficiently rigorous, which also gave the opponents of ISA the chance to criticize the benefits of the system. The many researches and trials gave the possibility to gain a better insight on the acceptance and behavior of the drivers. Some investigators mention the acceptance versus effectiveness paradox; the more effective ISA is on road safety (e.g. restricting ISA), the less accepted it is by the users. This could make the implementation of ISA more difficult.

Additionally, ISA is one of those systems where the acceptance gets higher if the driver is allowed to test the system. Frequently it was said that you could talk about ISA as long as you wanted but the best way to convince somebody of its benefits is to let him or her drive with ISA.

All in all we can conclude that the test-phase of ISA is over and that implementation strategies should be developed, although some barriers were found. One of the major issues is the development of a speed limit database. National and regional initiatives are made and on European level – with the action plan on ITS – governments will be stimulated to develop a national digital road database. Some stakeholders also indicated that some legal issues need to be resolved, especially when there would be malfunctions of ISA (certainly if a restrictive ISA is used). But it is also noted that these issues are not of such an order that they would hinder the introduction of ISA; it seems that problems are more connected with organizational difficulties and challenges than with legal risks and constraints.

If ISA is to be introduced, it would be more beneficial if governments would be involved in its implementation strategies. This could be done by supporting or creating a (technical) framework that would enable the use of ISA, or even actively promote ISA by giving subsidies or performing some other supporting actions. We also noticed that some local governments are taking the initiatives and are not waiting for the decisions that are to be made by higher (national) governments. Policy-makers have a key-role in the implementation of ISA, as mainly they are the problem owners, so they should do things, like:

- To communicate about ISA, to create more observability and political awareness of ISA (e.g. like the initiative on ISAweb.eu).
- To look for niches and implement ISA where it can be successful.
- To make a case for mandatory implementation because ISA is a preventive innovation and those who need ISA the most would never voluntary adopt it.

Initiatives taken by private companies to allow information about the speed limits or a warning if exceeding the limit in navigation systems can only be seen as a positive evolution in the use of ISA, but we conclude that after a test-period of 25 years the time has come to allow a broader public to experience the benefits of ISA. Only then an answer to the question how people react when using ISA in the real world would be given and maybe then we will know what the long-term effects would be. This would open new possibilities in the research field of intelligent speed adaptation.

References

S. Almqvist, M. Nygard, *Dynamic Speed Adaptation: A Field Trial with Automatic Speed Adaptation in an Urban Area* (Department of Traffic Planning and Engineering, University of Lund, Lund, 1997)

R. Argiolu, J. Van der Pas, N. Dragutinovic, G. Hegeman, V. Marchau, The future of advanced driver assistance systems, Reporting the results of an expert survey, in *Proceeding for TRAIL in motion,* Rotterdam, 2006

AVV, *Evaluatie Intelligent SnelheidsAanpassing (ISA): Het Effect Op Het Rijgedrag in Tilburg* (AVV, Nieuwegein, 2001)

C. Berg, G. Thesen, S. Brosvik Bayer, An analysis of attitudes and driving behavior among young drivers. *Proceedings of 15th ITS World Conference*, New York, 2008.

T. Biding, G. Lind, *Intelligent Speed Adaptation (ISA), Results of Large-scale Trials in Borlange, Lidkoping, Lund and Umea During the Periode 1999–2002* (Vägverket, Borlange, 2002)

S. Broekx, I.L. Panis, Potential network effects of intelligent speed adaptation, in *Proceeding of 10th International Conference on Urban Transport and the Environment in the 21st Century,* Dresden, 2004

K.A. Brookhuis, D. De Waard, Limiting speed, towards an intelligent speed adapter (ISA). Transp Res F **2**(2), 81–90 (1999)

K.A. Brookhuis, D. De Waard, Intelligent speed adaptor, in *Proceeding of European chapter of the human factors and ergonomics society annual conference*, Bochum, 1997

O. Carsten, M. Fowkes, *External Vehicle Speed Control. Executive Summary of Project Results* (University of Leeds, Leeds, 2001)

O. Carsten, F. Tate, Intelligent speed adaptation: accident savings and cost-benefit analysis. Accid Anal Prev **37**(3), 407–416 (2005)

O. Carsten, M. Fowkes, F. Lai, K. Chorlton, S. Jamson, F. Tate, B. Simpkin, *ISA-UK. Executive Summary of Project Results* (Department for Transport, Leeds, 2008)

Commission of The European Communities, *Proposal for a Directive of the European Parliament and of the Council* (European Commission, Brussels, 2008a)

Commission of The European Communities, *Action Plan for the Deployment of Intelligent Transport Systems in Europe* (European Commission, Brussels, 2008b)

S. Comte, New systems: New behaviour? Transp Res F **3**(2), 95–111 (2000)

S. Comte, S. Jamson, Traditional and innovative speed-reducing measures for curves: an investigation of driver behaviour using driving simulator. Saf Sci **36**, 137–150 (2000)

J. De Mol, S. Vlassenroot, *Krachtlijnen voor het leveren van snelheidsinformatie in functie van het toekomstig opstellen van een snelheidsdatabank* (Instituut voor Duurzame Mobiliteit, Universiteit Gent, Gent, 2006)

J. De Mol, M. Broeckaert, B. Van Hoorebeeck, W. Toebat, J. Pelckmans, *Naar een draagvlak voor een voertuigtechnische snelheidsbeheersing binnen een intrinsiek veilige verkeersomgeving* (Centre for sustainable development/Ghent University, BIVV, Ghent, 2001)

D. De Waard, K.A. Brookhuis, Behavioural adaptation to warning and tutoring messages. Results from an on-the-road and simulator test. Heavy Vehicle Systems, International Journal of Vehicle Design **4**, 222–234 (1997)

J. Ehrlich (ed.), *Carnet de Route Du LAVIA. Limiteur s'adaptant à la vitesse autorisée* (Inrets, Versailles, 2006)

European Commission, *Laying Down the Framework for the Deployment of Intelligent Transport Systems in the Field of Road Transport and for Interfaces with Other Transport Modes* (European Commission, Brussels, 2008)

E. Falk, A. Varhelyi, M. Draskoczy, Field trials with ISA in Hungary, in *Proceedings of ITS in Europe,* Budapest, 2004

Finish Road Administration (2006). Digiroad-project. http://www.digiroad.fi. Accessed 21 July 2009

M. Flament (ed.), *ActMAP Final Report. D1.2, 117v10* (ERTICO, Brussels, 2005)

F. Goodwin, F. Achterberg, J. Beckmann, *Intelligent Speed Assistance – Myths and Reality ETSC Position on ISA* (ETSC, Brussels, 2006)

L. Harms, B. Klarborg, H. Lahrmann, N. Agerholm, E. Jensen, N. Tradisauskas, Effects of ISA on the driving speed of young volunteers: a controlled study of the impact information and incentives on speed. *Proceeding for the 6th European Congress on Intelligent Transport Systems and Services*, Aalborg, 2007

M. Hjälmdahl, *In-Vehicle Speed Adaptation. On the Effectiveness of a Voluntary System* (Department of Technology and Society Traffic Engineering Lund Institute of Technology, Lund, 2004)

J. Hogema, H. Schuurman, C. Tampere, ISA effect assessment: from driving behaviour to traffic flow, in *Proceedings of ICTCT Workshop*. Nagoya, 2000

ISAweb.eu (2004) Intelligent Speed Adaptation. http://www.isaweb.eu. Accessed 3 August 2009

S. Jamson, Would those who need ISA, use it? Investigating the relationship between drivers' speed choice and their use of a voluntary ISA system. Transp Res F **9**, 195–206 (2006)

H. Keith, Intelligent speed adaptation. (Transport for London, 2008). http://www.tfl.gov.uk/corporate/projectsandschemes/7893.aspx. Accessed 3 June 2009

H. Lahrmann, J.R. Madsen, T. Boroch, Intelligent speed adaptation: development of GPS based System and Field Trial of the System with 24 test Drivers, in *Proceedings of the 8th World Congress on Intelligent Transport Systems*, Sydney, 2001

H. Lahrmann, N. Agerholm, N. Tradisauskas, J. Juhl, L. Harms, Spar paa farten. An intelligent speed adaptation project in Denmark based on pay as you drive principles, in *Proceedings of 6th European Congress on Intelligent Transport Systems and Services*, Aalborg, 2007

R. Liu, J. Tate, Network effects of intelligent speed adaptation systems. Transportation **31**(3), 297–325 (2004)

R. Liu, J. Tate, R. Boddy, Simulation modelling on the network effects of EVSC. Deliverable 11.3. of external vehicle Speed Control Project. Institute for Transport Studies, Leeds, 1999

V. Marchau, *Technology Assessment of Automated Vehicle Guidance* (Faculty of Technology, Policy and management, school of Systems Engineering Policy Analysis and Management Delft University of Technology, Delft, 2000)

V. Marchau, R. van der Heijden, E. Molin, Desirability of advanced driver assistance from road safety perspective: the case of ISA. Saf Sci **43**(1), 11–27 (2005)

V. Marchau, W. Walker, R. van Duin, An adaptive approach to implementing innovative urban transport solutions. Transport Policy. **15**(6), 405–412 (2009)

P. Morsink, C. Goldenbeld, N. Dragutinovic, V. Marchau, L. Walta, K.A. Brookhuis, *Speed Support Through the Intelligent Vehicle: Perspective, Estimated Effects and Implementation Aspects* (SWOV Institute for Road Safety Research, Leidschendam, 2006)

B. Nielsen, H. Lahrmann, Safe young drivers: Experiments with Intelligent Speed Adaptation, in *Proceedings of ITS at the Crossroads of European Transport*, Brussels, 2005

NVDB, The Swedish National Road Database. Proceedings of ITS-congress. Turino.OECD/ECMT. Final report of the working group on speed management. OECD/ECMT Joint Transport Research Committee, Paris, 2006

O. Carsten. European research on ISA: Where are we now and what remains to be done, in *Proceeding for ICTCT Workshop*, Nagoya, 2002

M. Paatalo, H. Peltola, M. Kallio, Intelligent speed adaptation – effects on driving behaviour, in *Proceedings of Traffic safety on Three Continents Conference*, Moscow, 2001

H. Persson, M. Towliat, S. Almqvist, R. Risser, M. Magdenburg, *Speed Limiters of Cars: A Field Study of Driving Speeds, Driver Behaviour and Traffic Conflicts and Comments by Drivers in Town and City Traffic* (Department of Traffic Planning and Engineering, University of Lund, Lund, 1993)

Piao et al. (2005)

M.A. Regan, T.J. Triggs, K.L. Young, N. Tomasevic, E. Mitsopoulos, K. Stephan et al., *On-Road Evaluation of Intelligent Speed Adaptation, Following Distance Warning and Seatbelt Reminder Systems: Final Results of the TAC SafeCar Project* (Monash University Accident Research Centre, Victoria, 2006)

F. Saad, G. Malaterre, Régulation de la vitesse: aide au contrôle de la vitesse. *Synthèse*, Rapport ONSER, 1982

SARTRE. European drivers and road risk. Part 1. Report on principal results. INRETS, 2004

SpeedAlert, Evolution of SpeedAlert concepts, deployment recommendations and requiremenuts for standardisation, version 2.0, ERTICO, Brussels, 2005

SWECO, WP 6 PROSPER. ISA implementation strategies. Legal and policy aspects on road speed management in Europe. PROSPER, Brussels, 2005

F. Tate, O. Carsten, *ISA-UK. Implementation Scenarios* (Department for Transport, Leeds, 2008)

Transek AB, SWECO VBB (2005). ISA in Stockholm. Results from trials and possibilities for implementation, Trafikkontoret, Stockholm, 2005

J. Van der Pas, V. Marchau, W. Walker, D. Agusdinata, Exploratory modeling to support multi-criteria analysis to cope with the uncertainties in implementing ISA, in *Proceeding of 14th World Congress on Intelligent Transport Systems*, Beijing, 2007

K. Van Wees, *Intelligente Voertuigen, Veiligheidsregulering en Aansprakelijkheid* (Technology, Policy and management, University of Delft, Delft, 2004)

A. Vàrhelyi, T. Mäkinen, The effects of in-car speed limiters – field studies. Transp Res C **9**, 191–211 (2001)

A. Varhelyi, M. Hjalmdahl, C. Hyden, M. Draskoczy, Effects of an active accelerator pedal on driver behaviour and traffic safety after long-term use in urban areas. Accid Anal Prev **36**, 729–737 (2004)

A. Várhelyi, S. Comte, T. Mäkinen, *Evaluation of In-car Speed Limiters: Final Report* (VTT Communities and Infrastructure, Helsinki, 1998)

S. Vlassenroot, S. Broekx, J.D. Mol, L.I. Panis, T. Brijs, G. Wets, Driving with intelligent speed adaptation: final results of the Belgian ISA-trial. Transp Res A **41**(3), 267–279 (2007)

Vlassenroot, S., Brookhuis, K.A., Marchau, V., Witlox, F. (2008). Measuring acceptance and acceptability of ITS, in Theoretical background in the development of a unified concept. Proceeding of TRAIL Research School Conference. Delft

Vlassenroot, S., De Mol, J., Dedene, N., Witlox, F., Marchau, V. (2008). Developments on speed limit databases in Flanders: A first prospective, in *Proceedings of the 15th World Congress on ITS*. New York.

H. Wallén Warner, *Factors Influencing Drivers' Speeding Behaviour* (Acta Universitatis Upsaliensis, Uppsala, 2006)

H. Wallén Warner, L. Aberg, The long-term effects of an ISA speed-warning device on drivers' speeding behaviour. Transp Res F **11**(2), 96–107 (2008)

K.L. Young, M.A. Regan, E. Misopoulos, N. Haworth, *Acceptability of In-Vehicle Intelligent Transport Systems to Young Novice Drivers in New South Wales* (Monash University Accident Research Centre (MUARC), Victoria, 2003)

S. Myhrberg, ISA in Stockholm Results from trials and possibilities for implementation, SWECO, Stockholm, 2006

J.H. Hogema, A.M. Rook, Intelligent speed adaptation: the effects of an active gas pedal on driver behaviour and acceptance. On request of the Dutch Ministry of Transport, Public Works and Water Management, Transport Research Centre AVV and the European Commission. TNO report TM- 04-D011. TNO Human Factors Research Institute, Soesterberg, 2004

A.M. Rook, J.H. Hogema, Effects of human-machine interface design for intelligent speed adaptation on driving behavior and acceptance. Human Performance; Simulation and Visualization (1937), 79–86 (2005)

A. Varhelyi, M. Hjalmdahl, C. Hyden, M. Draskoczy, Effects of an active accelerator pedal on driver behaviour and traffic safety after long-term use in urban areas. Accident Analysis and Prevention **36**(5), 729–737 (2004)

H. Peltola, J.R. Tapio, Intelligent Speed Adaptation – recording ISA in Finland. Presentation at the Via Nordica 2004 Conference on 7–10 June 2004 in Kopenhagen, Denmark, 2004

E. Adell, A. Varhelyi, Driver comprehension and acceptance of the active accelerator pedal after long-term use. Transportation Research Part F: Traffic Psychology and Behaviour **11**(1), 37–51 (2008)

V.A.W.J. Agusdinata, W.E. Marchau, Walker, Adaptive policy approach to implementing Intelligent speed adaptation, IET Intelligent Transport Systems **1**(3), 186–198 (2007)

Organisation For Economic Co-Operation and Development (OECD), Speed Management. Paris. ISBN: 92-821-0377-3 (2006)

J. Piao, M. McDonald, A. Henry, T. Vaa, O. Tveit, An assessment of user acceptance of intelligent speed adaptation systems, IEEE Intelligent transport Systems conference, Vienna, Austria, (2005)

Chapter 12
Watch Out! Something Precious is Moving

Anna Anund, Andreas Tapani, and Eleni Chalkia

12.1 Perception in the Framework of the Driving Task

Analysis of crash data has led to a number of ways of looking at crash causation. One such is the Haddon matrix (Haddon 1972).

In the matrix, the contributions of human, vehicle/equipment and environmental factors (both physical and socio economical) in injuring and death related accidents, as well as countermeasures, are described according to three phases: pre-crash, crash and post-crash. The major component in crash causation is the "human factor" (as opposed, for example, to mechanical failure and weather conditions), involved in around 90% of the crashes (Glendon et al. 2006).

Michon (1985) structured the driver task into three levels: strategic, tactical and operative. At the strategic level, the general planning of a journey is handled; for example, route planning and preparation before leaving. At the tactical level, the driver has to perform manoeuvres allowing him/her to e.g. make a turn or accept gaps to lead in front or lag behind vehicles. Finally, at the operative level, the driver has to execute simple actions that are automatic, for example, changing the gear, turning the steering wheel. At the strategic level time is not a critical aspect for success. Lack of time will be more and more important as the task is handled automatically.

Perception is an issue of importance when it comes to driving, especially on the tactical and operative level. Perception is most often defined as the process of attaining awareness or understanding of sensory information. Perception comes naturally and effortlessly and is involved in all daily living activities, from eating to driving a car (Blake and Sekuler 2006). Perception is essential in mental and physical daily activities and the result is depending on what we see, hear, feel, touch, smell and taste. From the drivers perspective the degree of safety will depend also on the driver's state. For example, a sleepy driver will need more time for

A. Anund (✉) and A. Tapani
Swedish National Road and Transport Research Institute (VTI), Linköping, Sweden
e-mail: anna.anund@vti.se

E. Chalkia
Centre of Research and Technology Hellas/Hellenic Institute of Transport, Thessaloniki, Greece

E. Bekiaris et al. (eds.), *Infrastructure and Safety in a Collaborative World*,
DOI 10.1007/978-3-642-18372-0_12, © Springer-Verlag Berlin Heidelberg 2011

perception, both regarding the use of different senses and the time needed for the cognitive demand or to act on the operative level.

12.2 Perception Enhancement Implementation Scenario

One critical situation, where a driver needs to have a high degree of awareness and perception, is when passing by a school bus that is stopped for embarkation or disembarkation of children. In many countries the most dangerous situation for children going by school transport is when the children are running out behind or in front of the school bus, after disembarkation (Anund et al. 2003). The most common crash is when an oncoming vehicle hits the children at high speed.

In this chapter, this scenario is used as an example of possible perception enhancement, in order to improve safety and security for the most vulnerable road users – the children. The scenario demands driver information and actions on both a tactical and operative level.

One of the scenarios within IN-SAFETY was focused on the idea of perception enhancement by providing the driver with information beforehand about a school bus stopped for children departure or exiting. The aim of this idea was to reduce the speed of the passing vehicle and thus contribute to increase driver awareness.

The technical solution was based on vehicle-to-vehicle (V2V) communication. An on-board unit communicates with roadside equipment, in this case the school buses. The driver receives a message on a LCD (Liquid Crystal Display) inside the car, about 300 m before the school bus appears. The pictogram used was a school bus sign, with a warning sign above. The scenario was related to inappropriate speed near unprotected road users, especially children behind/in front of school buses. From a technical point of view, data about the characteristics and the location of the vehicle ahead is needed for system operation. A radio transmitter and receiver can be used to handle the communication between both vehicles (i.e. in the form of a Dedicated Short-Range Communications, DSRC) (Fig. 12.1).

Fig. 12.1 The messages in the LCD displayed to the driver 300 m before the bus, when the bus was then not yet possible to be seen by the oncoming driver

The hypothesis was that information given beforehand will contribute to a perception enhancement and an action in terms of reduced speed and an increase in lateral distance to the school bus, when passing by it.

In order to test the effectiveness of this perception enhancement strategy, an experiment with 20 participants was performed in VTI's moving base driving simulator. The participants drove twice; once after having worked during the night (no sleep) and driving just after getting off the night shift (5–8 a.m.; drowsy state), and once after a night's sleep (9 a.m.–4 p.m.; alert state). The design was a within subjects design. The order between the no sleep condition and the night sleep condition was balanced for gender. The road scenario consisted of a total of 9 laps, of 9.4 km each with rumble strips systematically present on some sections. The scenario included interaction with other vehicles, like catching up a slower vehicle with and without visible oncoming vehicles, but also car following (Anund et al. 2009a).

A school bus parked on the right hand side was passed twice. The first time with no information at all beforehand, while the second time with in-vehicle information beforehand. The in-vehicle information was received in the end of the first hour, while drivers drove subsequently both after a good night sleep and after no sleep. The in-vehicle information is the one in focus here.

12.3 Test Results

The effect of information beforehand on the presence of a school bus was significant; speed was reduced by 11 km/h ($F(1,15) = 6.531$; $p = 0.022$), but the effect of condition (4 km/h) was not ($F(1,15) = 0.301$; $p = 0.591$) and here no significant interactions were found (Fig. 12.2).

12.4 Conclusion

The results of the perception enhancement in terms of information beforehand supported the hypothesis that an in-vehicle warning results in lower speeds when passing the bus compared to a no prior warning situation. This was true both when the drivers were alert and sleep deprived. Especially, the speed profile shortly before passing the parked bus was more favourable in terms of traffic safety when the drivers had received the information before passing. There is, consequently, a potential to use in-vehicle information to reduce speeds during temporary safety critical events. Possible adaptation effects of such a warning strategy should be examined in a long term study.

12.5 Further Research

Even though protecting the children – one of the most vulnerable transport system's users – is of great importance for all societies, bus transport to school is a highly underinvested area in many EU-countries. The EU funded project

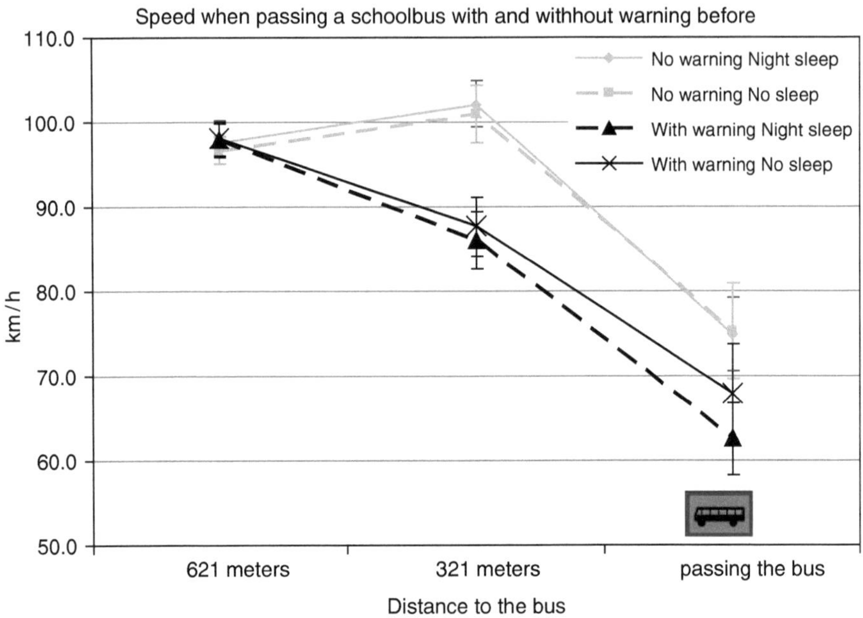

Fig. 12.2 Average speed (±SE) when passing a stopped school bus, as well as 621 and 321 m before that, with and without in-vehicle warning about the bus, and with and without prior night sleep of the test subject

SAFEWAY2SCHOOL (Grand Agreement: 233967) that started at September 2009 and continues until August 2012, aims to design, develop, integrate and evaluate technologies for providing a holistic and safe transportation service for children, from their home door to the school door and vice versa, encompassing tools, services and training for all key actors in the relevant transportation chain. These include optimal route planning and rerouting for school buses, to maximisation of safety through on-board safety applications (i.e. speed control and seat belts), "intelligent" bus stops, effective warning and information systems for bus drivers, children, parents and the surrounding traffic; as well as training schemes for all actors. The SAFEWAY2SCHOOL concept is based on a holistic approach from a door to door perspective, see Fig. 12.3.

The SAFEWAY2SCHOOL project aims to progress beyond the state of the art by means of the aforementioned holistic approach; considering procedures and processes to describe a living and dynamic safety structure for improvements. Chains of interdependencies and processes will be analysed and investigated, to identify critical parameters influencing road safety.

The target groups of the project captivate all the stakeholders that participate at the school transportation including school bus drivers, students/children from 6 to 16 years old, with and without disabilities, the families of the children, infrastructure operators (i.e. bus stops or bus fleet operators), car manufacturers (OEM's), authorities (legislators, municipal and school authorities) and finally all drivers of

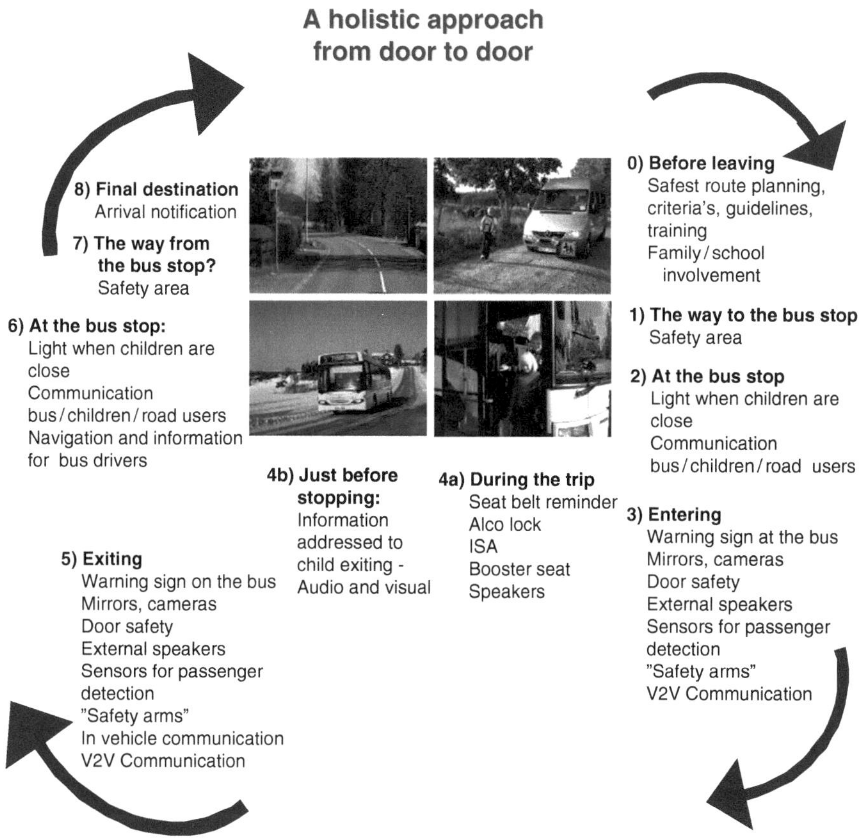

Fig. 12.3 The SAFEWAY2SCHOOL approach form door to door perspective
Source: Anund et al. 2009b

the surrounding traffic vehicles. These stakeholders, despite the fact that they have a common goal, which is the safe school transportation for the children; they have different needs and requirements that must be fulfilled in order to achieve this goal. During SAFEWAY2SCHOOL project, interviews and focus groups have been realised with the all kinds of different stakeholders in order to capture their thoughts and wishes and implement them to the development of the project.

The results of the survey of the stakeholders needs and requirements where various and very helpful, defining very clearly the problems of the school transportation. These outcomes where connected with the SAFEWAY2SCHOOL system requirements and that lead to the extraction of the Use Cases of the project. There are seven main Categories of Use Cases that are the following:

- Category 1: Routing and rerouting
- Category 2: Surrounding traffic information
- Category 3: On board systems

- Category 4: Intelligent bus stop
- Category 5: Notification
- Category 6: Training
- Category 7: Inventory tool

The SAFEWAY2SCHOOL Use Cases are illustrated at Fig. 12.4.

Thus, the SAFEWAY2SCHOOL project has to merge several technological innovative requirements in order to realise all the aforementioned Use Cases. Routing algorithms for the determination of the safest route for the school buses will be implemented, as well as algorithms for the location of the safe school bus stop. An automatic vehicle location (AVL) system will monitor the school bus and will allow the stakeholders to have the ability to check its position at any time. Dynamic rerouting will be also available in case of traffic or operational events that will redesign the route of the school bus according to the new traffic circumstances. All the aforementioned of course will be done under the presence of the discrete eyes of the parents and the family that will have the ability to be notified via different gateways when their child has entered the bus, when it has arrived at the school, as well as the situation of the child at any point of the transportation after special request.

To this end and to achieve the needs of the various stakeholders, there are different systems that must be integrated; like systems for boarding/deboarding recognition and identification, on board systems for safe transportation of children,

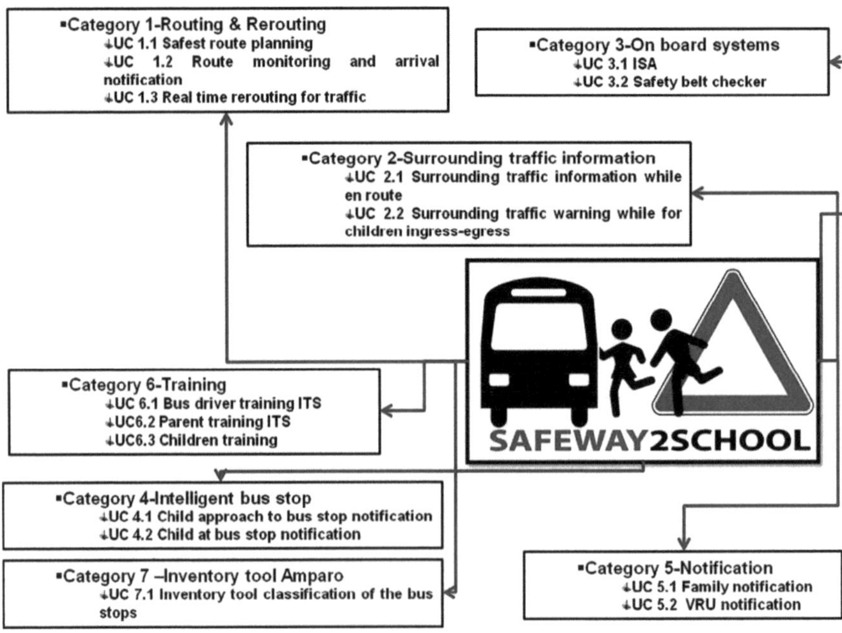

Fig. 12.4 The SAFEWAY2SCHOOL approach use cases
Source: Aigner-Breuss et al. 2010

systems for bus stop identification/localisation, systems for children localisation and monitoring, intelligent bus stop systems and other.

SAFEWAY2SCHOOL proposes a novel Vulnerable Road User-centred approach, by proposing safety solutions that place the needs and requirements of the stakeholders at the centre of the design process, at any stage of the school transportation. Safety and comfort concerns will be present in the whole project, and HMI issues will play a key role in ensuring safe school transportation. As a whole, SAFEWAY2SCHOOL's HMI concept will merge the adaptive driver-centred approach to a VRU-oriented perspective; that is, VRUs (and special users, in this case) are included in the overall design as ever present actors, and not only in critical traffic scenarios. This approach will set requirements for strong HMI flexibility, particularly as to the driver/supervisor's role.

References

E. Aigner-Breuss, M. Pilgerstorfer, A. Anund, T. Dukic, E. Chalkia, C. Ferrarini, R. Montanari, J. Wacowska, D. Jankowska, F. Diederichs, A. Pauzie, Comparison and analysis of user and stakeholder needs across different countries. EU project SAFEWAY2SCHOOL, D1.2 May 2010

A. Anund, J. Larsson, T. Falkmer, *Skolskjutsbarns inblandning i olyckor 1994-2001* (VTI, Linköping, 2003)

A. Anund, G. Kecklund, A. Tapani, A. Kircher, T. Åkerstedt, The effects of driving situation on sleepiness indicators after sleep loss: a driving simulator study. Ind. Health **47**, 1–9 (2009a)

A. Anund, T. Dukic, E. Chalkia, Project presentation. EU project SAFEWAY2SCHOOL, D10.1 September 2009b

R. Blake, R. Sekuler, *Perception*, 5th edn. (McGraw-Hill Education - Europe McGraw Hill Higher Education, New York, 2006)

A. Glendon, S. Clarke, E. Mckenna, *Human Safety and Risk Management*, 2nd edn. (Taylor and Francis, London, 2006)

W.J. Haddon, A logical framework for categorizing highway safety phenomena and activity. J. Trauma **12**, 193–207 (1972)

J. Michon, A critical view of driver behaviour models: what do we know, what should we do? in *Human Behaviour and Traffic Safety*, ed. by L. Evans, R. Schwing (Plenum, New York, 1985), pp. 485–520

Part IV
Self-Explanatory Road Environments

Chapter 13
A Message for You

Evaluation of Messages for Variable Message Signs to Enhance Comprehensibility

Karin Siebenhandl, Michael Smuc, and Florian Windhager

13.1 Research Aim

Motorists on European roads have to cope with ever more complex traffic environments and signs, with many of the latter now also supported by telematics technology. Since motorists now tend to travel farther than ever before, frequently crossing several countries in a single trip, confusion resulting from national differences in traffic signalling standards poses a risk to traffic safety. Chapter 5 gives an overview of requirements for harmonisation and shows examples for the diversification of Vienna Convention Symbols in several European countries. The legibility and comprehensibility of the messages conveyed thus becomes increasingly important as the number of variable message signs (VMS) in European countries rapidly continues to grow.

While developing comprehensible pictograms and minimizing the use of differing local languages formed one element of the research project, verbal messages and written text still cannot be substituted completely (i.e. they are needed to display additional information). Messages displayed on motorway VMS, which are passed at high speed, have to be particularly legible and accessible. Chapter 14 focuses on verbal messages or verbal message elements i.e. "Europeanisms", which have the potential to be harmonised EU-wide.

Any pictograms and verbal messages used should be comprehensible and easily recognised from a distance. This becomes even more relevant in impaired visibility conditions (such as darkness, rain or fog) or in the event of individual, for example visual, impairments (Garvey 2002).

Pictogram and typeface research, development and evaluation must, consequently, pay attention to (A) the attainability of maximum comprehensibility and legibility, (B) the different requirements of pictograms and fonts intended for printed signs or dynamic VMS matrix displays (Staplin et al. 1998; Siebenhandl et al. 2007a, b),

K. Siebenhandl (✉), M. Smuc, and F. Windhager
University of Krems (DUK), Krems, Austria
e-mail: karin.siebenhandl@donau-uni.ac.at

E. Bekiaris et al. (eds.), *Infrastructure and Safety in a Collaborative World*,
DOI 10.1007/978-3-642-18372-0_13, © Springer-Verlag Berlin Heidelberg 2011

and (C) the comparative analysis of symbols and signs in combination with verbal messages.

The objective of the implementation scenarios presented in this chapter are to make road environments more self-explanatory, by proposing a set of homogenized and comprehensive pictograms, which can be used to largely substitute verbal messages on VMS systems as far as possible, thereby reducing the complexity of information and minimizing the use of local languages.

13.2 Displaying Information on the Roadside

Traffic signs are the most commonly used means of controlling traffic. They convey messages using words and/or symbols and are placed at specific positions on roads to regulate, warn or guide road users. "Traffic signs are most effective when they comply with the following requirements: fulfil a need, command attention, convey a clear and simple message, command respect of the road users and give adequate time for proper response" (Pignataro 1973).

Given the fact that the use of consistent and easily understandable codes on the roadside can reduce motorists' workload (see Theeuwes and Godthelp 1995), there is an obvious need for solutions and techniques which overcome the numerous cultural differences encountered in the use of road signs.

The widespread use of pictograms and symbols as a substitute for language dependent verbal information is therefore recommended (Pline 1992). A "pictogram" is per definition a "visually perceptible figure referring to a 'real object' by resemblance" (Krampen 1965), which attains the effect of a public information symbol and can be considered to be backed by convention.

According to Luoma and Rämä, pictograms have many advantages over commonly used text passages: "For example, pictograms are more legible for a given size and hence cost. They are more easily recognized when their information is degraded due to poor condition of the sign, poor eyesight of the observer or poor environmental visibility; when motorists are familiar with both pictograms and text messages they can extract information more quickly from the former than the latter; words and abbreviations in foreign languages are not as well understood as the pictograms; and motorists who are poor readers and who therefore have difficulty understanding text messages are able to comprehend pictograms" (Luoma and Rämä 2001).

VMS display traffic signs electronically to inform road users of special activities or situations. They are commonly used to warn of traffic congestion, accidents, incidents, roadwork zones, speed limits or weather conditions. In urban areas, VMS systems are used in parking guidance and information systems to navigate motorists to available parking spaces. In roadwork zones, they can be used to guide vehicles to alternative routes, limit travel speed, indicate the duration and location of the work or simply provide information on the current traffic conditions. These signs can also be adjusted dynamically to display warnings and messages about changes in travel conditions (e.g. accidents, temporary restrictions, poor weather).

Balz (2003) lists the following areas of application for VMS and, in particular, dynamic traffic information signs (Balz 2003):

- To substitute signs for route guidance in the event of roadwork
- As an alternative to classic signs for route guidance
- To assist individual motorists in switching to public Transport
- To provide additional information for a specific stretch of road (e.g. problem areas)
- To supply preventive traffic information (e.g. regarding roadwork) and indicate to motorists that they should adapt their driving to the traffic conditions

Given the fact that the new generation of freely programmable VMS systems can display not only various individual information elements (e.g. symbols/pictograms), but also a combination of such elements (including verbal information) and even animated versions of elements, a new approach to displaying information is required (Balz 2003).

13.3 Evaluation of Displayed Information

The present study has been elaborated as part of the EU funded IN-SAFETY project. The objective of the research is to make road environments more self-explanatory by proposing a set of homogenized, comprehensive pictograms to substitute verbal messages on VMS systems. These should reduce the complexity of information and minimize the use of local languages. To achieve this objective, an easily readable font that could be used to supplement pictorial messages, as required, was also designed and tested.

The following key points were used to evaluate the different characteristics of traffic signs:

- Comprehensibility of displayed information elements
- Effect of animation on comprehensibility
- Amount of information relating to the number of information elements used
- Legibility of different type fonts used to supplement pictograms

13.4 Methods and Materials

A set of existing and proposed variants for 457 pictograms served as background material for the study. In line with the Vienna Convention on Road Signs and Signals, this set included three categories of normal road signs: regulatory signs, danger warning signs and informative signs (see Table 13.1).

Table 13.1 Final list of pictograms submitted for evaluation (Simlinger et al. 2008)

1	**Regulations**
1-1	**Lane allocations**
1-1-1	Lane control signals
1-1-2	Lane indication
1-2	**Carriageway guidance**
1-2-1	Closure ahead
1-2-1-1	Closure ahead: Road (similar meaning to 1-2-2)
1-2-1-2	Closure ahead: Pass/mountain road
1-2-1-3	Closure ahead: Tunnel
1-2-1-4	Closure ahead: Bridge
1-2-1-5	Closure ahead: X exit
1-2-2	Take next exit (similar meaning to 1-2-1-1)
1-2-3	Lane closure ahead
1-3	**Speed control**
1-3-3	Speed limit 10 km/h
1-3-4	Speed limit 20 km/h
1-3-5	Speed limit 30 km/h
1-3-6	Speed limit 40 km/h
1-3-7	Speed limit 50 km/h
1-3-8	Speed limit 60 km/h
1-3-9	Speed limit 70 km/h
1-3-10	Speed limit 80 km/h
1-3-11	Speed limit 90 km/h
1-3-12	Speed limit 100 km/h
1-3-13	Speed limit 110 km/h
1-3-14	Speed limit 120 km/h
1-3-15	Speed limit 130 km/h
1-4	**Regulations**
1-4-1	Regulations of use/dedicated lanes for target groups
1-4-1-1	Dedicated lanes: Buses
1-4-1-2	Dedicated lanes: Lorries
1-4-1-3	Car sharing lane/HOV lane
1-4-1-4	Dedicated lanes: Taxis
1-4-1-5	Dedicated lanes: Emergency vehicles
1-4-2	Smog/inversion weather/ environmental zone
1-4-3	No lorries at night
1-4-4	No lorries over x tonnes
1-4-5	Temporary prohibition: Dangerous goods
1-4-6	End of (temporary) restrictions/ limitations
1-4-7	Use/don't use hard shoulder (see 1-1-1)
1-4-9	No entry for vehicles having a mass exceeding x tonnes on one axle
1-4-10	Prohibited vehicular traffic in both directions
1-4-11	No entry
1-4-12	Overtaking prohibited
1-4-13	End of prohibition of overtaking
1-4-14	Overtaking prohibited for goods vehicles
1-4-15	End of prohibition of overtaking by goods vehicles
1-4-16	Driving less than x metres apart prohibited
1-4-17	Direction to be followed
1-4-18	Direction to be followed
2	**Danger warning**
2-1	**Danger warning (general)**
2-2	**Immediate warning on weather conditions**
2-2-1	Flooded road
2-2-2	Fog
2-2-3	Freezing fog
2-2-4	Snow/ice
2-2-5	Crosswind
2-2-6	Road surface temperature
2-2-7	Slippery road
2-3	**Immediate warning on traffic status – close ahead**
2-3-1	Traffic congestion/queue
2-3-2	Accident (has happened)
2-3-3	Vehicle broken down
2-3-4	Wrong way driver
2-3-5	Pedestrians on the road
2-3-6	Horses on the road
2-3-7	Cattle on the road
2-3-8	Deer on the road
2-3-8-1	Elk/reindeer on the road
2-3-11	Objects/obstacles on road
2-3-12	Two way traffic
2-3-13	Road uneven
2-3-14	Light signals
2-3-15	Road works
2-3-16	Swing bridge
3	**Informative**
3-1	**Advance warning**
3-1-1	Traffic status (see 2-3)
3-1-2	Weather condition (see 2-2)
3-1-3	Speed camera
3-2	**(Implicit) advice**
3-2-1	Rerouting
3-2-2	Last exit before
3-2-2-1	Last exit before toll station
3-2-2-1-1	Toll road ahead

(*continued*)

Table 13.1 (continued)

3-2-2-2	Last exit before pass/mountain road	3-3-2-18	Hospital
		3-3-2-19	Restaurant
3-2-2-3	Last exit before tunnel	3-3-2-20	WC/toilet
3-2-2-4	Last exit before temporarily closed tunnel	3-3-3	Parking space available
		3-3-4	Emergency phone
3-2-2-5	Last exit before bridge	3-3-5	Emergency phone number
3-2-3	Exit after next exit closed	3-3-6	Snow chains mounting area
3-2-4	Fog speed control	3-3-6-1	Snow chains compulsory
3-2-5	Filling station	3-3-7	Length/distance
3-3	**Driver comfort**	**4**	**Miscellaneous**
3-3-1	Temporarily free lane ahead	4-1	Direction (see 1-4-17/1-4-18)
3-3-2	Services	4-2	Follow (see 1-4-17/1-4-18)
3-3-2-1	Parking facilities	4-3	Reachable
3-3-2-2	Park and ride	4-4	Fines doubled
3-3-2-3	Tram	4-5	Switch off engine if congestion persists
3-3-2-4	Ferry boat		
3-3-2-5	Sport events	4-6	Switch on hazard warning lights
3-3-2-6	Fair	4-7	Motorway entry ramp/junction
3-3-2-7	Picnic/rest area	4-8	Motorway exit
3-3-2-8	Children's play area/playground	4-9	Height control
3-3-2-9	Internet	4-10	Truck-to-rail terminal
3-3-2-10	Caravan site	4-11	Motorail station
3-3-2-11	Mobile home	4-12	City centre
3-3-2-12	Information	4-13	Compulsory direction for lorries to check point
3-3-2-13	Camping site		
3-3-2-14	Refreshments or cafeteria	4-14	Peage/toll (see 3-2-2-1-1)
3-3-2-15	Hotel or motel	4-15	Underground trains depart every x minutes
3-3-2-16	Drinking water		
3-3-2-17	Full accessibility/toilets accessible		

A total of 2,977 symbol/pictogram variants were subsequently developed in an iterative testing and redesign process, in line with the ISO 9186 "Test methods for judged comprehensibility and for comprehension" quality standard (ISO 2001). These methods are used to evaluate the validity of redesigned and new pictograms. Designs were drafted and submitted for testing and evaluation, resulting in a reduction in variants and an increase in insights on how to improve the remaining pictograms.

In line with ISO 9186:2001, the results for each participating country were listed and tabulated; this list of responses was also used to resolve anomalies in the results from the different countries. The responses for each test were then categorized by three independent assessors into seven standard categories. Based on these tables, the mean and median response values were calculated for each variant and each country. The total scores were averages, computed from the values of all participating countries. The variant with the highest overall score is considered as the most comprehensible variant.

13.5 Empirical Studies

Four empirical studies were coordinated by the Danube University Krems and conducted in the Czech Republic, Hungary, Spain and Austria with a total of 2,667 participants.

13.5.1 Comprehensibility Judgement Test

The (first) comprehensibility judgement test was conducted in all four countries (Siebenhandl et al. 2007a, b). The aim of this test was to select the best of a wide range of variants for further analysis, "to determine the variants judged highest on comprehensibility" (ISO 2001). Studies by Zwaga (1989) support the validity of this procedure for identifying promising variants within a larger set.

The test material used in the comprehensibility judgement test was presented in the form of test booklets. As shown in Fig. 13.1 below the name of the symbol/ pictogram, its function, and field of application appeared in the centre of each page. Each participant was asked to judge the comprehensibility of each variant by indicating the percentage of the population he/she would expect to understand its meaning.

In total, 243 variants of 33 pictograms were submitted for testing. Fifty-six variants were judged as highly comprehensible. In accordance with the ISO evaluation procedure, this sample was then considered for submission to the subsequent series of tests (Fig. 13.2).

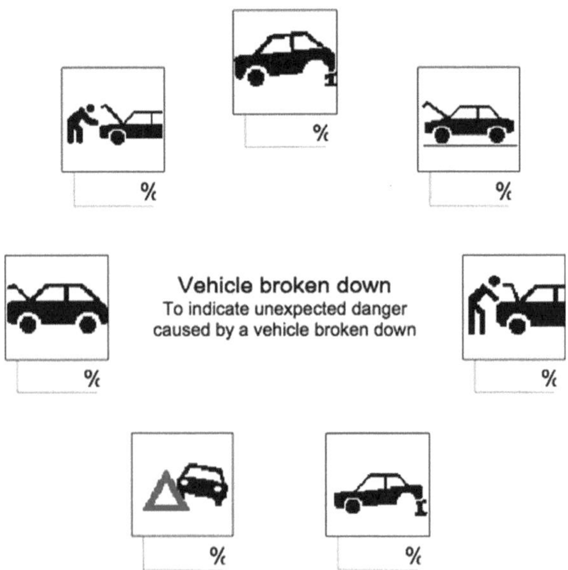

Fig. 13.1 Comprehensibility judgement test sample page (Brugger 2006)

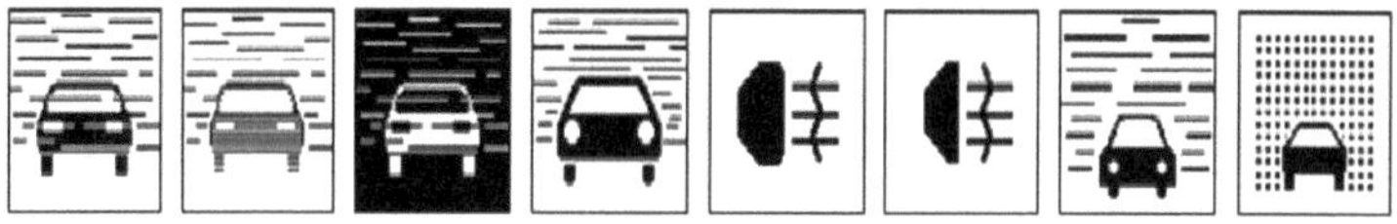

Variants sorted by mean estimate		1	2	3	4	5	6	7	8
Austria	Median	70,0	60,0	50,0	40,0	20,0	20,0	20,0	3,5
	Mean	58,1	54,6	46,8	38,2	34,3	33,2	28,2	10,5
Czech	Median	50,0	50,0	30,0	30,0	40,0	35,0	20,0	10,0
	Mean	50,8	52,0	34,9	36,2	39,3	38,1	25,5	12,2
Hungary	Median	50,0	52,5	30,0	45,0	30,0	30,0	35,0	20,0
	Mean	51,4	53,5	37,1	39,0	37,3	38,0	37,6	31,9
Spain	Median	70,0	67,5	60,0	50,0	30,0	20,0	42,5	20,0
	Mean	60,5	57,9	53,8	45,4	33,9	31,8	42,9	27,3
Total	Median	60,0	57,5	42,5	41,3	30,0	26,3	29,4	13,4
	Mean	55,2	54,5	43,2	39,7	36,2	35,3	33,6	20,5

Fig. 13.2 Evaluated variants for "fog" (Brugger 2006)

13.5.2 Comprehension Test

The comprehensibility test was followed by a comprehension test in three countries (Austria, Czech Republic and Hungary) with 604 participants (Siebenhandl et al. 2007c). The aim of this test was to evaluate the comprehension rate of the pictograms. A total of 84 variants were tested for 33 pictograms.

The test material used in the comprehension test was again provided in the form of test booklets. The signs to be tested were printed in the centre of each page, with a line provided below for the tester's response. Each participant had to judge a total of 21 symbols.

The comprehension test began with a verbal explanation of the project and the aim of the tests, after which the participants were shown the test booklet. The verbal explanation included information on the general context in which the participants could expect to see the symbol (Fig. 13.3).

> We are studying the comprehensibility of symbols used on highways. Please write down under the symbol what you think it means and which action you would take in response.

Based on the test results, a total of 22 pictograms were determined to be comprehensible enough to be recommended for use. These included informative pictograms indicating elemental objects ("ferry boat", "mobile home" or "deer on road") and pictograms used in public places, e.g. for a "children's playground" (see Fig. 13.4). The answers provided by the test participants and categorized as "wrong" were most useful and significant in terms of the (re)design process and reassessing existing hypotheses.

As hypothesized, pictograms of a more symbolic nature proved to be less comprehensible. For example, three variants of the pictogram "city centre" were tested all showing circles with different line widths, but without displaying a city name (see Fig. 13.5). All three variants achieved a low comprehensibility score

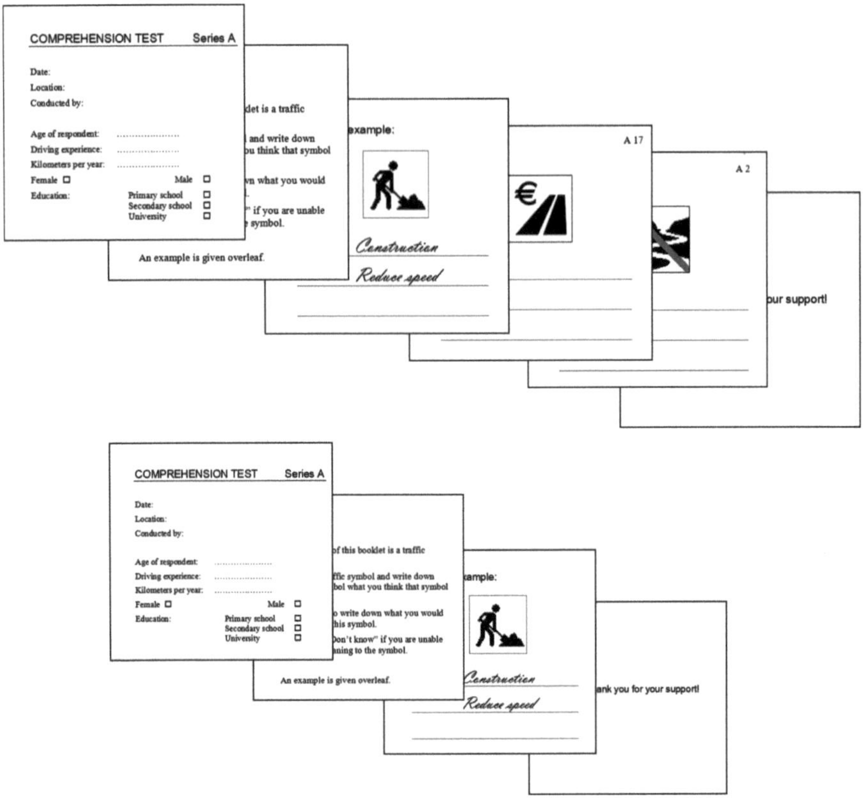

Fig. 13.3 Preparation of the test booklets (Siebenhandl et al. 2007d)

(around 10%). This leads to the assumption that unfamiliar symbolic pictograms are difficult to comprehend, especially when shown without a context (in this case the name of the city).

Pictograms indicating complex topics (such as "obstacles on the road" or "oncoming illegal traffic") achieved the lowest comprehension rates in the test series (Figs. 13.6 and 13.7).

13.5.3 Comprehension Test Using Animated Pictograms

The comprehension test described above concluded the initial static information optimisation process. To also address the challenges posed by freely programmable VMS systems (which can display traffic information using animated pictograms), the project then went on to examine the design of such animated pictograms.

Fig. 13.4 Pictogram samples: "children's playground", "mobile home", "ferry boat", "deer on road" (Simlinger et al. 2008)

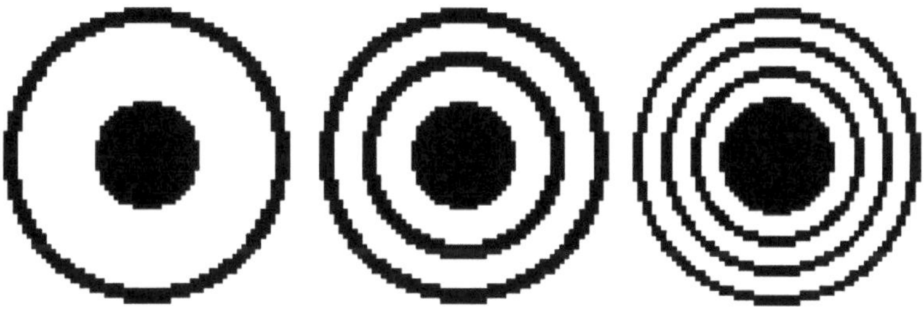

Fig. 13.5 Pictogram samples: three variants of "city centre" (Siebenhandl et al. 2007d)

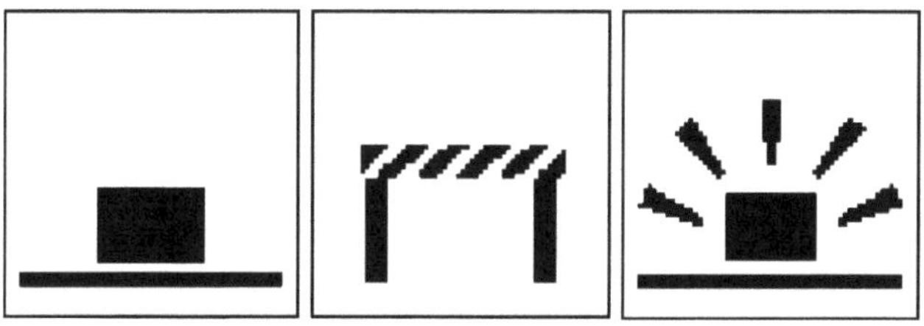

Fig. 13.6 Pictogram samples: three variants of "obstacles on the road" (Siebenhandl et al. 2007d)

Fig. 13.7 Pictogram samples: four variants of "oncoming illegal traffic" (Siebenhandl et al. 2007d)

Fig. 13.8 Pictogram samples: "oncoming illegal traffic" (Siebenhandl et al. 2007e)

The intention here was to investigate the potential of animated information for faster recognition and better understanding. It was hypothesized that pictograms indicating objects/subjects in motion (such as "driver going the wrong way") and those indicating activities (e.g. "switch off engine") might attain higher comprehensibility scores.

Two types of animation were studied:

- Static pictograms with superimposed flashing danger or prohibition elements
- Animated pictograms, consisting of several picture frames shown in rapid succession to create a film to transmit messages, which cannot be delivered using static symbols

The aim of this experiment was to identify the influence animation might have on the level of comprehension of the pictograms evaluated in the previous experiments.

Animated versions of 13 specific symbols/pictograms were prepared and tested using an animated pictogram comprehension test (CAT) in two countries (Austria and Czech Republic) with 308 participants (Siebenhandl et al. 2007e). Whilst the preceding tests were carried out using "paper and pencil", the pictograms in the CAT were projected on a screen. Participants were instructed to imagine that they were driving along a highway, where they would be faced with a number of (graphic) signs. They were requested to indicate what they thought these signs meant. Their reactions to the signs were also noted.

Only a few of the tested symbols/pictograms showed an improvement in comprehensibility score over the results of the static pictogram comprehension test. In three cases (see Figs. 13.8–13.10), the test results hit the defined benchmark and showed a significant improvement over their static counterparts. This leads to the assumption that in rare cases (those indicating activities or showing subjects in

Fig. 13.9 Pictogram samples: "vehicle broken down" (Siebenhandl et al. 2007e)

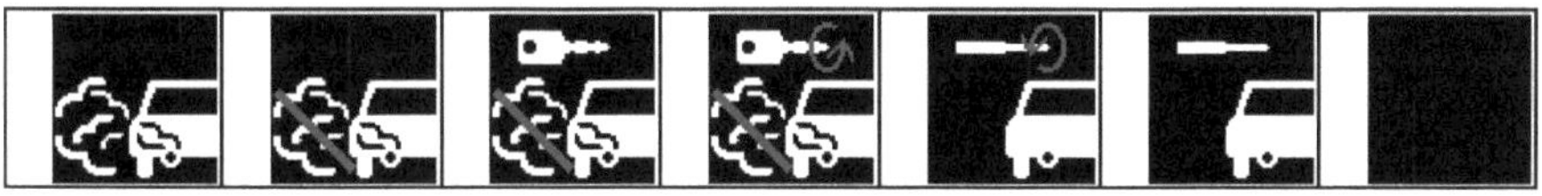

Fig. 13.10 Pictogram samples: "switch off engine" (Siebenhandl et al. 2007e)

Fig. 13.11 Sample of combined pictograms (Siebenhandl et al. 2007f)

motion), animation can facilitate the correct comprehension of symbols/pictograms. Consequently, the study does not support the general use of animated symbols/pictograms.

13.5.4 Content Structure Test

In a final step, a range of different symbols/pictograms (both animated and static), Vienna Convention traffic signs and text information were combined and tested with respect to their content structure (CST) (Siebenhandl et al. 2007f). The aim of this test was to evaluate the comprehensibility of the combined information by exploring the structure of the content. Special attention was given to the impact of the number of information elements involved. Selected combinations of pictograms and/or text were projected onto a screen at a simulated driving speed of 100 km/h. Forty-four information element combinations were evaluated by a total of 291 participants in two countries (Austria and Czech Republic). The participants were instructed to imagine that they were taking a trip to three different destinations, where they would encounter graphic symbols/pictograms. They were asked to indicate the meaning of the combined symbols and their reaction in response to these symbols (Figs. 13.11 and 13.12).

Fig. 13.12 Sample of combined pictograms (Siebenhandl et al. 2007f)

13.5.5 Results

The main results of these SER related studies can be summarised per element tested, as follows.

13.5.5.1 Comprehensibility of Displayed Information Elements

With regard to the many conditions governing the fast recognition and correct understanding of symbols/pictograms, the test results indicate correlations between the pictograms and information that has already been learned. As hypothesized, clear and simple symbols are understood more quickly than their more detailed counterparts. Pictograms of a more symbolic nature proved less easy to understand, leading to the assumption that symbolic pictograms, which are shown without context (e.g. "city centre"), are difficult to comprehend.

Traffic signs usually have an advantage over newly introduced (information) pictograms. Since the meaning of some signs (both symbol and pictogram based) seems very difficult to understand, the results of our study indicated that these should be advertised prior to use, to enable consumers to learn them.

Three classes of symbols/pictograms were defined in the Table 13.2.

13.5.5.2 Effect of Animation on Comprehensibility

The evaluation showed that animated content heightens alertness. In rare cases, it can also facilitate the correct understanding of symbols/pictograms. However, animation more frequently causes irritation and distracts motorists from other potentially more important information. Consequently, the study does not support the general use of animated symbols/pictograms.

13.5.5.3 Amount of Information Related to the Number of Information Elements Used

The evaluation of the number of information elements used showed that, in practice, a maximum of four information elements can be used in optimal

Table 13.2 Classes of information elements (Simlinger et al. 2008)

Class 1	All Vienna Convention signs required for messages on VMS, with the exception of those which achieved correct understanding scores of less than 88% in the comprehension tests It can basically be assumed that most – if not all – Vienna Convention signs are well understood, since drivers have to learn them at driving school Symbols/pictograms not regulated by the Vienna Convention, which yielded correct understanding scores of 88% and above in the comprehension tests
Class 2	Vienna Convention signs which yielded correct understanding scores of between 77 and 88% in the comprehension tests Symbols/pictograms not regulated by the Vienna Convention, which have been accepted either after achieving convincing scores when tested for judged comprehensibility or which yielded comprehension scores of between 77 and 88%
Class 3	Vienna Convention signs which yielded correct understanding scores of between 66 and 77% in the comprehension tests Symbols/pictograms not regulated by the Vienna Convention, which yielded comprehension scores of between 66 and 77%. Some of these symbols/pictograms have only been considered with the provision that they are subsequently taught in driving schools and are advertised widely to induce a learning process among motorists should they come into force

environmental conditions without placing an additional workload on the motorist. This however is in contradiction with relevant results of the project "TRaffic OPtimisation by the Integration of Information and Control" (TROPIC 1998), which recommends that "only short messages of up to four units of information" should be displayed on VMS on roads with speed limits above 110 km/h.

13.6 Presenting Verbal Information

13.6.1 Scope

While developing comprehensible pictograms and, thus, minimizing the use of different local languages leads in general to more self-explanatory roads, verbal messages and written text cannot, however, be substituted completely.

Traditional traffic signs – and the content of modern VMS systems – often only gain their full information value and precise meaning through the inclusion of additional written remarks or verbal extensions (such as place names, chronological restrictions or other context related specifications). This makes the legibility of textual information vitally important (Dewar et al. 1997); in particular, the messages displayed on motorway VMS's, which are frequently passed at high speed and consequently have to offer maximum legibility and ease of access.

The three most influential types of European traffic fonts (of the 28 types identified) were subjected to an extensive evaluation: the so-called "Transport" font ("Transport D" and "Transport 360"; GB), the "RWS" font ("RWS Ee VL" and "ANWB Ee"; NL) and the "DIN" standard font ("DIN-Mittelschrift" and "MITT2R"; DE, AT). Each of these three fonts was tested in two forms: a "normal", printed version for static signs and a "VMS" version, for rasterised matrix displays.

The aim of an empirical study was to compare the legibility of the new prototype "TERN" typeface (Trans European Road Network) with that of the three most influential European traffic fonts. In a subsequent step, typeface design experts compared the different fonts, on a character by character basis, to determine their disadvantages/advantages in terms of legibility. Equipped with this knowledge, the "TERN" prototype typeface was subjected to a new design process and an optimised version was produced.

13.6.2 Experiment

The configuration of the equipment used in the typeface experiment was similar to the one used in other visual tests, with the test person standing in front of a screen and reading aloud what he/she is able to perceive, while an observer records and checks the answers. Participants in this experiment were shown different combinations of characters in varying typefaces on a 15″ notebook screen. Six letters were shown on each page, and each series of tests comprised 50–100 pages, which had to be read by participants at three different distances. These distances correspond to different levels of visual acuity and, therefore, to different levels of "visual impairment". For example, the letter "e" in small letters was 7.25 mm high for a viewing distance of 5.5 m (visual acuity 1.0). This corresponds to a visual acuity of 0.65 at 7.4 m viewing distance and 0.5 at 8.3 m. Font sizes were calibrated to ensure fair testing (see Figs. 13.13 and 13.14).

In this experiment, a total of 150 participants from varying demographic backgrounds and with different levels of driving experience were asked to read aloud the randomly displayed characters.

Fig. 13.13 Calibration of typefaces for normal displays

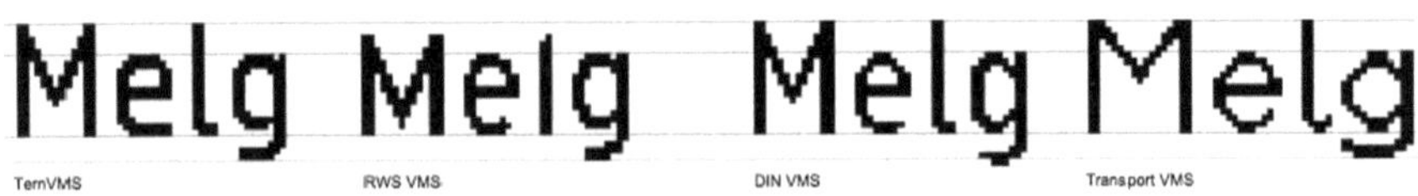

Fig. 13.14 Calibration of 24-pixel typefaces for VMS displays

13.6.3 Results

A comparison of the one newly designed and three existing traffic sign fonts mentioned above showed some remarkable differences with regard to their general legibility. An empirically based legibility ranking was established for the extended testing conditions of *impaired visibility* and *dual purpose display mode* (*normal* and *VMS*): for typefaces displayed on VMS systems, a general decline in legibility was revealed from the *Transport* font to the *TERN*, *DIN* and *RWS* fonts (see Figs. 13.15 and 13.16). The results were similar for normal displays, with the *Transport* and *TERN* fonts both proving more legible than the other two fonts.

In addition to a general analysis of legibility, the ability of the fonts to compensate for typical reading errors caused by the "usual suspects" (Spencer 1969), i.e. characters which are commonly confused (e.g. the number 8 and the letter B), was studied (see Fig. 13.17). A comparison of the error rates for specific characters in all four fonts delivered many concrete design and optimisation recommendations for the new TERN font (see Fig. 13.18).

As expected, the results indicate a general preference for wide characters. Although wide characters perform better in terms of legibility, road operators often opt for more condensed fonts for practical reasons. The optimised version of the TERN font (shown in Fig. 13.18 for the upper case letter "D") both reflects the survey results and addresses this demand for redesign.

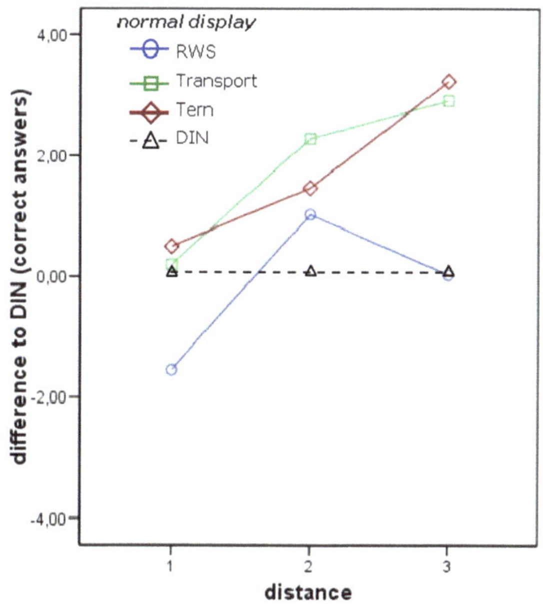

Fig. 13.15 Normal display; overall comparison of three test fonts using the *DIN* font as reference font; frequency of correct answers (test fonts minus *DIN*); distance 1 = 5.5 m, distance 2 = 7.4 m and distance 3 = 8.3 m (Smuc et al. 2007)

Fig. 13.16 Variable message sign display; overall comparison of three test fonts; frequency of correct answers (test fonts minus *DIN*) (Smuc et al. 2007)

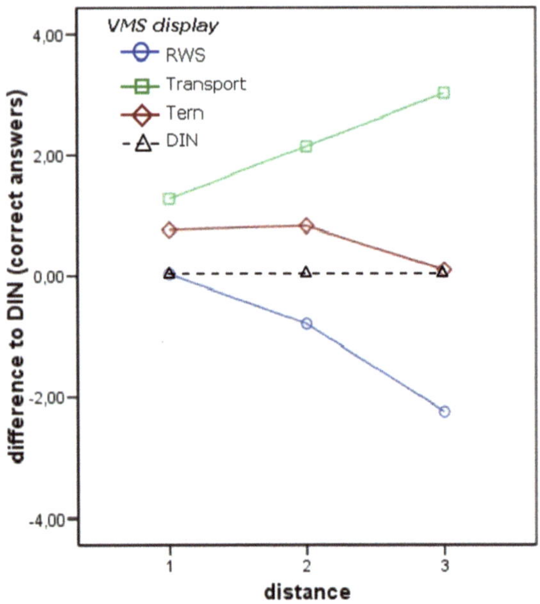

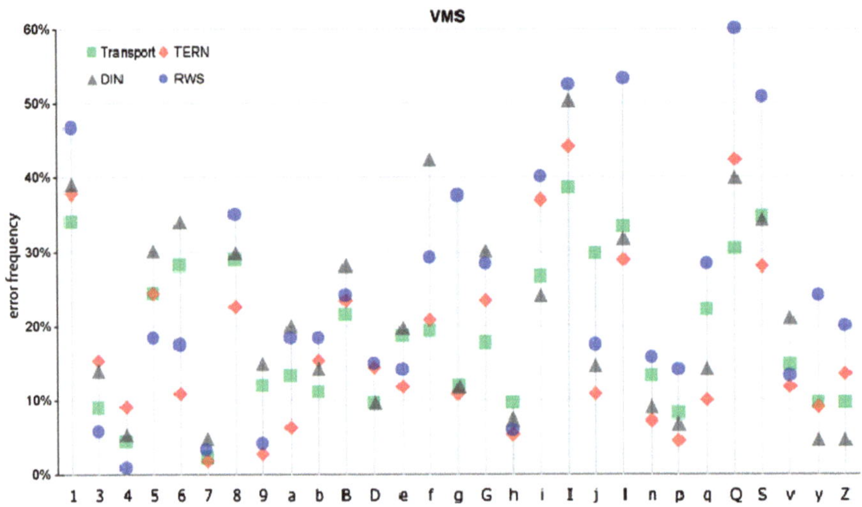

Fig. 13.17 "Usual suspect's" error frequency for the TERN (first draft) and RWS, Transport and DIN fonts (Smuc et al. 2007)

Transport VMS	*RWS VMS*	*DIN VMS*	*TERN VMS*	*New TERN VMS*

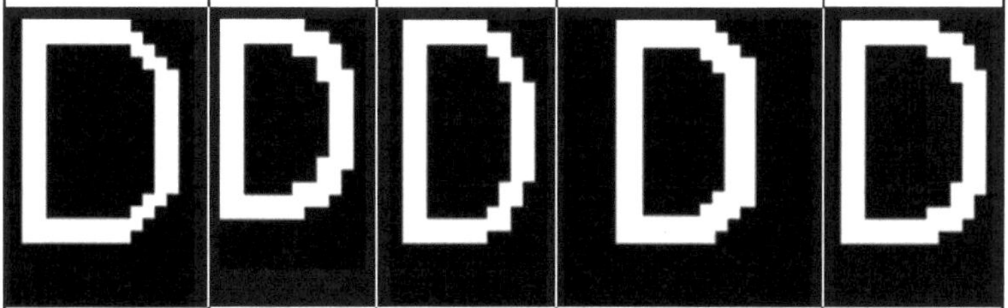

10.37%	13.33%	13.93%	14.41%	
Analysis	With this character, there is a clear preference for a wide font. It would also seem preferable not to design the rounded edge like part of a circle; its bend should instead show a more horizontal motion or be straight in the centre.			

Fig. 13.18 Average percentage of subjects who confused this character with another (including errors resulting from the use of upper and lower cases) (Smuc et al. 2007)

13.7 Evaluating SER Scenarios

Within IN-SAFETY project, a wide range of symbols/pictograms and a new font design were tested for understanding and early recognition in an experimental, iterative evaluation process. The pictograms which achieved the highest comprehensibility scores and the new TERN font design were published in the "Proposal on unified pictograms, keywords, bilingual verbal messages and typefaces for VMS in the Trans European Road Network (TERN)" (Simlinger et al. 2008).

Integrating user participation into the design process proved a useful method of identifying shortcomings in the existing pictogram/font designs, generating new ideas/hypotheses and reassessing existing assumptions. The interdisciplinary involvement of designers, psychologists and traffic engineers was effective and improved the process.

When it comes to enhancing road safety information systems, the study has led to the following conclusions:

- There is a strong need for a harmonisation of pictograms and traffic information signs.
- The more complex the pictogram/symbol, the less understandable it is.
- A maximum of four information elements (pictograms/symbols) should be displayed even in optimal conditions.

- Animated content in general heightens alertness, but only facilitates the correct comprehension of symbols/pictograms in rare cases.
- An evaluation of the typefaces used should be mandatory in the development of new traffic systems and environments. The need for a uniform typeface standard should be emphasised, given the significant differences in the legibility of the existing typefaces.
- Further research should include in situ investigations (e.g. "on the road" observation of driver behaviour), assess legibility beyond a character level (words) and examine the impact of systematically varying specific typographic design elements (graphemic cues).

The experiments carried out in this study also revealed a number of issues which should be addressed in further research. Firstly, the symbols/pictograms which failed to reach the defined benchmark for inclusion in the list of recommended symbols/pictograms should be redesigned, refined and retested. Some of them (e.g. "Rerouting arrow") have complex meanings and require further investigations.

Secondly, some of the methodological issues faced in this study could be improved. As recommended in ISO 9186:2001, total scores were averages computed from the values of all participating countries, a method which does not allow the assessment of intercultural differences. The evaluation of combinations of pictograms (through the content structure test) was not regulated in the ISO standard, and the method used did not permit a detailed evaluation of the differences between the combined images.

Thirdly, advanced test techniques should be developed to assess the effectiveness of composite messages. Considering how important it is for motorists to react correctly to the information displayed, advanced test techniques could combine message interpretation and driver performance by employing a driving simulator in a first step. The results and subsequent conclusions should be replicated in real life test conditions on the road, to measure the impact on driving behaviour. This would, for instance, provide researchers with the opportunity to determine whether an increase in driving speed influences the comprehensibility of pictograms on VMS displays.

References

W. Balz, Dynamische Verkehrsinformationstafeln – Einsatzmöglichkeiten und –grenzen. Strassenverkehrstechnik **47**(5), 250–253 (2003)

Ch. Brugger, Comprehensibility judgement test. In-Safety Deliverable, 2006

D. Dewar, K. Kline, F. Scheiber, A. Swanson, *Symbol Signing Design for Older Drivers* (Turner-Fairbank Highway Research Center, McLean, 1997)

P.M. Garvey, *Synthesis on the Legibility of Variable Message Signing (VMS) for Readers with Vision Loss* (Access Board, Washington, DC, 2002)

ISO, International Standardization Organization, *ISO 9186, Graphical Symbols – Test Methods for Judged Comprehensibility and for Comprehension* (ISO, Geneva, 2001)

M. Krampen, *Signs and Symbols in Graphic Communication*, Design Quarterly 62 (Walker Art Center, Minneapolis, 1965), p. 12
J. Luoma, P. Rämä, Comprehension of pictograms for variable message signs. Traffic Eng. Contr. **42**(2001), 53–58 (2001)
L.J. Pignataro, *Traffic Engineering: Theory and Practice* (Prentice-Hall, Englewood Cliffs, 1973)
J.L. Pline (ed.), Traffic engineering handbook, in *Institute of Transportation Engineers*, 4th edn. (Prentice-Hall, Englewood Cliffs, 1992)
K. Siebenhandl, S. Egger, P. Simlinger, J. Weinberger, J. Vasek, Results of the test of content structure of pictorial and verbal messages on VMS conducted in Austria and Czech Republic. Report In-Safety, 50671, 07/12/2007, 2007a
K. Siebenhandl, H. Risku, Ch. Brugger, P. Simlinger, Evaluating the comprehensibility of visualized information for the trans-European roadnetwork (TERN), in *Proceedings 20th International Technical Conference on the Enhanced Safety of Vehicles*, Lyon, 2007b
K. Siebenhandl, H. Risku, Ch. Brugger, P. Simlinger, Evaluating the comprehensibility of visualized information for the Trans European Road Network (TERN) as part of the EU Project IN-SAFETY: INfrastructure and SAFETY, in *Proceedings of the International Conference: RSS 2007*, Road Safety& Simulation, Rome, 2007c
K. Siebenhandl, Ch. Brugger, P. Simlinger, S. Egger, P. Holló, J. Weinberger, J. Vasek, Results of the comprehension tests on pictograms conducted in Austria, Czech Republic and Hungary. IN-SAFETY Report, 2007d
K. Siebenhandl, Ch. Brugger, P. Simlinger, S. Egger, J. Weinberger, J. Vasek, Results of the comprehension tests on animated pictograms conducted in Austria and Czech Republic. IN-SAFETY Report, 2007e
K. Siebenhandl, Ch. Brugger, P. Simlinger, S. Egger, J. Weinberger, J. Vasek, Results on content structure of pictorial messages on VMS conducted in Austria and Czech Republic. IN-SAFETY Report, 2007f
P. Simlinger, S. Egger, C. Galinski, Proposal on unified pictograms, keywords, bilingual verbal messages and typefaces for VMS in the TERN. Deliverable 2.3, C.N. 506716, In-Safety project, January 2008, http://www.insafety-eu.org/documents/IN-SAFETY_Deliverable_2.3.pdf
M. Smuc, F. Windhager, K. Siebenhandl, S. Egger, Impaired visibility typeface test report. IN-SAFETY Report, 2007
H. Spencer, The visible word, Lund Humphries in association with the Royal College of Art, 1969
L. Staplin, K. Ball, D. Park, L.E. Decina, K.H. Lococo, K.W. Gish, B. Kotwal, *Synthesis of Human Factors Research on Older Drivers and Highway Safety* (Turner-Fairbank Highway Research Center, McLean, 1998)
J. Theeuwes, H. Godthelp, Self-explaining roads. Safety Sci. **19**, 217–225 (1995)
TROPIC, TRaffic OPtimisation by the Intergation of information and Control, trial phase: text and combined message reference manual, 4.3 Major findings of WP12, 1998, p. 13
H.J. Zwaga, Comprehensibility estimates of public information symbols; their validity and use, in *Proceedings of the Human Factors Society 33rd Annual Meeting*, The Human Factors Society, Santa Monica, 1989, pp. 979–983.

Chapter 14
A Sign Equals Thousand Words

Consistency of Traffic/Road Signs and Verbal Messages

Christian Galinski

14.1 Variations Among Road/Traffic Signs and Verbal Messages in Europe

The Convention on Road Signs and Signals (1968, commonly called "Vienna Convention"), which allows certain national variations of road/traffic signs, is implemented by means of national laws and/or regulations or (public as well as industry) standards with minor or larger modifications. Some of these legal and technical standards not only constitute an obstacle to pan-EU traffic information/control systems or to the interconnection of such systems, but they reduce above all user-friendliness with regard to self-explanatory roads (SER) and forgiving roads (FOR), thus augmenting the potential for accidents caused by distracted drivers.

If a driver crosses a country border (sometimes even the border of an autonomous region) on the highway in the EU today, s/he is more often than not suddenly confronted not only with a different language, but also with:

- Design variations of familiar road/traffic signs
- Different positions of road/traffic signs
- Sometimes new road/traffic signs (to her/him)
- Unfamiliar additional signs (added to familiar or not familiar road/traffic signs)
- New VMS boards (to her/him), etc.

Some of which may be justified, while others are not or not fully. In any case, they require a learning process, which may distract the driver from concentrating on road and traffic – thus increasing the possibility of accidents.

In practice many countries have registered national exceptions and peculiarities to the Vienna Convention, which leads to a multitude of variations of representations of meaning in traffic/road signage. In order to cope with these variations, a systematic and comprehensive methodology has been developed, based on

C. Galinski
International Information Centre for Terminology (Infoterm), Vienna, Austria
e-mail: cgalinski@infoterm.org

E. Bekiaris et al. (eds.), *Infrastructure and Safety in a Collaborative World*,
DOI 10.1007/978-3-642-18372-0_14,

existing international terminology standards. Hereafter, this methodology is referred to as ***"Europeanisms"***, i.e., verbal messages or verbal message elements, which have the potential to be harmonized EU-wide.

Some of these Europeanisms can either replace or verbalize a graphic symbol. Thus, some graphic or pictogrammatic symbols (e.g. for "bus") can also become the object of harmonization. In this respect, it must be emphasized that Europe-wide harmonization does not necessarily conflict with "multilingualism", since the individual languages will still be used in most verbal messages – especially those of a more complex nature, whether in navigation systems or in other types of car-driver information and communication systems.

14.2 Methodology for Key Meanings and Bi-/Multilingual Messages

In view of:

- The need for more integration of road management systems (necessitating also a higher degree of content interoperability)
- Latest ICT (Information and Communication Technology) developments also in traffic telematics, increasingly requiring the integration of different forms of multimodality, as well as
- The fact that Europe is multilingual and needs multilingual approaches in the development of application systems

a methodology for establishing a sound basis for a systematization of representations of meanings in road/traffic signage was developed, based on experiences with terminology standardization over the last decades. This methodology is geared towards identifying candidates for European-wide harmonization, in order to reduce unnecessary variations, on the one hand, and towards obtaining more flexibility for developing road/traffic signs implemented on existing and future technologies, on the other hand.

Verbal representations (in the form of terms, facts, statements, indications, requests, demands, etc.) can occur in combination with or in addition to or independently from traffic signs. A total of about 40 traffic sign samples were selected, comprising about 50 verbal representations. All verbal representations (i.e., text messages in static traffic signs and variable message signs – including variants in the same language) of the same meaning were recorded in one record each in all 23 official EU working languages of the 25 EU countries up to 2007 – including Germanic, Romance, Slavic languages and Greek, with the respective character sets. In addition to linguistic information, graphical information was recorded in all cases where applicable. If no equivalent existed in a given language, translations were proposed – duly taking into account state-of-the-art localization methods and intercultural aspects. These data were recorded and maintained in a state-of-the-art terminology management system (TMS), adapted for that purpose.

Excluded from the investigation was the use of:

- Lights (traffic lights, flashing lights, car lights, etc.)
- "Arrows", which obviously need to be harmonized (within traffic systems as well as between traffic systems and other environments, such as train stations, airports, etc.)
- Geographical names

Concerning arrows, the elaboration of an application-oriented arrow design methodology, in coordination with other environments where arrows are used (airports, hospitals, train stations, etc.), was recommended.

The methodology outlined hereunder, is comprised of, in addition to the systematically compiled ***basic concepts***:

- An investigation into the categorization, classification and typology of road/traffic signs and messages
- The formulation of the concept of "Europeanisms"
- An approach to extend VMSs towards in-vehicle information and communication
- The IN-SAFETY data model for verbal messages
- A proposal for a systematic distributed database management scheme
- Standardization issues

and conclusions as well as draft recommendations drafted in the respective reports.

14.3 Theoretical Foundation and Basic Concepts

In order to arrive at a generic data model (both in terms of human communication semantics and the formal semantics of computer science), a number of fundamental concepts in the field of terminology had to be re-defined to meet the requirements of SER enhancement implementation scenarios. This generic data model seamlessly interfaces with data models in product data management (e.g. for traffic signs regarded as physical products, which have to be technically described, produced, traded, etc.) and other areas of content management. In IN-SAFETY these concepts were anticipated as in Table 14.1.

14.4 Categorization, Classification and Typology of Road/Traffic Signs and Messages

Differences in categorizations, classification or typologies of road/traffic signs pose – as well-known in the field of computerized terminology – serious problems in the processing and maintenance of data related to road/traffic signage. Variation at higher granularity, such as of the road/traffic signs themselves or their "morphological" elements, are possibly easier to be dealt with.

Table 14.1 Modification of fundamental concepts across different domains towards SER enhancement

Concept	Explanation and/or examples	Remarks
Meaning (expressed by presentation of meaning)	Different presentations of meaning: • On large panels down to smallest screens • By voice or other acoustic means • By haptic means • Via multimedia means are communicating meaning to the driver (or to the car acting on behalf of the driver)	Several fields of science have to be aligned and arrive at joint solutions to problems arising from the ever growing variation in the presentations of meaning due to the development and application of the ICTs
Semantics (in linguistics)	Study of the ways in which words, phrases, and sentences can have meaning	Has to be extended to comprise also non-verbal representations on traffic signs in general and on VMS in particular
Syntax (in linguistics)	Study of the rules, or "patterned relations" that govern the way the words in a sentence come together	Has to be extended to comprise also non-verbal representations on traffic signs in general and on VMSs in particular
Sign (in general) and symbol	The majority of traffic signs and some VMSs are complex signs, consisting also of different kinds of symbols (such as letter symbols, graphic or pictogrammatic symbols e.g. for "bus", "car", "horse")	In addition to a great variety of possible fixed or flexible combinations, the use of ICTs allows for different modalities (such as written and spoken) and languages
Object (also called subject, e.g. in ISO 7239)	Simple or complex situations (referring e.g. to traffic, environment, weather, geographical or other information)	Due to the development of the ICTs, more and more complex situations can be processed
Concept (called referent, e.g. in ISO 7239)	Expected driver's behaviour due to traffic or other situations	More and more complex tests have to be developed in order to be sure of the expected driver's behaviour
Modality	So far, comprised mainly of written and spoken representations of meaning	Has to be extended towards haptic and other kinds of non-verbal and non-visual symbols (in principle any modality can occur in the communication between driver and car and – to some degree – on traffic signs)
Representations (of meaning or concepts)	• Displayed messages of traffic signs, VMSs, additional signs, etc. • Morphologic elements of traffic signs or additional signs	Some of these representations can occur in different modalities and languages

(*continued*)

Table 14.1 (continued)

Concept	Explanation and/or examples	Remarks
	• Official names of traffic signs, VMSs, additional signs, morphological elements, etc. • "Popular" names of the above • Official rules or explanations of the above • Messages to the driver for a desired behaviour in verbal or other form (e.g. for in-vehicle information/ communication)	
Representation autonomy (data modelling in traffic sign databases)	Principle allowing each representation of the concept to be documented with all necessary data categories	Concept representations in traffic sign databases can be written or spoken, verbal, alphanumeric, graphic or pictogrammatic, haptic, acoustic, etc. or any combination thereof
Data modelling (in information system design)	Development of a model for formal semantics from the software developers point of view	Both a semiotic model from the point of view of content entities and from the point of view of formal semantics are needed, which must fit together

The earliest road signs gave directions; for example, the Romans erected stone columns throughout their empire giving the distance to Rome. In the Middle-Ages, multi-directional signs at intersections became common, giving directions to cities and towns. In modern times, traffic signs became more important with the development of automobiles. The basic patterns of most traffic signs were set at the 1908 International Road Congress in Rome.

Since language differences can create barriers to understanding, international signs, using symbols in place of words, have been developed in Europe and adopted in most countries of the world. Shape, size, colours and sign elements have been harmonized to quite an extent on international level. The ***Vienna Convention on Road Signs and Signals*** (see Vienna Convention), entering into force on 21 May 1977, defines eight categories of signs (according to Annex 1, Sections A–H):

A. ***Danger warning signs*** (Section A)
B. ***Regulatory signs***: signs intended to inform road-users of special obligations, restrictions or prohibitions with which they must comply; they are subdivided into:
   (i) ***Priority signs*** (Section B): signs indicating the order in which vehicles should pass intersection points
   (ii) ***Prohibitory or restrictive signs*** (Section C)

(iii) ***Mandatory signs*** (Section D)
(iv) ***Special regulation signs*** (Section E)

C. ***Informative signs***: signs intended to guide road-users while they are travelling or to provide them with other information which may be useful; they are subdivided into:
(i) ***Information, facilities or service signs*** (Section F)
(ii) ***Direction, position or indication signs*** (Section G)
(iii) ***Additional signs*** (Section H)

However, individual countries (or even regions/provinces/states) may categorize road signs in different ways, e.g. in Germany (s. StVO: Straßenverkehrsordnung/ German Convention on Road Signs and Signals), they are categorised as:

- Pictograms ("Sinnbilder")
- Danger waning signs ("Gefahrenzeichen")
- Signs giving orders ("Vorschriftzeichen" – including priority signs, prohibitory signs, mandatory signs and special regulation signs)
- Information and direction signs ("Richtzeichen" – except traffic management signs)
- Traffic management signs ("Verkehrslenkungstafeln")
- Traffic installations ("Verkehrseinrichtungen")
- Additional signs ("Zusatzzeichen")

or in the United States of America, where they are categorised as:

- Regulatory signs (covering warning signs and guide signs)
- Route marker signs (covering expressway signs and freeway signs)
- Informational signs (such as recreational and cultural interest signs)
- Emergency management signs (encompassing temporary traffic control signs, school signs, railroad and light rail signs, bicycle signs)

Concerning the composition of road/traffic signs and their verbal and non-verbal messages:

- ***Pictograms*** (German: "Sinnbilder") have been identified as:
  - Pictogrammatic–'morphologic' elements of traffic signs
  - Additional signs to traffic signs
  - In combination to other (verbal or graphic) additional signs
- ***Additional signs*** (German: "Zusatzzeichen") can consist of:
  - Alphanumeric symbols
  - Graphic symbols (e.g. arrows, etc.)
  - Pictogrammatic symbols (e.g. "truck")
  - Combinations thereof and with traffic signs

There are ***traffic signs with integrated content, including***:

- Pictograms or graphic symbols (as semiotic–'morphologic' elements)
- Alphanumeric information
- Combinations thereof

and others being ***supplemented by additional signs***, which contain:

- Pictogrammatic symbols, or
- Graphic symbols, or
- Alphanumeric information, or
- A combination thereof

Textual (i.e., alphanumeric, namely verbal or quasi-verbal) information being the central part of a traffic sign or being integrated in regular traffic signs or in their additional signs are:

- Emergency, Police, WC, etc. (+ TEL symbol + distance indication...)
- Names: London, Paris, etc.
- "EXIT", "STOP", give way 50 yds, etc.
- "One-way", "...-zone", beginning/end of ..., etc.
- "Slippery road" (if raining, if freezing, if dirty, etc.)
- Time indications: "Sundays and holidays", from 20 h to 06 h, etc.
- H (=bus stop in Germany), U12 (temporary or permanent re-routing)
- 100 m (meaning for a distance of 100 m or in 100 m – e.g. railway crossing)
- Other measurements, such as:
  - 5.5 t (gross weight), 8 t (axle weight), etc.
  - 2 m (width), 3.8 m (height), 10 m (length, distance, ...)
- (Speed:) 80 (=80 km/h) + time (period) indication
- (Degrees:) 10% (gradient road, dangerous hill), 0° (temperature), etc.

which may or may not be combined with graphic or pictogrammatic symbols.

So far, there are at most only steps towards a systematization of messages on signs. Ballardin et al. (2005) indicates that there may indeed be further potential for harmonizing and systematizing messages on additional signs by a combination of existing symbols (some of which could be taken from the signing of airports or train stations, etc.). This would, however, require a thorough investigation of the syntax of the messages to be conveyed.

14.5 Designations (and Different Kinds or Levels of Naming)

All road/traffic signs have names, and some may also be expressed in terms of expected drivers' behaviour. In multilingual databases these names and phrases (or micro-statements) cannot simply be translated, but rather need to be "localized".

The traffic signs (comprising integrated "morphologic" elements or not) and their additional signs (comprising integrated "morphologic" elements or not) may have:

- Simple (more or less self-explanatory) designations, such as: curve, warning, "STOP"

- Simple designations, such as 10% (here: road gradient), which, however, more often than not may mean something like "Steep downgrade – You should shift to a lower gear. The degree of the slope is shown"
- 'Difficult' legal designations (used in law) vs. popular names (used for instance in driving schools)

and may need a new short/concise and easy to understand name and/or explanation in real traffic situations – and especially for in-vehicle information and communication purposes.

In this context, verbal can mean both written verbal and spoken verbal, which in actual use could be literally different, since 'noise' (in the meaning of visual interferences) in written communication may be different from 'noise' (in the meaning of acoustic interferences) in spoken communication. Both have to be considered as 'equivalent/synonym' from the semantic point-of-view, even if their 'linguistic outer form' is quite different (potentially) in any language or language combination. Furthermore, the legal designation in one language may be perceived as 'difficult' by people of that language community, but quite simple and easy to understand in another language by people of that community. Popular names may exist in some languages, but not in others. This has a strong impact on data modelling and information design.

As a side-effect of these investigations, it became clearer how verbal messages on static sign boards and VMS displays could be optimized by duly considering:

- Harmonization
- Comprehension
- Multilinguality

and taking into account:

- Cultural aspects
- Localization methods
- Road equipment standards and national regulations
- Future necessities of car navigation systems

Some of the criteria for terminological optimization are:

- Transparency (morphological/semantic motivation)
- Consistency (i.e., consistent use of terms in all types of verbal messages)
- Appropriateness, namely:
  - Familiar to the reader (localization)
  - Don't cause confusion or insecurity
  - Have no negative connotations (neutral, politically correct)
- Linguistic economy (e.g. avoiding long terms)
- Derivability (e.g. *medicinal plant vs. herb* → *herbal, herbalist, ...*)
- Linguistic correctness
- Preference for native language (whenever appropriate)

14.6 Linguistic Variation in Europe and Bilingual Traffic Signs

Given the 28 EEA (European Economic Area – i.e., 25 EU member countries and 3 EFTA countries), there were at the start of the IN-SAFETY Project:

- 22 EU (European Union) Members' official working languages (incl. Irish, and Luxembourgish)
- Plus additional 2 EEA official languages (Icelandic and Norwegian), amounting to 35 official language situations (in combination with road/traffic sign variants = ***locales***), if 4 variants each of German (incl. Schwyzerdütsch) and French, and 2 variants each of English, Italian, Dutch, Swedish and Greek are included
- Not to mention additional official *regional* languages, such as Catalan, Basque, Valencian, etc.

Not all of the above are realized in road/traffic signage – but potentially all of them could be or become used in road/traffic signs. But for increasingly integrated or interoperating road management and other systems the data model and system specifications must cope with all languages, their linguistic features and all sorts of representation of meaning.

Bi-/multilingual traffic signs and messages are due for:

- Language policy reasons
- Historical reasons
- Change of traffic signage at borders

Bilingual signing in Wales and elsewhere has caused traffic engineers to inquire into the safety ramifications of providing verbal messages in multiple languages. As a result, some countries have opted to limit bilingual signing to dual-name signs near places of cultural importance (e.g. New Zealand), or to use it only in narrowly circumscribed areas, such as near borders or in designated language zones (e.g. the NAFTA countries).

Karhunen (1998) evaluated bilingual VMSs in Finnish and Swedish. Every message could be seen for 2 s in each language. The signs are empty for half a second between the messages. The results showed that drivers in general consider the display time of 2 s as being long enough. Additionally, 75% of the sample recalled and accepted the message signs. Concerning the cognitive demand, there is no difference between displaying variable message signs alternatively compared to displaying them simultaneously. Nonetheless, elderly drivers consider VMSs as more demanding than regular static message signs.

The TROPIC project Final Report 9 (1999) yielded results about message signs with or without redundant pictograms and translations. Bilingual messages can reduce drivers' ability to recall the message in their own language. Regarding visual distraction, it does not make a difference if bilingual messages are displayed consecutively or simultaneously. So finally, it is acceptable to display bilingual messages by turns of 2 s each. Redundant pictograms seem to be useful, because the most comprehensible message style is text with a pictogrammatic element.

Jamson et al. (among others 2001 and 2005) evaluated the effect of various bilingual VMS configurations on driver behavior. Results showed that both mono- and bilingual drivers can read (mono- and bilingual) message signs with no compensatory effects in their mean vehicle speed. But four-line bilingual message signs led to a decrease in mean speed. Different arrangements of signs, in this case the sequencing of signs, had no impact on factors associated with driving performance. Further investigations tackled the question of an additional separation line between the two used languages. But such a line made no difference with regard to driver performance. Finally, a four-line bilingual VMS with two lines of text in each language seems to be the best compromise, because it can be read by mono- and bilingual drivers as good as a monolingual two-line sign. In Jamson et al. (2001), four-line bilingual VMSs, and especially four-line monolingual VMSs, led to a decreased mean vehicle speed, which indicates an increase of the time needed to read the sign. But still four-line bilingual VMSs with two lines of text in each language are read almost in the same manner by both mono- and bilingual drivers as two-line monolingual signs. In Jamson et al. (2005) drivers significantly reduce their vehicle speed and increase their headway in front of a leading vehicle in order to read four-line monolingual and four-line bilingual message signs. This implies that they are also reading the irrelevant text on the bilingual VMS. The reading of one- and two-line monolingual signs and two-line bilingual signs led to no disruption in their driving behavior.

Clark (2005) conducted a study in which the message text was presented in turns of 4 s each in English and French. English speakers found the mono- and bilingual formats equally easy to read, the results from the drivers with French as the first language varied. The best results were obtained using a four-line display with two lines for both languages each.

Anttila et al. (2000) recorded the eye movement behavior of their subjects while driving. This makes the comparison of the visual demand imposed by different VMS configurations possible. The data shows that a VMS displaying alternating bilingual messages is no more visually demanding than a VMS displaying the same messages simultaneously. So far the comprehensibility of combinations of bi-/multilingual verbal messages with pictograms has not been tested.

The cited studies regarding bilingual VMSs lead to the following conclusions:

- It is no more demanding to display variable message signs alternatively than to display them simultaneously
- Elderly drivers consider VMSs as more demanding
- Messages should consist of less than six units of written text, if the displayed information is supposed to be recalled
- Four-line bilingual VMSs, comprised by two lines of text in each language displayed in turns, have been tested and validated as the most acceptable solution

The combination of bi-/multilingual messages and pictograms may require additional testing features.

14.7 Different Scripts and Transcription

Transliteration is an additional linguistic feature in connection with bilingual road/traffic signs. Given the fact that Greek is written in Greek characters and other countries using a non-Latin script joined the EU, some thought must be given on how to use transcription of words in these languages/scripts into Latin. For this purpose the confusing differences in simplified transcription schemes, target language oriented transcriptions (e.g. Russian in Latin letters for French readers), common transliterations (e.g. in newspapers) vs. standardized ones, must definitely be harmonized. In addition to pertinent ISO standards the ALA-LC Romanization Tables can be consulted on the Library of Congress Website.

The Vienna Convention (1968) stipulates under "Informative Signs" Art. 14: "2. The inscription of words on informative signs (ii) of Art. 5, para. 1 (c), in countries not using the Latin alphabet shall be both in the national language and in the form of a transliteration into the Latin alphabet reproducing as closely as possible the pronunciation in the national language. 3. In countries not using the Latin alphabet, the words in Latin characters may be entered either on the same sign as the words in the national language or on a repeat sign. 4. A sign shall not bear inscriptions in more than two languages." This stipulation should in essence also be applicable to the transliteration of verbal messages in Greek on traffic signs in Greece.

The Greek transliteration system, ELOT 743 (1982), might be most appropriate, but there are issues with certain diacritics either on Greek letters or on Latin letters used for transliteration. For present VMS displays the most simple/simplified transliteration has still to be used, which may mean deprecation of some – possibly disambiguating – features (such as diacritics on Greek letters and on Latin letters for Greek in transliterated form). This could also lead to difficulties in future automatic pronunciation of messages to car drivers, whether from the Greek original spelling or from the transliteration, to have a correct pronunciation produced.

In the field of terminology it has been recognized that a situation, which seems to be governed by many language pair relations, can best be dealt with by means of a multilingual approach to data modelling and system design.

14.8 Data Model

If personalization features of interactive in-vehicle systems more or less fully comply with political, historical as well as other reasons for bilingual signage, preference should be given to in-vehicle representation of the message in a personalized form. This would also apply in the case that future VMS message displays become (fully) freely programmable. Thus, any information for the driver – in principle, maybe not necessarily in practice – could be communicated through:

- ***Multilingual information*** (comprising also language variants)
- ***Multimodal information*** (comprising also modes beyond written and spoken)
- ***Multimedia*** (going beyond a combination of visual, audio and video presentations)

This necessitates a sophisticated – but not necessarily complicated – data model, which probably can also accommodate requirements stemming from personalization, eAccessibility (or eInclusion for people with special needs). Such a data model would relieve technical devices from constraints and make technical communication very flexible, on the one hand, and may even support multi-channel output via many different types of devices, on the other hand. However, every kind of representation – whether written or spoken verbal representation, graphic/pictogrammatic or multimedia presentation, etc. – has its own inherent constraints (first of all in terms of human perception), which have to be taken into account in the data model.

On the basis of these considerations, IN-SAFETY specific "structured content" (at the level of lexical semantics) required the adaptation of the present terminology data model (instead of concept/entry – language section – term) as in Fig. 14.1.

The ***Concept Entry*** is a container for all information that pertains to a single traffic sign concept; therefore, this container should be repeatable for each concept entry being part of the data collection. It usually contains, for example, the graphic sign assigned to a concept, descriptive information like a sign classification, and administrative information concerning the concept. It can contain one or more locale sections depending on whether the data collection covers one or more national traffic sign regulations.

The ***Locale Section*** contains all the traffic sign information (of a given concept entry) that is used in a given country or geographical region (location). Usually it contains, for example, sign names, explanations or driver instructions, etc. associated with that location. The locale section must be repeated for every national or geographic region treated in the relevant concept entry.

The ***Representation Section*** contains information about the representation of the traffic sign. If more than one representation is possible in a given locale, the

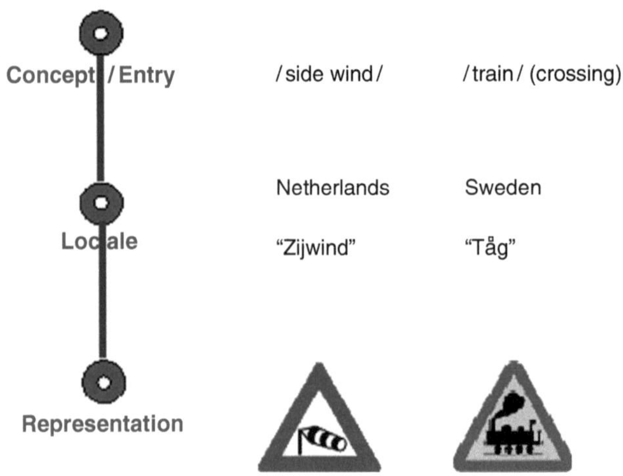

Fig. 14.1 The road/traffic sign meta-model for IN-SAFETY

representation section must be repeated. Usually the representation section contains a single representation of the traffic sign, as well as any other information (e.g. status, source, etc), associated with that representation.

14.9 Design for a Cluster of Repositories for IN-SAFETY Messages (CRIM)

Structured content, such as traffic sign information and VMS in different languages and with a considerable degree of variation, could be stored and maintained in distributed databases (repositories), forming a cluster of repositories for IN-SAFETY messages (CRIM). For the sake of ***content interoperability***, a ***federated database system*** seems to be the most appropriate. The conceptual framework of rules for CRIM should be conceived following the principles of comprehensive content management:

- Single source and resource sharing
- Based on metadata methodology (and XML-based)
- A metadata, micro-datamodel and meta model repository
- Workflow management of distributed (i.e., web-based) cooperative content creation and maintenance
- Workflow hierarchy according to language or other aspects

This would support, if not guarantee ***content interoperability*** across languages and national conventions. Whether it needs only *one* lead repository for all messages (plus attributes and all related data), or – more likely – a set of lead repositories according to major types of content (all modelled according to metadata methodology) needs to be discussed.

14.10 Conclusions

In conclusion, we can summarise that:

- The complex situation described above requires a terminological approach to data modelling.
- This approach should be based on pertinent standards and standardization activities.
- Every European country has or can have road/traffic sign variants, both referring to verbal messages as well as to non-verbal signs and symbols; a locale, therefore, is a particular road/traffic sign variant or information on an additional sign or a particular combination of these with one or more verbal messages in the official language.

14.11 Europeanisms as Key Meanings

A number of verbal messages or verbal message elements have been selected, which could be considered as candidates for "Europeanisms" – i.e., verbal messages or verbal message elements, which have the potential to be harmonized EU-wide.

Verbal messages or verbal message elements can occur:

- In traffic signs: STOP, 60 (= max 60 km/h), 5 t, 15%, etc.
- In additional signs: EXCEPT, GRATIS, h (= hour), min (= minute), etc.
- On informative signs: WC, P, etc.
- In a verbalization of:
  - A traffic sign or additional sign: STOP, OIL (slippery road), etc.;
  - A graphic symbol: BUS, TRAM, etc.;
- In larger verbal messages:
  - Based on traffic signs, additional signs or graphical symbols
  - Based on informative signs
  - Informing the driver on traffic conditions, weather conditions, traffic broadcasts, etc.

Systematization would, on the one hand, enhance the comprehensibility and perceivability, thus increasing transparency and robustness and, on the other hand, reduce variation of existing signs and messages, while assisting dynamic development or ad-hoc creation of new verbal messages without deprecating comprehensibility and perceivability.

The harmonization based on systematization would help:

- The driver to become familiar with road/traffic signage in a different country within a few minutes
- Road and traffic administrations to develop flexible (multilingual) data model for content management
- The system developers to develop (multilingual and highly personalisable) data models for car-driver communication
- Administrations and system developers to implement requirements for people with special needs

and, overall, to make systems generically multilingual and highly flexible for new developments and new requirements. *This would also greatly increase the infrastructure safety for road managers and traffic telematics system developers.*

In this context, multilinguality will not be a barrier, but would even facilitate the development of navigation systems or other types of car-driver communication systems in different languages of Europe. In principle, also all non-linguistic representations may have to be rendered in verbal (written or spoken) form in some situation or for some special user groups. Non-linguistic representations here refer not only to graphic signs and symbols, but also to other non-linguistic representations, such as haptic information (e.g. vibration of the steering wheel in case of speed limit violation).

In general, fundamental/most important short verbal messages and verbal message elements in written form for the sake of harmonization may:

- Be written in upper case letters (exception: measures, units, quantities are recommended to be written in lower case letters, as well as i = information)
- Not exceed 5–6 letters whenever possible (exceptions: EXCEPT, CONTROL, etc.)
- (If applicable) be based only on metric measures, units and quantities in Europe
- Avoid misunderstanding (H = Halt vs. H = Hospital; min = minute vs. min = minimum; m = metre vs. m = mile, etc.) either by harmonization or by means of layout, etc.

These fundamental/most important (short) verbal messages and verbal message elements in written form should be recognizable as "icons" rather than being read. Therefore, a European-wide harmonization would reduce the learning necessity and increase the *iconic* character of these (short) verbal messages and verbal message elements. Needless to say that, for the spoken form, a different – however coordinated – systematization/methodology has to be developed, which will heavily rely on the phonological system of the respective language.

Altogether, about 40 candidates for such Europeanisms were identified within IN-SAFETY, some of which – such as telephone, bus, tram, etc. – are also represented by a graphic symbol. The systematic approach chosen made it necessary to take as well the respective graphic symbols into account. Europe-wide harmonization does not necessarily conflict with "multilingualism", since the individual languages will still be used in most messages – especially those of a more complex nature, whether in navigation systems or in other types of car-driver information and communication systems. In addition, within the project, 20 completely new or newly designed traffic signs have been proposed and tested, some of which again represent verbal messages or verbal message elements (as outlined above).

These 60 verbal messages or graphical symbols constituted the short list of a total of originally more than 100 candidates. They were entered into a comprehensive XLS file together with pertinent RDS-TMC "events" selected from EN-ISO 14819-2:2003 "Traffic and Traveller Information (TTI) – TTI messages via traffic message coding" and DATEX (Data Exchange Service). All related traffic signs or graphic symbols had been added.

As this XLS file proved to be by far too complex when asking experts from 34 language/country (or language/region) combinations for feedback, simplified tables were constructed. These tables were sent to nearly all EEA countries, including Norway, but excluding Malta (where British rules and signage was said to prevail), Iceland (not interested) and Vatican (= Holy Sea). Input was provided more or less complete by 22 language/country (or language/region) combinations, 8 did not respond at all, 2 did not provide input in time (Table 14.2).

As a result:

- 35 Candidates for Europeanisms were identified for consideration in a Europe-wide harmonization process
- 9 Candidates were assumed as immature to be harmonized all across Europe

Table 14.2 Examples for the collected feedback to the simplified tables

No.	Potential Europeanism	Name/meaning in English	Name/meaning in your language
25	@	Internet (access available)	cs – /Internet/ de – Internet da – Internet/@ = snabel a et – /Internet/ es – Internet (acceso disponible) ca – Internet (accès disponible) eu – Internet gl – Internet fi – Internet fr – /Internet/ el – /Ιντερνέτ/ ga – Idirlíon en – Internet it – Internet hu – /Internet/ lt – Internetas (galima naudotis internetu) lv – /Internet/ nl – @: Internet no – Internett (tilgangsmulighet) pl – /internet/ pt – Internet (acesso disponível) sk – Internet (internetové pripojenie) sl – Internet sv – Internet cy – /Rhyngrwyd/
32	BUS CZ  ES    	Buses (in combination with: - For busses, or - Bus stop, or - Busses allowed, or - Only for buses)	cs – bus de – Bus da – Busser (kun for busser) et – /buss/ es – autobús ca – autobus eu – autobusa gl – Carril reservado para autobuses fi – Linja-auto fr – Voie réservée aux véhicules des services réguliers de transport en commun el – /λεωφορείο, leoforeio/ ga – Lána Bus en – Bus Lane it – Riservato bus hu – /autóbusz /busz / lt – Autobusas (leidžiama tik autobusams) lv – /autobuss/ nl – Autobus no – Holdeplass for buss pl – / szyna / pt – Autocarros (permitido apenas para autocarros) sk – BUS (len pre autobusy)

(continued)

Table 14.2 (continued)

No.	Potential Europeanism	Name/meaning in English	Name/meaning in your language
	FR 		sl – avtobus sv – Buss cy – /bws/
	IT  		
	BE 		
	NL 		
	NO 		

Countries: *BE* Belgium; *CZ* Czech Republic; *FR* France; *IT* Italy; *NL* The Netherlands; *NO* Norway
Languages: *ca* Catalan; *cs* Czech; *cy* Welsh; *da* Danish; *de* German; *el* Greek; *en* English; *es* Spanish; *et* Estonian; *eu* Basque; *fi* Finnish; *fr* French; *ga* Irish; *gl* Galician; *hu* Hungarian; *it* Italian; *lt* Lithuanian; *lv* Latvian; *nl* Dutch; *no* Norwegian; *pl* Polish; *pt* Portuguese; *sk* Slovak; *sl* Slovene; *sv* Swedish

Recommendation 1 recommends (short) verbal messages and fundamental verbal message elements to be harmonized or considered for Europe-wide harmonization.

Table 14.3 Examples for harmonization that can be easily implemented

Letter sign	Meaning represented	Existing standards	Recommendations and comments
@	Internet access available		Should be harmonized (also on maps, etc.)
RADAR	Radar (control)		Should be harmonized
RADIO	Traffic broadcast (+frequency)		Should be considered
TRAIN	Train (crossing...)	Symbol is standardized acc. to VC	Should be considered

Recommendation 2 recommends (short) verbal messages and fundamental verbal message elements, which need further investigation, especially if occurring in clusters of related messages, before they can be considered for Europe-wide harmonization.

Table 14.4 Examples for potential harmonization that needs further study

Letter sign	Meaning represented	Remarks
FOG	Fog warning	
GAS	Gasoline, petrol, benzin	LPG = Liquified/Liquefied Propane/Petroleum Gas is used for gas (not: gasoline) driven vehicles
HOV	High occupancy vehicles	In some countries called "Car sharing lane" or "Car pools"
CASH	Dues/fees to be paid in cash	
FAIR	(Trade) fair	

Recommendation 3 recommends a systematization of the syntax of certain verbal message and symbol clusters.

Based on the above comparative analysis, a set of recommendation was issued. The most relevant and important are briefly presented below (Tables 14.3 and 14.4).

Certain combination clusters may have several complexity dimensions – e.g. co-occurring signs, graphical symbols and verbal messages/verbal message elements – and should be further investigated with the aim to systematize and simplify (across language boundaries). Information on "PARKING" for example belongs to more than one clusters (Table 14.5).

Quantities and units occur (less in than around) traffic signs – e.g. in additional signs and verbal messages for information purposes. They occur in combination with numbers (integers or decimals – sometimes negative) and other kinds of information, such as:

Table 14.5 Examples for highly combinable elements in road/traffic sings

Base symbol	Extension level 1	Extension level 2 plus indication of	Extension level 3 plus indication of
P	For cars (&number)	Distance	Time
	For buses (&number)	Direction	WC
	For trucks (&number)		Handicapped WC
	For caravans (&number)		Drinks, restaurant...
	For other kind of vehicle (&number)		Other
P+(R)	+R (park-and-ride)	Distance	Time
	+BUS	Direction	Frequency
	+METRO		Capacity
	+TRAIN		Name (of location)
	+TRAM		
	+Other		Other

Recommendation 4 recommends the harmonized use of quantities and units.

- Distances:
  - (Length:) = Stretch of road from here to...: /arrow up/ 500 m /arrow up/,/ arrow up/ 5 km /arrow up/, etc.
  - Distance (= in m): 100 m (e.g. railway crossing), give way 50 ys, etc.
- Other measurements:
  - 5,5 t (gross weight), 8 t (axle weight), etc.
  - 2 m (width), 3.8 m (height), 10 m (length, distance, ...)
- (Speed:) 80 (= 80 km/h) + time (period) indication
- (Degree:) 10% (gradient road, dangerous hill), 0° (= 0°C. temperature), etc.

They may indicate:

- A fact: e.g. 10% gradient road
- A minimum: e.g. minimum number of passengers 2+ for HOV
- A maximum: e.g. 60 (= speed limit 60 km/h)
- a range of:
  - Time: from 10 h to 15 h
  - Time: "Sundays and holidays, 20:00–06:00", etc.
  - Distance: between 2 and 5 km from here
- A frequency: every 15 min (e.g. trains leaving from P + R)

It is recommended to follow this systematic approach and harmonize as many verbal messages and verbal message elements (as well as traffic signs and graphic symbols) as possible, as a first step. As a second step, further verbal messages and signs/symbols should be considered for harmonization.

14.12 Formulating SER by Europeanisms

The development of traffic information systems, road management systems, and other related systems, which increasingly have to communicate with the car-driver system, is heading for more integration or interoperability, at least in multilingual Europe. The methodology outlined in this chapter shows a way, how to reduce the proliferation of unnecessary variation in the representation of meaning, while at the same time creating more flexibility for new re-combinations of existing and development of new representations of meaning, without overburdening the driver. This can be achieved by harmonizing existing road/traffic signs and road/traffic sign elements step-by-step. This harmonization will also result to more investment security for their research and development efforts.

References

American Library Association; Library of Congress (ed.), *ALA-LC Romanization Tables: Transliteration Schemes for Non-Roman Scripts*, (1997) Some of the tables have been revised in the meantime, see: http://www.loc.gov/catdir/cpso/roman.html

V. Anttila, J. Luoma, P. Rämä, Visual demand of bilingual message signs displaying alternating text messages. Transp. Res. Part F Traffic Psychol. Behav. **3**(2) 65–74 (2000)

D. Ballardin, D. Bruno, E. Rovida, Some observations about the semanticas and syntax of road signs. TEC **6**, 267–269 (2005)

P. Bilak, Stereotypes on the street, in http://www.icograda.org/feature/current/articles64.htm from *Dot Dot Dot, Graphic Design/Visual Culture Magazine*

R. Clark, Bilingual signs pass road test in Canada. The national website of Wales, 2005. Accessed 4 Jan 2007, see: http://www.walesonline.co.uk/news/wales-news/tm_objectid=15595653&method=full&siteid=50082&headline=bilingual-road-signs-pass-test-name_page.html

H. Erke, Psychologische und grafische Aspekte der Konzeption von Piktogrammen, 20 Aug 2003 (unpublished manuscript)

EURESCOM Report on P923, Multilingual web sites: best practice, guidelines and architectures, D1 Guidelines for building multilingual web sites – Sept 2000, http://www.eurescom.eu/Public/projectresults/P900-series/923d1.asp

Federal Ministry of Transport, Building and Housing, *German Road Traffic Regulations* (inofficial translation of the German Straßenverkehrsordnung). (BAST, s.a, Bergisch-Gladbach) unpublished

R. Hawkins (ed.), *Study of the Standards-Related Information Requirements of Users in the Information Society. Final Report to CEN/ISSS, 14 February 2000* (SPRU, Brussels, 2000)

E. Hovy et al. (eds.), Multilingual information management: current levels and future abilities. A report commissioned by the US National Science Foundation and also delivered to the European Commission's Language Engineering Office and the US Defence Advanced Research Projects Agency, Apr 1999, http://www.cs.cmu.edu/~ref/mlim/index.html

S.L. Jamson, F.N. Tate, A.H. Jamson, Bilingual variable message signs: a study of information presentation and driver distraction, in *1st International Driving Symposium on Human Factors in Driver Assessment, Training and Vehicle Design*, Aspen, CO, 2001, pp. 153–158

S.L. Jamson, F.N. Tate, A.H. Jamson, Evaluating the effects of bilingual traffic signs on driver performance and safety. Ergonomics **48**(15), 1734–1748 (2005)

M. Karhunen, Bilingual variable message signs – drivers' opinion and visual load, 1998

S. Matteini, Multilinguality and the Internet. European Parliament, 2001 (Research Directorate A. STOA – Scientifical and technological options assessment. Briefing Note No. 2/2001), http://www.sustainable-design.ie/sustain/EPbrief-InternetMultilinguality.pdf
Ovum (ed.), *Repositories and XML: Technology Choices for Meta-Data Management* (Ovum, London, 1999)
Pricewaterhouse Coopers (comp.), Cultural diversity market study. Final Report, Luxembourg, 2001
TROPIC (ed.), TRaffic OPtimisation by the Integration of Information and Control. Tropic Final Report, EC Contract RO-96-SC.303/2, 1999
UN/ECE (ed.), Convention on Road Signs and Signals. Vienna, 1968 (Vienna Convention)
UNESCO (ed.), in *Recommendation Concerning the Promotion and Use of Multilingualism and Universal Access to Cyberspace, adopted by the UNESCO General Conference*, UNESCO, Paris, 2003 (32 C /Resolution 41)
VAMOS Consortium (ed.), White Book for Variable Message Signs Application, Oct 1991

Pertinent Standards

EN 12767:2000, Passive Safety of Support Structures for Road Equipment – Requirements and Test Methods (CEN, Brussels, 2000)
EN 12899-1:2001, Fixed, Vertical Road Traffic Signs – Part 1: Fixed Signs (CEN, Brussels, 2001)
EN 12966-1:2005, Road Vertical Signs – Variable Message Traffic Signs – Part 1: Product Standard (CEN, Brussels, 2005)
EN 12966-2:2005, Road Vertical Signs – Variable Message Traffic Signs – Part 2: Initial Type Testing (CEN, Brussels, 2005)
EN 12966-3:2005, Road Vertical Signs – Variable Message Traffic Signs – Part 3: Factory Production Control (CEN, Brussels, 2005)
CEN/ISSS/CWA 13699:1999, Model for Metadata for Multimedia Information (CEN, Brussels, 1999)
CEN/ISSS/CWA 13873:2000, Information Technology – Multilingual European Subsets in ISO/IEC 10646–1 (CEN, Brussels, 2000)
CEN/ISSS/CWA 13989:2000, Description of Structure and Maintenance of the Web Based Observatory of European Work on Metadata (CEN, Brussels, 2000)
CEN/ISSS/CWA 14094:2001, European Culturally Specific ICT Requirements (CEN, Brussels, 2001)
ELOT (ed.), ELOT 743–1982 (Conversion of the Greek Alphabet into Latin) (ELOT, Athens, 1982)
ISO 80000 (Series), Quantities and Units (International Organization for Standardization, Geneva)
ISO 843:1997, Information and Documentation – Conversion of Greek Characters into Latin Characters (International Organization for Standardization, Geneva, 1997)
ISO 7001:1990, Public Information Symbols (International Organization for Standardization, Geneva, 1990)
ISO/TR 7239:1984, Development and Principles for Application of Public Information Symbols (International Organization for Standardization, Geneva, 1984)
ISO 9186:2001, Graphical Symbols – Test Methods for Judged Comprehensibility and for Comprehension (International Organization for Standardization, Geneva, 2001)
ISO/DIS 10241–1:2009, Terminological Entries in Standards – Part 1: General Requirements and Examples of Presentation (International Organization for Standardization, Geneva, 2009)
ISO/DIS 26162, Systems to Manage Terminology, Knowledge and Content – Design, Implementation and Maintenance of Terminology Management Systems (under preparation)

ISO 12620:2009, Terminology and Other Language Resources – Specification of Data Categories and Management of a Data Category Registry for Language Resources (International Organization for Standardization, Geneva, 2009), s.a. ISO Data Category Registry, http://www.isocat.org
ISO 16642:2003, Computer Applications in Terminology – Terminological Markup Framework (TMF) (International Organization for Standardization, Geneva, 2003)
ISO 15836:2009, Information and Documentation – The Dublin Core Metadata Element Set (International Organization for Standardization, Geneva, 2009)

Chapter 15
As You Like *IT*

Evangelos Bekiaris, Evangelia Gaitanidou, Maria Panou, Konstantinos Kalogirou, and Pavlos Spanidis

15.1 The Need for Personalization in Traffic Info Provision

Apart from the benefits that the growing penetration of Intelligent Transport Systems (ITS) in the automotive market imply (such as their impact on driving and road safety enhancement), there are also behavioural issues that are being met more and more often as the penetration rate of ITS raises. The most common behavioural issue that is raised among developers and researchers is the distraction of the driver. In particular, regarding VMS, studies have been conducted and are currently being undertaken, examining the behavioural effects of these systems to the drivers. As indicated in (Erke et al. 2007), there is a number of different reasons due to which drivers may be distracted by VMS. The most prominent reason is that significant attention should be allocated to the VMS in order to read and understand the message and to decide upon the consecutive action that should be taken. This results in conflicting attention demands by the VMS and the driving task, which usually lead to reduction of the amount of attention dedicated to one of the two actions that could be lethal. The attention demands that the combination of VMS and driving are imposing, can be located at different levels of the driving task (i.e. operational level: increased visual load, tactical level: attempt to gain time by reducing speed, etc.). There is thus the danger of reduction of the level of traffic safety. This effect becomes even more intense, when there are also other, in-vehicle devices providing warnings and/or information to the driver.

Introducing a secondary task to an easy primary activity (i.e. driving on a highway with low traffic) has been stated to impair primary task performance. On the other hand, the introduction of a secondary task to a hard primary task (i.e. urban scenarios of high traffic and conflicts level) may result in significant decrements to primary task performance (Gopher 1990) (Fig. 15.1).

E. Bekiaris (✉), E. Gaitanidou, M. Panou, K. Kalogirou, and P. Spanidis
Centre for Research and Technology Hellas/Hellenic Institute of Transport (CERTH/HIT), Thessaloniki, Greece
e-mail: abek@certh.gr

E. Bekiaris et al. (eds.), *Infrastructure and Safety in a Collaborative World*,
DOI 10.1007/978-3-642-18372-0_15,

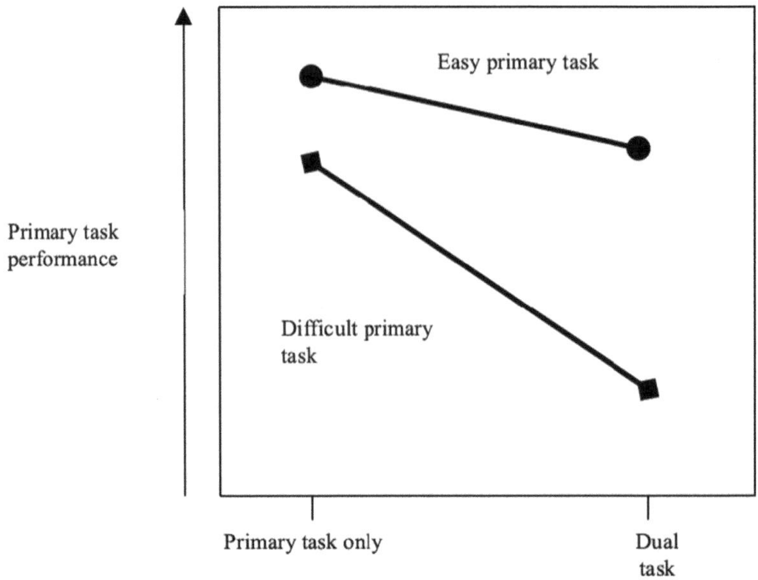

Fig. 15.1 Theoretical dual task performance (Gopher 1990)

Distraction is indeed among the critical parameters that affect the driving performance according to the DRIVABILITY model (Bekiaris et al. 2003). DRIVABILITY is a dynamic driver behaviour model that defines the ability to drive of a specific person in a specific environment and under specific circumstances. Distraction is directly related to the level of attention of a driver, having immediate consequences at his/her risk awareness level. Furthermore, distraction is highly correlated with the workload due to the inability to detect the needed info in the VMS sign. Evidence from studies of perceptual workload and visuocortical processing imply that perceptual workload can amend attention focus early in visuocortical processing. Furthermore, increasing the workload of the foveal targets decreases the amount of residual attention capacity available for allocation to task-irrelevant parafoveal locations (Handy et al. 2001).

One way to reduce the distraction and workload caused by the VMS to the driver would be the use personalised cooperative systems, by means of designing and developing an in-vehicle system, able to display to the driver, inside the vehicle, information provided by VMS/VDS in a personalised way. More specifically, information should be presented according to his/her own needs, residual abilities, preferences and goals (i.e. with adapted in-vehicle HMI and according to his/her route, scheduled in the route guidance system). The main advantage is that the driver, not only would receive the information of the VMS on the in-vehicle device, but also, he/she would receive only the information of his/her interest.

Alternative ways of in-vehicle personalised provision of information are visually, acoustically or haptically, as well as exclusion of information which is out of the field of interests and needs of the driver, etc., as pre-defined by each individual driver. As an example, information about specific parking places for wheelchair users will not be displayed to all drivers, while wheelchair users will receive this information at priority while driving. Moreover, there will be the possibility to display the information provided by the VMS also on the in-vehicle device, in a language independent manner, either by substituting text with adequate pictograms or by displaying textual messages, audible or visual, in the preferred language of the driver.

The design, development and testing of such an application is analysed below.

15.2 System Architecture

The first step in designing the system is to define its architecture, both by means of theoretical as well as practical design. Originally, three architectures have been proposed, using different technological solutions. These were correspondingly based upon:

- Wireless LAN
- GPS with GPRS
- Bluetooth

After examining the potential of each one of these architectures, it has been concluded that the use of Bluetooth technology would not be appropriate for the intended system, as it is facing certain limitations. The main problem is the transmission speed. Bluetooth offers communication in a range of 10–150 m. Interactions at higher distances cannot be supported. For example, if the driver is driving with 120 km/h at a motorway and the Bluetooth range assumed is at 100 m, the establishment of the connection between parties and the transmission of a message to the car's PC must be completed in 3 s. Another pitfall is its limited bandwidth. The Bluetooth can transmit and receive data at a mere 720 kbps (http://ntrg.cs.tcd.ie/undergrad/4ba2.05/group3/index.html).

Furthermore, the "pairing" process between Bluetooth devices needs some reasonable time (approximately 1–2 s). Thus, the vehicle would be restricted to move with no more than 120 km/h, otherwise the driver might see the message of a previous VMS, while passing the next one. Finally, a Bluetooth device must be integrated in the VMS, to establish communication with vehicles. Old types of VMSs may not support it and the integration of Bluetooth hardware will come at a cost.

Baring into account the above mentioned issues, it was decided to implement the remaining two of the proposed architectures. The detailed design and implementation of the two architectures that are applied in the Greek pilot are described in the following chapters.

15.2.1 Wireless LAN Architecture

The Wireless LAN architecture is focused on the VMS device. The VMS is connected to an industrial PC, which has an integrated WLAN module. This architecture can be seen below (Fig. 15.2).

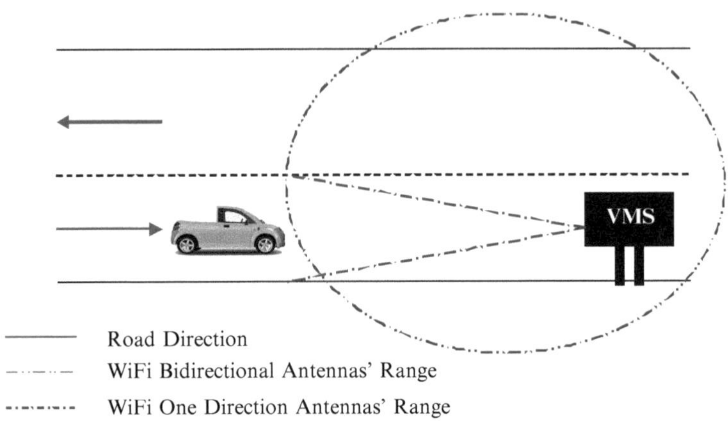

Fig. 15.2 The bidirectional range of the WiFi signal

The WLAN architecture consists of two Agents, each one serving the corresponding side of the application and they are totally independent (http://www.wi-fi.org).

15.2.1.1 Server Agent Architecture

In detail, the Server Agent, which is running on the server of the application (the PC on the VMS), provides the client/vehicle with the desired information. This information can be e.g. the road conditions and/or the existence of an emergency situation.

The server application is continuously running on the server. It can be said that it is in a "sleep" mode and it awakes only when it receives a message from the client (on the vehicle), in order to serve it.

It has also been decided that the transmitted information should be of the minimum amount. Therefore, the server has been designed to send only a string with an integer stored on it and also to include the preferred language of the user. The integer corresponds to the identification number of the set of information on the system's database (http://jade.tilab.com).

15.2.1.2 Client Agent Architecture

The Client Agent is an application which is loaded in the start up of the Operating System (OS) of the vehicle and is continuously running. It is silently searching for the server/VMS in the background of the OS. Therefore, each time that the vehicle reaches a VMS, the Client Agent is ready to establish a connection and process the received information. Finally, it displays the appropriate information on the screen of the driver.

When the client application is loaded, it is minimized, in order not to distract the driver with its appearance on the in-vehicle screen and it is ready to pop-up and display the information to the driver. Another part of the client application is the database of the system, which contains the required information of the application.

In more detail, the client application receives a message from the server whose content is an integer. This integer corresponds to the ID field of the database. Therefore, depending on the ID number that the client received, it retrieves the appropriate information from the database.

Using the retrieved information, the client loads the corresponding files (pictogram and sound file) and passes them to the appropriate fields of the code.

The visual warnings are displayed on the in-vehicle screen, on top of all other applications, for about 1 min. This is considered as an appropriate period of time because it lasts long enough to give time to the driver to finish something that he/she is doing (e.g. an overtaking manoeuvre) and short enough not to be characterized as annoying. After this period of time, the client (pop-up window) minimizes itself and waits until the next VMS, the next valid information.

In the case where there are two VMSs in a very close distance and one signal overlaps the other, the client is able to display the most vital information. The categorization of the message importance is done using priority levels on each stored warning. Three priority levels are distinguished: (a) high priority, (b) middle priority and (c) low priority level, each one corresponding to the integers 1, 2 and 3 respectively (http://jade.tilab.com).

15.2.2 GPS with GPRS Architecture

The GPS with GPRS architecture is VMS independent. This means that the system is totally dependent on the Traffic Management Centre, which acts as the server on this architecture. The GPS with GPRS architecture is illustrated in Fig. 15.3.

The GPS with GPRS architecture also consists of two, totally independent, Agents, each one serving the corresponding side of the application.

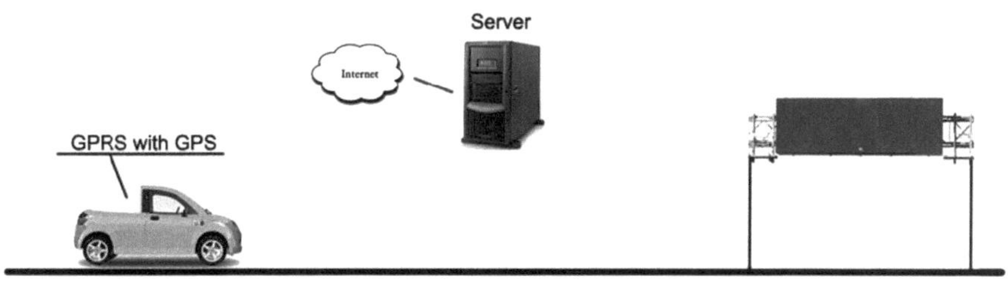

Fig. 15.3 GPS with GPRS architecture

15.2.2.1 Server Agent Architecture

In contrast to the WLAN architecture, the server of this architecture is a PC, resided on the Traffic Management Centre. Again, the Server Agent is running on the server and provides the client/vehicle with the desired information.

Once more, the server application is continuously running on the server PC. The Server Agent has been designed with a static IP, which is set in the code lines of the client application.

The Server Agent of the GPS with GPRS architecture follows exactly the same principles of design with the Server Agent of the WLAN architecture. The difference between the two is that this Server Agent is also connected to a database, which resides on the TMC PC. This database contains the coordinates (latitude and longitude) and the direction of the road that each VMS is associated with.

In addition to the GPS coordinates, a moving object also has a decimal value from 0° to 360°, which provides the direction of the object in degrees. This value has been used to separate the left from the right direction of the road and by extension the left from the right VMS (whenever they might exist in opposite directions) (http://jade.tilab.com).

15.2.2.2 Client Agent Architecture

The Client Agent of the GPS with GPRS architecture follows the same principles with the Client Agent of the WLAN architecture.

In this case, the vehicle does not communicate with the VMS but with the TMC. During its trip, the vehicle sends periodically its coordinates. These are: (a) the latitude, (b) the longitude and (c) the direction of the vehicle.

Once the Server Agent receives the message with the coordinates, it constructs a bounding box around the vehicle. If the coordinates of a VMS (found in the server database) are included in this bounding box, it means that a VMS has been detected in the area of the vehicle. Then the server/TMC sends to the client/vehicle the information corresponding to that VMS (Fig. 15.4).

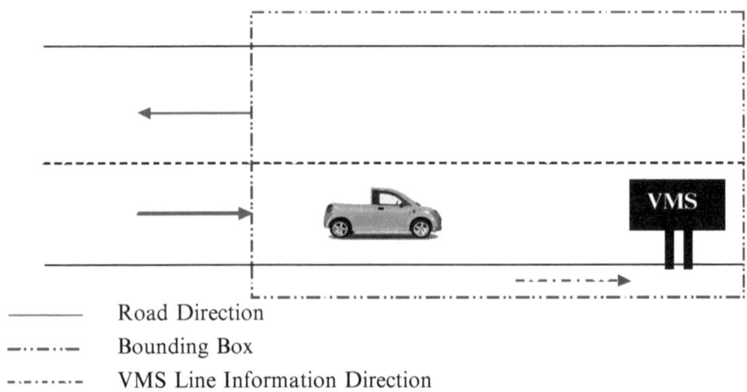

Fig. 15.4 GPS with GPRS bounding box

The TMC keeps informing the driver for the content of the VMS, as long as the VMS is detected in the bounding box of the vehicle (http://jade.tilab.com).

15.3 Human–Machine Interaction (HMI)

The information of a VMS is transmitted with one of the above architectures to the in-vehicle PC. Extensive research has been undertaken within IN-SAFETY, in order to define/redefine the appropriate standard pictograms, which will substitute the verbal messages on the VMS and a relevant list is available, after testing their user-friendliness and acceptability with numerous users in different countries. For the substitution of the VMS message by pictograms, two different technologies have been applied. The database of pictograms can reside at the car's PC locally, or this database can reside at the TMC (remotely). This service constitutes the Virtual VMS service (VVMS), provided locally or remotely. In both cases, the text message provided by the TMC is converted into pictograms. Furthermore, these selected pictograms or earcons are displayed by the in-vehicle system. Within IN-SAFETY, a long list of pictograms has been proposed for all types of messages (see Chap. 13). A selection among these has been used for the needs of the pilot testing of the VVMS, as seen in Table 15.1.

Besides, user friendly messages to be provided by the in-vehicle GUI have been examined and implemented. From their evaluation, the following main conclusions can be drawn:

- For each message displayed, it is recommended to add an initiation sound to capture the driver attention.
- If a text (string) message is displayed, it is recommended also to provide the same message audibly, in the driver's language.

15.4 On-Site Implementation

A variety of incremental and iterative tests took place during the development of the application (all architectures). The scope of these tests was to examine the three above mentioned architectures in real-conditions and to identify their weaknesses and bugs.

15.4.1 Greek Site Implementation

Both WLAN and the GPRS architectures have been realized in the Greek pilot-site (Kalogirou et al. 2008).

Table 15.1 Examples of pictograms on the in-vehicle screen used to communicate VMS messages (Simlinger et al. 2008)

Pictogram	Message	Pictogram	Message
	Speed limit 90 km/h		Speed limit 120 km/h
	Next exit closed		Traffic congestion/Queue ahead
	Accident (ahead)		Car break down ahead

15.4.1.1 WLAN Architecture

As already been mentioned, the VMS had to be connected to an industrial PC providing all the required processing power and storage devices to support the Server side of the application. The WiFi Linksys Access point and the PC were directly attached to the Compex switch.

The Access point has also attached the bidirectional antennas of Linksys on it in order to succeed a good range for testing.

An industrial PC has been used in the client's side, which is also part of the vehicle's equipment. This PC has also attached the WiFi PCI adapter on it in order to communicate with the Server. Figure 15.5 shows the bidirectional antenna of the client PC attached on the right side of the vehicle.

After several examinations/tests, it had been proven that the application works properly, although some weaknesses were spotted. Figures 15.6 and 15.7 show how the application layout appears on the 7 in. LCD touch-screen of the vehicle.

Fig. 15.5 The bidirectional antenna on the vehicle

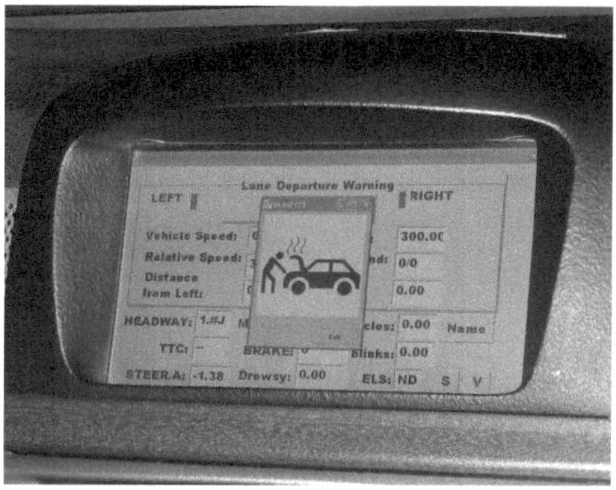

Fig. 15.6 The application's layout on the in-vehicle screen

Fig. 15.7 The pop-up window

The maximum range of the client with the server that has been succeeded was approximately 300 m in line of sight, using the antennas on both sides (Server and Client side). This distance reaches the theoretical maximum distance of the devices with the antennas attached on them.

15.4.1.2 GPS with GPRS Architecture

In this architecture the industrial PC of the vehicle communicates with the TMC via the GPRS technology. The only installation that had been done on the vehicle was the connection of the in-vehicle PC with the GPS and the GPRS devices. The GPS device had been placed in the back windscreen in order to be in line of sight with the satellites.

After a number of tests, it has been proven that the GPS with GPRS architecture works as expected. The layout of the application is exactly the same as with the previous architecture, as the mechanism, that provides the warning signs on the in-vehicle screen, is exactly the same.

15.5 Greek Pilot Results

15.5.1 Test Scenaria

Of the above, the WLAN and GPS/GPRS architectures were tested with real users in Greece, in 'Attiki Odos', a major highway in Athens. Two IN-SAFETY scenarios defined in (De Brucker et al. 2006) within four situations were tested, which are described in Table 15.2.

The measurements utilized for the evaluation of the scenarios were pre- and post-test subjective user reports. The results from the performed pilot demonstrate

Table 15.2 An overview of the scenarios and methodological details of the Greek Pilot (Anund et al. 2008)

Scenario description	Type of accidents/ incidents addressed	Technology used	Driver groups
IN-SAFETY scenario 4: Virtual rumble strips warning to prevent from lane departure	Head-on, head-tail, and lateral accidents	CCD camera, LDW	Ten commuters
IN-SAFETY scenario 4: Personalised adaptation of ADAS warnings based on driver behaviour	Head on, head-tail, and lateral accidents	ACC, CAS, LDW, Intelligent Agents	Ten commuters
IN-SAFETY scenario 1: In-vehicle speed limit warnings	Accidents due to inappropriate speed	In-vehicle HMI, infrastructure sensors	Ten commuters

that an in-vehicle system, as the one proposed and tested in Athens, can provide a solution that is easy to use and much more cost-efficient when compared to changes in the traditional infrastructure.

15.5.2 Test Participants

A total of 40 participants completed the Pilots, 4 of which were female and 36 male. The participants ranged from the age of 23 to 45, with the average age of 32. Each scenario was tested by ten participants who were experienced drivers with 4–24 years of driving experience (an average of 15 years) and the majority of them possessed driving license category B (with two of the participants having a license category A/B, two D, and one G). Half of the participants reported having a vehicle equipped with various driving technologies and the reported kilometres per year driven were in the range of 12,000–45,000 km/year, with an average of 18,700 km/year. All of the participants had a high level of educational background. Finally, the participants in the Pilots reported being in good health and with normal hearing and normal or corrected-to-normal visual acuity.

15.5.3 Pilot Design

The first phase of the pilot included the *no system scenario*. In this phase the participants drove in a specific route without the use of any system. They were receiving incidence and traffic information only from the standard on-road equipment (VMS, traffic signs), while also being warned about speed limits by roadside signs (static or electronic). This was used as the reference scenario for the next two, in order to compare with the actual situation and assess the effect of the use of the IN-SAFETY systems.

After this, two scenarios were tested: the *in-vehicle incidence information/ warnings scenario* and the *in-vehicle speed limit warnings scenario*.

In the *in-vehicle incidence information/ warnings scenario*, the system informs the driver via an auditory and a visual signal of the VMS content when that is not visible. A driver can therefore be informed of a situation occurring on the road when the vehicle is away from the VMS, avoid exposure to a message that is irrelevant to him/her, and/or be informed in a personalized HMI. The participants were asked before driving to report their overall judgment of the proposed IN-SAFETY solution using nine different pairs of items. Their judgments were based on a 5-point scale, with 1 being the highest opinion rating and 5 the lowest point in the scale. The participants were asked the same question after driving (post-pilot questionnaire), also in comparison with the *no-system scenario*. The same procedure was done for the two architectures using: GPS/GPRS or WiFi.

The same technology has been used also for the *in-vehicle speed limit warnings scenario,* which is based on a system that informs the driver of the existence of speed limitations in a given road. The system informs the driver only when he/she exceeds the given speed limit of that particular road. Before the pilot of this scenario, participants were asked to report their overall judgment of the proposed IN-SAFETY solution using nine different pairs of items. Their judgments were based on a 5-point scale, with 1 being the highest opinion rating and 5 the lowest point in the scale (as defined in Gemou et al. 2007). The participants were also asked the same question in the post-pilot questionnaire, also in comparison with the *no-system scenario.*

15.5.4 Results

15.5.4.1 In-Vehicle Incidence/Information Warning

For both architectures, participants' reports were in general more positive in the post-pilot questionnaires as compared to the pre-pilot questionnaires, with the reports for GPS/GPRS being more positive than those for WiFi. However, the WiFi reports were slightly better than those of pre-testing, see Fig. 15.8.

Thus, it seems that both architectures satisfied user's expectations.

The participants were asked to judge how pleasant they found the proposed IN-SAFETY system to be. Both of the architectures were viewed favorably in comparison between the pre- and post-questionnaires, with the GPS/GPRS being slightly higher that the WiFi.

The assisting feature of the system was also evaluated. Participants in the pre-testing thought that the system is going to be of assistance to them. After testing took place, this belief remained the same for both the architectures utilized, rating GPS/GPRS a little bit more positive compared to WiFi. The same results also hold true for the participants' opinion regarding desirability.

Overall, for both architectures, participants' reports were in general more positive in the post-pilot questionnaires as compared to the pre-pilot questionnaires, with the reports for GPS/GPRS being more positive that those for WiFi. One measure that participants were negative about was that of cost-effectiveness (as seen also in the pilots reported previously), namely they were not sure if the system was or was not really cost-effective in transportation.

Another reported disadvantage was related to the auditory warning signal (i.e. some reported the signal as being inaudible). It must be noted here that no problems were mentioned in relation to the timing of the warnings, thus showing that the warning signals were efficient and timely for all the drivers that tested the system. It must also be noted that during the testing of the WiFi architecture, remarks were made about the late timing of the warning signals.

In fact, the used WiFi equipment seemed to work in real time up to vehicle speeds of 80 km/h, beyond which the info was provided after the driver had passed

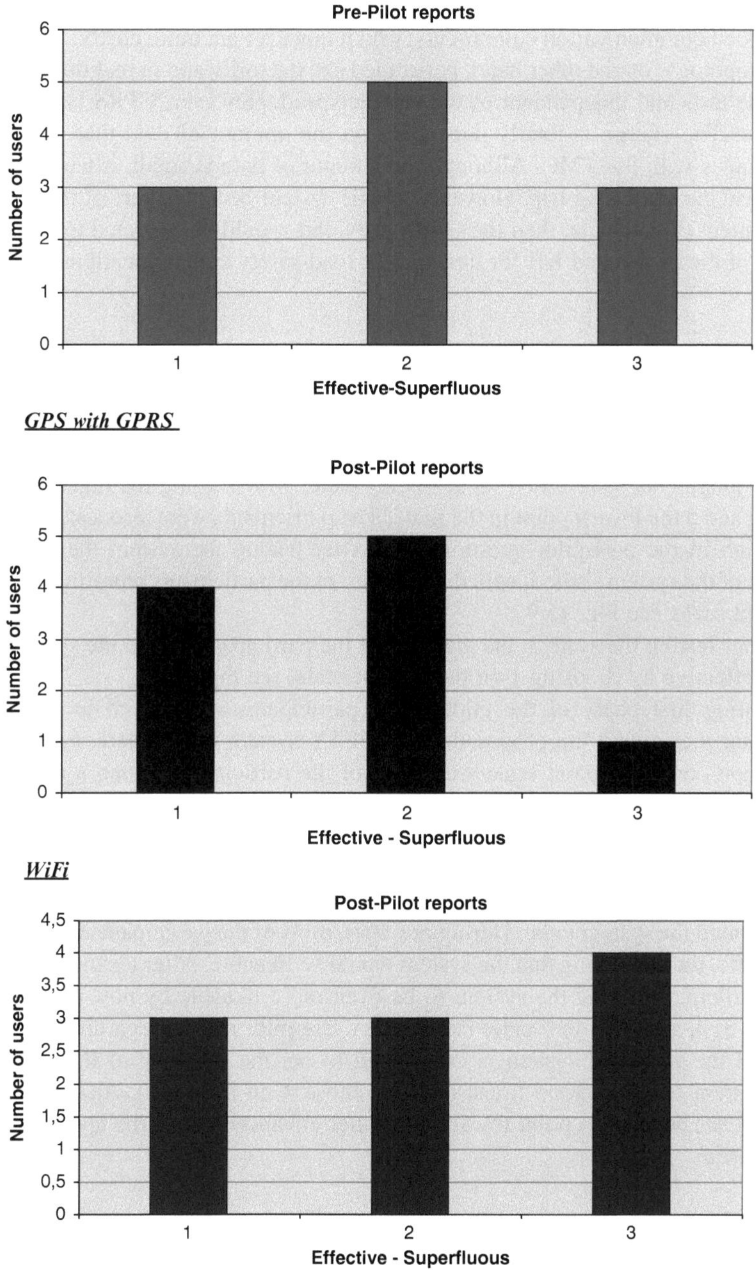

Fig. 15.8 Participants' pre- and post-pilot reports on how effective-superfluous they found the GPS/GPRS- and WiFi-based in-vehicle VMS warning/information system of IN-SAFETY to be (Anund et al. 2008)

the VMS spot (considered as too late). This weakness may be overcome using advanced communication equipments, which however are more costly. The GPRS/GPS topology, on the other hand, performed satisfactorily and in real time throughout the tests and independent of the vehicle speed. However, GPRS is not free of charge. The charge is totally dependent on the amount of data that the vehicle exchanges with the TMC. Although the amount of data is small, it would rapidly increase during a long trip. However, if this system becomes part of the standard equipment of a vehicle, then the GPRS providers would be expected to reduce the price of the exchanged KB for this kind of road-safety communications.

15.5.4.2 In-Vehicle Speed Limit Warning

Before the pilot of this scenario, participants were asked to report their overall judgment of the proposed IN-SAFETY solution using nine different pairs of items. Their judgments were based on a 5-point scale, with 1 being the highest opinion rating and 5 the lowest point in the scale. The participants were also asked the same question in the post-pilot questionnaire. After testing the system the usefulness rating of the system shifted, with the majority of the participants reporting a 1 on the 5-point scale, see Fig. 15.9.

After testing the system, the majority of the participants found the system to be very effective by choosing 1 on the 5-point scale, see Fig. 15.10.

During first phase of the pilot, when participants were asked to judge how pleasant they found the proposed IN-SAFETY system to be, there was a spread of reports on the 5-point scale with most of the participants giving a rating of 3. However, after testing of the system took place, the participants' reports shifted towards the rating of 1 and 2 on the 5-point scale, thus finding the system very pleasant.

Participants were also asked to make a judgment on how irritating or likeable they found the system to be. During pre-tests, most of the participants reported 3 on the scale, thus believing that the system would be likeable. After testing the system, the participants found the system to be even more likeable by now choosing the points 3, 4, and 5 on the scale. Participants' pre-pilot reports regarding how good or bad the proposed system is considered to be, the majority of them reported the system as being good (most of them chose 4 on the scale). Again, after test experience seemed to point towards a further enhancement of the appreciation of the system.

15.6 Towards Personalizing VMS Info On-Board

This chapter presented two architectures that have been used in a personalized in-vehicle driver information system, which would transmit any information displayed on VMSs, in a user friendly and personalized manner. These architectures

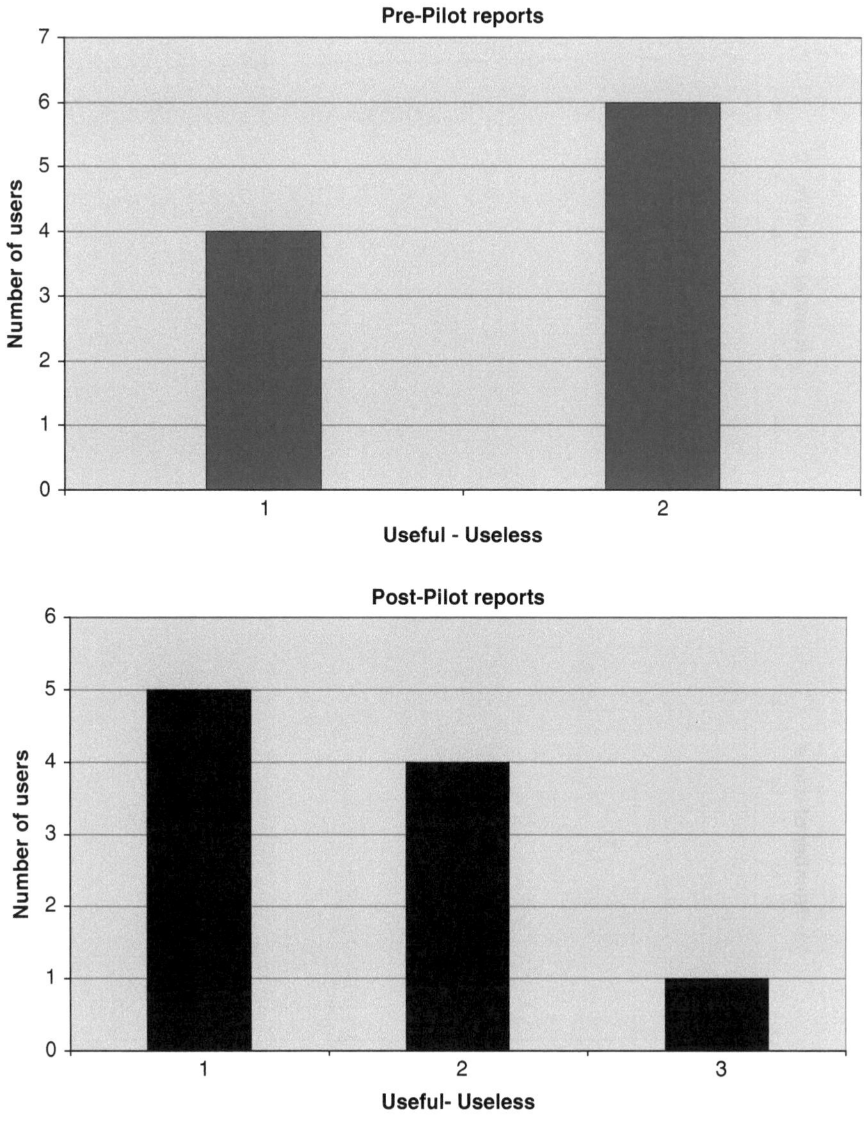

Fig. 15.9 Participants' pre- and post-pilot reports on the usefulness of the speed limit violation warning application of IN-SAFETY (Anund et al. 2008)

have been tested with real users during the IN-SAFETY pilot tests in the Greek pilot, in Athens.

The technological solutions that have been implemented are the WLAN and the GPS/GPRS. All have been used for the development of the system within IN-SAFETY. The tests showed that both architectures are working quite well and are meeting the original expectations. Regarding the WLAN, some minor

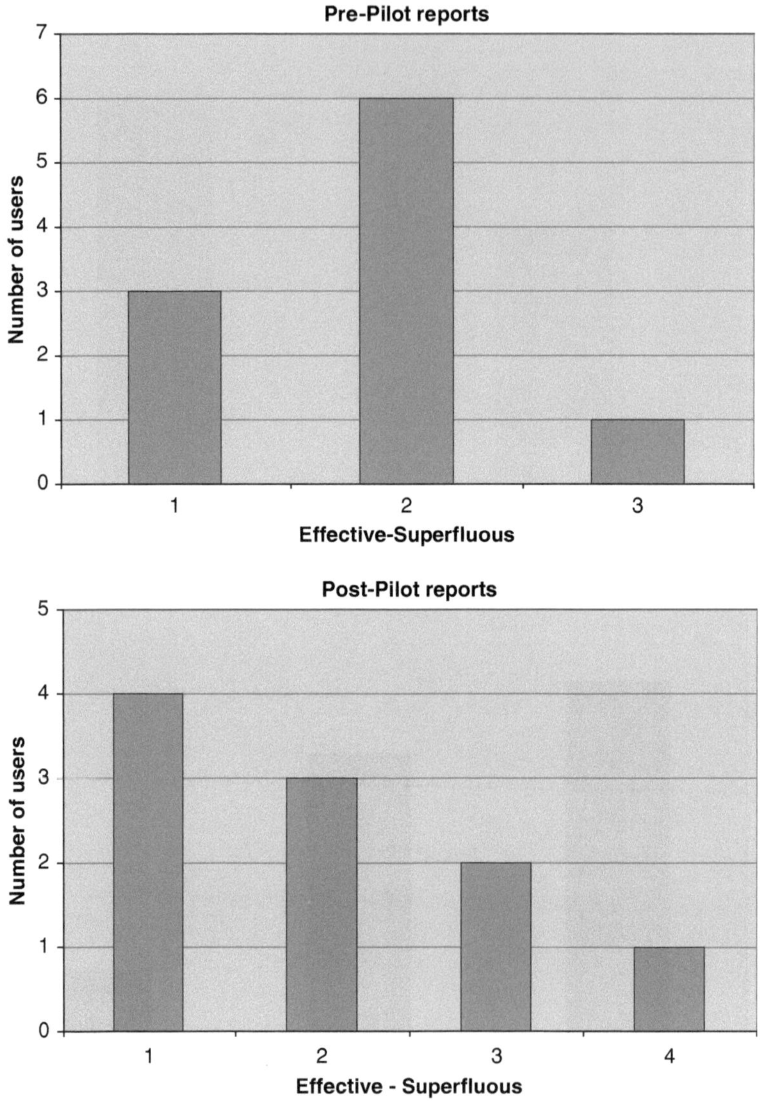

Fig. 15.10 Participants' pre- and post-pilot reports on how effective-superfluous they found the speed limit violation warning application of IN-SAFETY to be (Anund et al. 2008)

malfunctions of the system were identified requesting for high-end equipment. In the real application, these problems would not exist. As for the GPS/GPRS architecture, the only shortcoming is the cost of communication, which hopefully would be overcome in market applications by a relevant agreement between the road operators and the communication supplier.

The system assessment results, as derived from the pilot testing, using pre- and post-testing questionnaires that were answered by the participants, comparing driving the same route with and without the IN-SAFETY solutions, revealed a general acceptance of the system, which, in most cases, became even higher after testing the actual system.

Finally, it can be stated that the developed applications are working well and are very promising, since various future improvements are possible and could offer a lot more possibilities to the user. The main advantage of the system is that it provides the user with any available information in time, continuously updated, in his/her in-vehicle display. The primary concern in VMS provided info is the elimination of distraction from the driving task, which is achieved by minimizing the use of text, which by the tested system is substituted by adequate pictograms or sounds, and by offering the possibility to receive the information in the driver's preferred language.

References

A. Anund, A. Tapani, A. Kircher, C. Margberer, E. Bekiaris, A. Vatakis, P. Spanidis, K. Kalogirou, C. Liberto, S. Damiani, M. Raimondi, M. Wiethoff, T. Benz, IN-SAFETY D4.2, "Pilot results consolidation", Project Deliverable, 2008

E. Bekiaris, A. Amditis, M. Panou, DRIVABILITY: a new concept for modelling driving performance. Int. J. Cogn. Technol. Work. **5**(2), 152–161 (2003)

M. Gemou, E. Bekiaris, G. Wenzel, V. Vegiri, S. Damiani, A. Anund, C. Brugger, K. Sidiropoulou, IN-SAFETY D4.1, "Pilot plans", Project Deliverable, 2007

K. De Brucker, M. Wiethoff, IN-SAFETY, D2.1, "Implementation scenarios and concepts towards self-explaining roads", Project Deliverable, 2006

A. Erke, F. Sagberg, R. Hagman, Effects of route guidance variable message signs (VMS) on driver behaviour. Transp. Res. Part F Traffic Psychol. Behav. **10**(6), 447–457 (2007)

G. Gopher, *Attentional Allocation in Dual Task Environments. Attention and Performance III* (Elsevier, Amsterdam, 1990)

T.C. Handy, M. Solitani, G.R. Mangun, Perceptual load and visuocortical processing: event-related potentials reveal sensory-level selection. Psychol. Sci. **12**(3), 213–218 (2001)

K. Kalogirou, P. Spanidis, S. Damiani, M. Darin, F. Visintainer, IN-SAFETY D2.2, "Intelligent agents for driver information personalisation", Project Deliverable, 2008

P. Simlinger, S. Egger, C. Galinski et al., IN-SAFETY D2.3, "Proposal on unified pictograms, keywords, bilingual verbal messages and typefaces for VMS in the TERN", Project Deliverable, 2008

Part V
Final Evaluation

Chapter 16
Best Things First. The Application of Multi-Criteria Analysis to Derive Implementation Priorities for Innovative Road Safety Measures

Klaas De Brucker and Cathy Macharis

16.1 Generation of Implementation Scenarios

A number of potential alternative systems aimed at increasing road traffic safety by creating a more forgiving road (FOR) and a more self-explaining road (SER) environment have been identified in Chap. 4 of this book. These general categories were obtained by combining six typical errors, namely (1) excessive speed in unexpected sharp bends, (2) speeding in general, (3) violation of priority rules, (4) wrong use of the road, (5) failure when overtaking and (6) insufficient safety distance, with three dimensions, along which scenarios can be developed namely: (1) the vehicle, (2) the infrastructure and (3) the vehicle-infrastructure interface. By doing so, a total of 18 generic categories of potential alternatives were obtained. These should be considered as alternative venues for the design of innovative systems.

In this chapter, however, a limited set of more concrete scenarios will be selected for further in-depth study and final prioritization. The in-depth study took the form of extensive pilot studies[1] conducted by the experts, the results of many of which are described in this book.

The set of concrete scenarios, called "implementation scenarios", to be subjected to the above mentioned pilot studies was selected on the basis of the initial selection and prioritization process, as well as on extensive discussions with experts in two

[1]These pilot studies were conducted at the University of Stuttgart (Germany), the Research Centre Fiat (Turin, Italy), the Hellenic Institute of Transport (Thessaloniki, Greece) and the National Road and Transport Research Institute (Linköping, Sweden).

K. De Brucker (✉)
Faculty of Economics and Management, Hogeschool-Universiteit Brussel (HUB), Brussels, Belgium
e-mail: klaas.debrucker@hubrussel.be

C. Macharis
Department MOSI-Transport and Logistics, Vrije Universiteit Brussel (VUB), Brussels, Belgium

E. Bekiaris et al. (eds.), *Infrastructure and Safety in a Collaborative World*,
DOI 10.1007/978-3-642-18372-0_16, © Springer-Verlag Berlin Heidelberg 2011

special workshops,[2] and taking into account technical feasibilities. The idea that came out of these workshops was to firstly focus on the society's priorities, especially in terms of generic safety. Furthermore, users' comfort was also considered as important, since users will still be the consumers, and must be willing to pay. Such willingness to pay is critical to the manufacturers' investment risks. Manufacturers' objectives were also considered as important, but by this reasoning, seen as a third priority. Final prioritisation, which is the topic of this chapter, should, therefore, at least involve the alternatives that were ranked, in the initial prioritisation process, within the top three by the stakeholders "society/public policy makers" and "users". The alternatives ranked at the top from the society's point of view were related to speed warning and blind spot detection. The alternatives ranked at the top from a users' point of view were related to sharp bend warning and blind spot detection. Based on this information, combined with discussion among experts and taking into account technical feasibilities, it was decided to select a set of six implementation scenarios, as presented in Table 16.1. These scenarios were subject to pilot tests and it is these scenarios that need to be prioritised here in this chapter, as part of the final prioritisation process.

A more detailed description of each of the scenarios presented in Table 16.1 can be found in the extensive research report of the IN-SAFETY project (Macharis et al. 2008). There, each of these scenarios is illustrated by a set of graphical representations and real pictures.

For the purpose of evaluation of the scenarios described in Table 16.1, reference scenarios needed to be defined. In most cases the reference scenarios were equal to the current situations, with possible minor adaptations. Impacts were derived using an error-based approach or an empirical approach on the basis of pilot tests or a combination of both.

The implementation of the scenarios presented in Table 16.1 includes installation of infrastructure equipment (e.g., measuring equipment, variable message signs (VMS), roadside beacons, etc.) and vehicle equipment (e.g., radio receivers, devices for giving information or warnings to the driver, measurement equipment, vehicle-to-vehicle communication, etc.). For some scenarios, only vehicle equipment is required. As regards vehicle components, three possible implementation schemes are possible, namely: (1) *full scale implementation*, i.e., installation in all vehicles (both new and old ones) at the same time as the infrastructure equipment, with a market penetration rate of 100% from day 1, (2) *market trend*, i.e., installation in all new vehicles from day 1, (3) *voluntary*, which means that installation in new vehicles is voluntary. Implementation scheme no. 2 was considered the most realistic one and it is this scheme that was used in the final prioritisation process described here.

---

[2]The first workshop took place at the German Federal Highway Research Institute in Bergisch-Gladbach (Germany) on 4 September 2006 and the second one at the Danube University of Krems in Austria on 5 March 2007.

Table 16.1 Summary of a detailed scenario description, considered within the IN-SAFETY project

Name	Description	Type of system	Data needed for operation	Data collection for operation	Condition requirement	Reference scenario for evaluation	Main contributing factor for target accidents	Pilot studies
In-car variable message signs (VMS) info (dynamic legal speed limit motorways)	Self-explaining system Dynamic speed limit based on weather and traffic conditions	Roadside VMS Warning into vehicle	Current speed, environmental data, traffic volume/flow	Vehicle sensors, roadside sensors	Reliable detection systems, algorithms for calculating safe speed	(A) Current state (B) Roadside VMS, dynamic speed limit (no info into vehicle)	Inappropriate speed on motorways	Greek, Italian, German pilots
In-car school bus ahead warning	Self-explaining system Warning when school bus stops ahead	Warning from bus into vehicle	Vehicle location, school bus location	Vehicle equipment for vehicle-to-vehicle communication	Reliable detection systems, reliable radio transmitter and receiver	No in-vehicle warning	Not detecting school children after leaving or before entering a school bus	Swedish pilot
In-car curve speed warning (rural roads)	Self-explaining system Safe curve speed calculated based on curve geometry and weather conditions	Vehicle autonomous	Current speed, curve geometry, environmental data, vehicle characteristics	Digital maps, vehicle sensors	Reliable updated data basis for infrastructure conditions, algorithms for calculating safe speed	No in-vehicle warning	Inappropriate speed in curves on rural roads	–
In-car lane departure warnings (LDWA) (motorways)	Forgiving system Lane departure warnings based on lane markings + road side beacons in work zone	Warning into vehicle	Lane markings, speed, local conditions (e.g., roadworks)	Vehicle sensors (LDWA) Road side beacons (adaptive LDWA)	Lane markings, reliable detection systems	(A) No lane departure warning (B) Rumble strips no measures at road works	Lane departure on motorways	Swedish, German, Greek pilots
Overtaking assistant "blind spot vehicle detection" (more than 1 lane per direction)	Forgiving system Warning when overtaking while vehicle approaching from behind	Vehicle autonomous	Position and speed of vehicle approaching in blind spot, current speed	Vehicle sensors for detection of vehicle behind	Reliable detection systems	No overtaking assistance	Overseeing vehicle approaching from behind while overtaking	–
Overtaking assistant oncoming vehicle detection (1 lane per direction)	Forgiving system Warning when overtaking with oncoming traffic	Vehicle-to-vehicle communication	Location and speed of own vehicle and oncoming traffic	Vehicle sensors, equipment for vehicle-to-vehicle communication	Reliable detection and communication systems	No overtaking assistance	Overseeing oncoming traffic while overtaking	–

Source: Macharis et al. (2008)

16.2 Definition of a Set of Criteria for the Final Prioritisation of Scenarios

16.2.1 Definition of Criteria

The next phase in the MCA methodology is the construction of a set of criteria for the evaluation of the scenarios. This set, which is shown in Fig. 16.1, was elaborated after extensive discussions with experts and policy makers in several technical[3] and policy workshops[4] organised for this purpose. A draft for a criterion tree was presented by experts based on their experience with previous research projects related to road safety (e.g., the ADVISORS[5] project) (De Brucker et al. 2002 and Macharis et al. 2004) and by analysing the objectives and criteria that policy makers consider to be relevant in other, similar decision-making problems. This hierarchy of criteria was constructed according to the principles of the analytic hierarchy process (AHP) of Saaty (1977, 1986, 1988, 1995). However, it was the policy makers (and stakeholder representatives) who ultimately had the last word in the decision regarding the final structure of the criterion tree.

The top level of the decision tree shown in Fig. 16.1 represents the focus or overall objective, namely creating benefits by making the road environment more forgiving and more self-explanatory. At the second level, three groups of main stakeholders are considered, namely (1) the users, (2) society/authorities and (3) manufacturers. Within each group of stakeholders, a number of subcategories were identified, such as drivers, fleet owners and emergency centres (for the main category "users"), road managers and authorities (for the main category "authorities") and vehicle manufacturers, equipment manufacturers, system providers and content providers (for the main category "manufacturers"). As regards these subcategories, it turned out that it was not necessary to include them as separate groups, since the preferences of these subgroups were not substantially different from each other and since some of these subgroups were not organised in such a way so as to exert a substantial influence on policy making. At the third level, the criteria that these main stakeholders consider relevant are listed. At the lowest level, the scenarios that need to be prioritised are shown.

---

[3]A technical workshop with experts from consortium partners was organised at the Technical University of Delft on 6 and 7 February 2006.

[4]Two policy workshops were organised with representatives from policy makers, users and manufacturers. The first one took place at the premises of the Intertraffic Conference in Amsterdam on 6 April 2006 and the second one in Brussels during the annual meeting of POLIS on 24 October 2007. The aim of the former was to validate the set of criteria and to derive criterion weights. The aim of the latter was to validate the criterion weights and to present the pre-final results. Both of these workshops were organised by POLIS, which is a network organisation of leading European cities and regions.

[5]ADVISORS is the abbreviation for "Action for advanced Driver assistance and Vehicle control systems Implementation, Standardisation, Optimum use of the Road network and Safety". EC FP5 project GRD11999 10047.

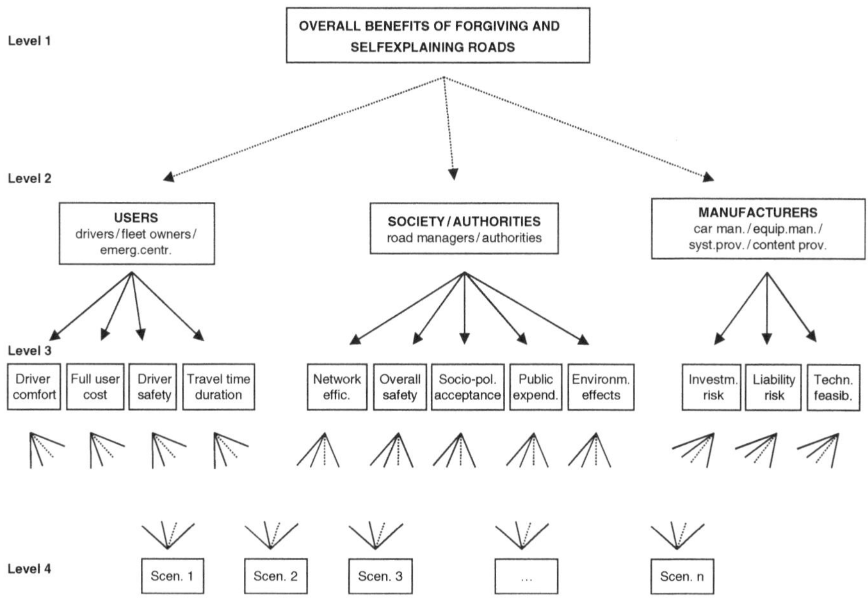

Fig. 16.1 Decision hierarchy for the prioritisation of FOR and SER scenarios
Source: the authors and IN-SAFETY project team, based on the AHP

It should be noted that the second stakeholder (at level 2) in Fig. 16.1 represents the point of view of public policy in general. The subsystem that is formed by this stakeholder and all its lower level elements is the most important one, since it represents the overall societal point of view. The two remaining subsystems, formed by the users (i.e., the demand side of the market), respectively the manufacturers (i.e., the supply side) and their lower level elements, are also important but in another context. Successful implementation of alternatives by public policy makers (i.e., the middle subsystem) is indeed only possible when the decisions made or the options chosen by these public policy makers are in accord, at least to a certain extent, with the interests of the other stakeholders. If this is the case, then the public policy objective will be facilitated by the actions taken by the other stakeholders and it will be easier for public policy makers to have their policies implemented. This way of using stakeholder management as facilitating (or hindering) public policy implementation is fully in line with the actual definition of the concept of "stakeholder" by Freeman (1984), who defined a stakeholder as any individual or group who can affect an organization's performance or who is affected by the achievement of this organization's objectives. The MCA that will be performed in the following sections, therefore, needs to be designed in such a way, so as to be able to investigate to which extent the solutions chosen within the second subsystem (public policy view) are compatible with the solutions preferred by the users and the manufacturers and whether this compatibility needs to be improved (or not), using a specific implementation path designed for this purpose.

In a perfect market (which is the standard assumption in neo-classical economics), however, the priorities derived at the demand side of the market would be expected to be fully consistent with the ones derived at the supply side, and government or public policy intervention (this is the middle subsystem in Fig. 16.1) would not be an important issue. It is thus assumed that what would be good for individual users would also be good for the society. This is definitely not the case here and several reasons can be identified for this. First, there are a number of external effects, such as effects on safety (including third part safety effects, such as effects on pedestrians and cyclists), environmental effects, etc. Second, infrastructure and also safety have the character of a public good, which can only be financed with governmental funds, to be provided by public policy makers. Third, there may be bounded rationality and consumer preferences may be inconsistent over time. When deciding on the type of goods to buy, consumers often have a preference for goods which result in an immediate, but temporary award, but which may result in a large cost or sacrifice in the future (e.g., road accidents). This future cost or sacrifice is often underestimated at the time the decision is made. This means that intervention in the market by public policy makers is highly necessary here. Fourth, the tools or systems analysed are highly innovative and the market still has to be developed. In such cases, government incentives or an active supply policy by government may be instrumental to stimulating and structuring the institutional structures of this evolving market. The decision problem which public policy makers are confronted with here is, therefore, not a simple but a complex one and the hierarchical structure developed here (Fig. 16.1) should be viewed as an attempt to order this complexity.

16.2.2 Definition of Criterion Weights

As described in Sect. 16.2.1, criterion weights were derived and validated during a couple of policy workshops with representatives from policy makers, users and manufacturers. For the workshop to elicit weights, the room was rearranged to facilitate a group decision room (GDR) session. A GDR consists of a network of computers running group systems software, which enables the participants in the session to express their opinion anonymously, and to be heard without having to draw the attention to themselves.

All the participants had to compare the importance of the criteria in pairs, using the pairwise comparison scale of the AHP, described in Chap. 3 of this book, Sect. 3.5.3. In order to synthesize the various pairwise comparisons given by each representative, the geometric mean was calculated, as suggested by Saaty (1995). Also the spread was calculated using the traditional statistical variables, such as mean, mode, highest and lowest score and standard deviation. More detailed information regarding the set up of these workshops and the results that came out of them, including the procedure followed for the calculation (and rounding) of the geometric mean, can be found in the extensive research report of the IN-SAFETY project (Macharis et al. 2008).

The final results of all these pairwise comparisons made by each representative of the various stakeholders and synthesised using the GDR software are shown in Tables 16.2–16.4. Part A of these tables shows the synthesis (i.e., the geometric mean) of the various pairwise comparisons and Part B contains the final relative priorities for the criteria (i.e., the criterion weights) calculated on the basis of these pairwise comparisons.

Tables 16.2–16.4 represent the relative priorities or weights[6] of the criteria, i.e., the priorities in terms of the overall objective of one specific stakeholder, respectively "users", "society/authorities" and "manufacturers".

The users gave the highest weight to the criterion "driver safety" (45.2%), followed by "full user cost" (28.2%). They gave less weight to "travel time duration" (17.9%) and to "driver comfort" (8.7%).

Table 16.2 Pairwise comparison matrix and relative priorities users' criteria

Stakeholder	Part A: Pairwise comparisons				Part B
Users	Driver comfort	Full user cost	Driver safety	Travel time duration	Relative priority
Driver comfort	1	1/3	1/5	1/2	0.087
Full user cost		1	1	1	0.282
Driver safety			1	4	0.452
Travel time duration				1	0.179

Inconsistency ratio: 0.06
Source: designed by the authors

Table 16.3 Pairwise comparison matrix and relative priorities society's criteria

Stakeholder	Part A: Pairwise comparisons					Part B
Society/ authorities	Network efficiency	Overall safety	Socio-political acceptance	Public expenditure	Environmental effects	Relative priority
Network efficiency	1	1/5	2	4	1	0.171
Overall safety		1	5	5	3	0.509
Socio-political acceptance			1	1	1/2	0.082
Public expenditure				1	1/3	0.068
Environmental effects					1	0.170

Inconsistency ratio: 0.04
Source: designed by the authors

[6]These weights were calculated using the computer programme ExpertChoice™, which makes it also possible to calculate the (in)consistency ratios. These ratios, which are shown at the bottom of each pairwise comparison table, express the degree to which the pairwise comparisons are internally (in)consistent from a mathematical point of view. They are calculated using the theory of eigenvectors described in chapter 3 of this book. An inconsistency lower than 10% is generally considered acceptable (Saaty 1995).

Table 16.4 Pairwise comparisons and relative priorities manufacturers' criteria

Stakeholder	Part A: Pairwise comparisons			Part B
Society/authorities	Investment risk	Liability risk	Technical feasibility	Relative priority
Investment risk	1	1/2	2	0.276
Liability risk		1	5	0.595
Technical feasibility			1	0.128

Inconsistency ratio: 0.01
Source: designed by the authors

From the societal point of view, the criterion "overall safety" turned out to be the most important criterion (50.9%). The criteria "network efficiency" and "environmental effects" received a lower weight, but nearly ex aequo (respectively 17.1 and 17.0%). The criteria "socio-political acceptance" and "public expenditure", received the lowest weight (respectively 8.2 and 6.8%). Manufacturers gave the highest weight to the criterion "liability risk" (59.5%), followed by "investment risk" (27.6%). Interestingly, the criterion "technical feasibility" received a much lower priority (12.8%).

16.3 Scoring of Scenarios on each Criterion

16.3.1 Introduction

After having defined a set of criteria and criteria weights, scenarios now need to be evaluated on each criterion separately. By doing so, a partial evaluation is obtained. This should be done by experts, on the basis of the information which is available to them. For some criteria hard data are available and these should of course be used. Some scenarios were subject to pilot studies which were conducted by consortium partners. The partial evaluations to be performed by experts in fact amount to assessing a cardinal value function for the criterion studied.

In the next section three possible ways to construct such a value function are being discussed. In the subsequent section, the capabilities of the pairwise comparison mechanism of the analytic hierarchy process (AHP) to implicitly derive such a value function are also briefly described.

16.3.2 Three Ways of Constructing a Cardinal Value Function

A. Case no. 1. The hard data describe a cardinal value function: no further operations are necessary.

   When the scores associated with the scenarios in terms of one specific attribute or criterion describe a cardinal value function, then these scores can be used for prioritisation without any further operations or calculations. A cardinal value function makes it possible to argue that a score of, e.g., 2*X*, is considered to be twice as good/bad as a score of *X* for each value of *X* (assuming that a natural zero point exists for this value scale). For instance, as regards the criterion or attribute "public expenditure", estimates of levels of expenditure expressed in euros are available, e.g., from the cost–benefit analysis described in Chap. 17 of this book. This information is expressed on a ratio scale.[7] It is quite natural to assume that a tool associated with a public expenditure of 200,000€ (such as tool B) is twice as bad compared to a tool with a public expenditure of only 100,000€ (such as tool A), in terms of the criterion *public* expenditure (i.e., in terms of a societal point of view). In this case, the scores obtained on the criterion or attribute "public expenditure" need no further computations. In fact, the value function transforming the attribute scale (z_j) into a value scale (v_j) is a linear one in this case (for the criterion j), as illustrated by the straight line in Fig. 16.2.

B. Case no. 2. The hard data is presented on a ratio scale, but it does not describe a cardinal value function.

The fact that the scores obtained e.g., from pilot studies are expressed on a ratio scale does not automatically imply that these scores describe a cardinal value function. The relation between the attribute scale (z_j) and the cardinal value scale (v_j) is not necessarily a linear one. For instance, as regards income for a private person, it is generally accepted that the marginal utility of income is decreasing. A job that pays twice the salary of another job is not necessarily considered to be twice as good in terms of the criterion salary. In case the salary is low, the job with the double salary may be considered close to two times better than the other job. In case the salary is high, it may be considered e.g., only 1.5 times better. The underlying value function transforming the attribute scale into a value scale is a concave one in this case. A value of 0.5 on attribute scale then corresponds to a score of e.g., 0.8 on value score (as shown in Fig. 16.2). It is mostly in cases where revenue (or other attributes, such as e.

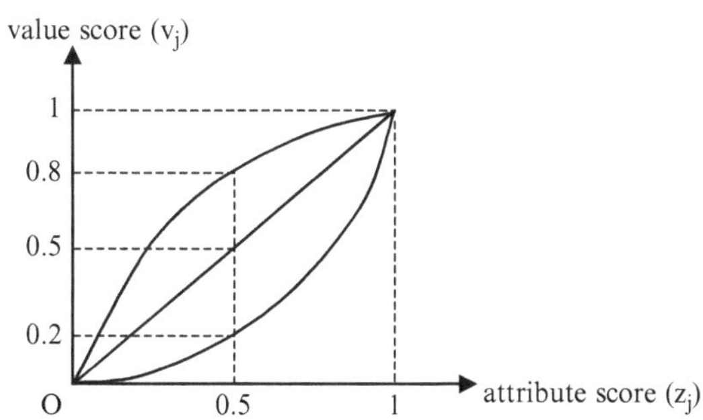

Fig. 16.2 Deriving a value function
Source: designed by the authors

[7]A ratio scale is a scale which not only consists of equidistant points, but which has also a meaningful zero point, e.g., revenue, cost and profit are ratio scales.

g., reduction of accidents) is studied in terms of society as a whole, that this relationship may be assumed to be a linear one.

C. Case no. 3: The hard data can only be expressed on an ordinal scale.

For some criteria it is not possible to obtain scores expressed on a ratio scale, but only scores on an ordinal scale. For instance, as regards the criterion "driver comfort" the highest or most sophisticated measurement scale that is possible may be an ordinal scale. This means that the scenarios can only be classified and ranked in terms of their contribution to this criterion ("driver comfort"), using explicitly or implicitly a scale, like e.g., – – –/– – /–/0/+/+ + /+ + +. Although this scale can be considered as a value function, it is purely an ordinal one. In this case, further operations in order to obtain a cardinal value function will be necessary. One technique to derive such a cardinal value function is the pairwise comparison mechanism of the AHP described below.

16.3.3 *The Use of the Pairwise Comparison Mechanism as a Tool to Construct a Cardinal Value Function*

The pairwise comparison mechanism of Saaty's AHP makes it possible to obtain a cardinal value function for the prioritisation of scenarios in terms of specific criteria. This will be especially useful in the cases B and C described above (in Sect. 16.3.2).

The use of this mechanism was described in detail in Chap. 3 of this book. More details can also be found in Saaty (1977, 1986, 1988, 1995). The relative priorities of the scenarios in terms of contribution to a specific criterion (g_j) (e.g., "driver comfort") are determined by comparing these scenarios in pairs, in terms of their contribution to that criterion. The values that may be used for making this comparison were presented in Chap. 3 (Table 3.4). These values ($Pg_j(a_i,a_{i'})$) represent the preference intensity for a specific pair of scenarios (e.g., the pair a_i - $a_{i'}$ in Table 3.3 in Chap. 3) in terms of the criterion studied (g_j). In the next section a concrete example of such a pairwise comparison matrix is given as applied to the IN-SAFETY project, for one specific criterion, namely "driver comfort" and for one specific expert.

16.3.4 *Scoring of Scenarios on each Criterion Separately*

The scoring of scenarios on the various criteria was performed by a number of experts coming from IN-SAFETY Consortium partners,[8] taking into account as

[8]Experts involved in the scoring of scenarios came from BASt (*Bundesanstalt für Strassenwesen,* i.e., the German Federal Highway Research Institute), CRF (*Centro Ricerche Fiat*, i.e., Fiat Research Centre), CERTH/HIT (Centre for Research and Technology Hellas/Hellenic Institute of Transport) and KfV (*Kuratorium für Verkehrssicherheit,* i.e., the Austrian Road Safety Board).

much as possible the hard data available to them, on the basis of results from pilot studies or other relevant studies or relying on their expert judgments. Within each partner, however, several internal experts were consulted to assist working on this scoring exercise. The whole scoring procedure was coordinated by the authors of this chapter, through a number of workshops and following a specific procedure comparable to traditional Delphi techniques.

Firstly, workshops[9] were organised with these experts, in order to explain the scoring procedure to them. Then, an extensive scoring document was submitted to the experts. This document contained a complete description of the scenarios to be scored (including the reference scenarios), a clear specification of the exact meaning of the criteria, as well as a clear explanation of the various ways of constructing a cardinal value function, including the role of the pairwise comparison mechanism of the AHP in this procedure. The main part of this scoring document was, however, a list of scoring tables, to be filled in by the experts. Experts had the choice to give their scores in a pairwise fashion, i.e., using the pairwise comparison mechanism or to give their scores using direct ratings. Most experts preferred to give their ratings in a pairwise fashion. An example of such a scoring table filled in by one expert for one specific criterion (namely "driver comfort") is given in Table 16.5 as an example. Such pairwise comparison matrices had to be filled in for each individual criterion (12 criteria in total) and for each individual expert.

In Table 16.5 the scenarios are compared in pairs, in terms of their contribution to one specific criterion, namely "driver comfort". The scenarios are listed horizontally on the first row of the matrix, as well as vertically in the first column of the matrix. The criterion in terms of which the pairwise comparison is made

Table 16.5 Example of a pairwise comparison matrix filled in by one expert[10]

Crit. Driver comfort	Scen. 1 VMS info into vehicle	Scen. 2 School bus ahead warning	Scen. 3 Safe curve speed warning	Scen. 4 Lane depart. warning	Scen. 5 Overtak. ass. with lane separation	Scen. 6 Overtak. ass. without lane separation
Scen. 1	1	7	1	1	5	1/2
Scen. 2	1/7	1	1/4	1/5	1/2	1/9
Scen. 3	1	4	1	1/2	3	1/3
Scen. 4	1	5	2	1	3	1/2
Scen. 5	1/5	2	1/3	1/3	1	1/8
Scen. 6	2	9	3	2	8	1

Overall inconsistency: 0.02
Source: KfV

[9]The first workshop took place at the *Donau Universität Krems* (Austria) on 5 March 2007. A second one took place at the Centre for Research and Technology Hellas/ Hellenic Institute of Transport (Thessaloniki, Greece) on 28 June 2007. By the time of this second workshop, some experts had already performed a preliminary scoring.

[10]The full name of the six scenarios can be found in the following table (Table 16.6). A detailed explanation of each scenario was given in Table 16.1.

(c.q. "driver comfort") is mentioned in the first cell in the first row of the matrix. The remaining cells contain the values of the actual pairwise comparisons made by the expert. For instance, the value 5 in the shaded cell of the upper right triangle of this matrix (on row 2) expresses that scenario no. 1 (which is the so-called "row element" or "row scenario") is considered to have a "higher importance" (value 5) over scenario no. 5 (which is the so-called "column element" or "column scenario") in terms of contribution to the criterion driver comfort. When the opposite comparison is made, i.e., when scenario no. 5 is compared with scenario no. 1, then this comparison should receive the reciprocal value of the initial comparison, i.e., the value 1/5 as shown in the shaded cell in the lower left triangle of the matrix (on row 6 of the matrix).

As soon as all the experts had completed their task, the scoring matrices were, then, collected by the coordinating team. The preliminary results of the scoring exercise were presented to the experts and other partners during another workshop.[11] Since the pilot tests were not finalised yet at that time, a revised scoring document was sent to the experts after this workshop. Finally, all the information provided by the experts (i.e., all scoring tables) was compared and analysed in depth by the coordinating team and then presented again to a committee of experts during a conclusive workshop.[12] By the end of this conclusive workshop, scores were

Table 16.6 Relative priorities of scenarios in terms of criteria by the user representatives

Scenario / Criterion (Weight)	Driver comfort (0.087)	Full user cost (0.282)	Driver safety (0.452)	Travel time sav. (0.179)	Overall (users) (1.000)
Scen. 1: In-car VMS dynamic speed limit	0.150	0.277	0.125	0.190	0.182
Scen. 2: School bus ahead warning	0.100	0.403	0.025	0.040	0.141
Scen. 3: In-car curve speed warning	0.150	0.043	0.250	0.090	0.154
Scen. 4: In-car lane departure warning (motorways)	0.150	0.124	0.100	0.040	0.100
Scen. 5: Overtaking assistant (blind spot vehicle detection) (>1 lane per direction)	0.150	0.115	0.175	0.040	0.132
Scen. 6: Overtaking assistant oncoming vehicle detection (1 lane per direction)	0.300	0.038	0.325	0.600	0.291

Overall inconsistency ratio: 0.06
Source: Own computation using ExpertChoice™

[11]This workshop took place at the Centre for Research and Technology Hellas/Hellenic Institute of Transport (Thessaloniki, Greece) on 28 June 2007.

[12]This conclusive workshop was organised by the coordinating team (i.e., the authors of this chapter) and took place in Brussels on 22 October 2007 at the premises of POLIS. To this workshop were invited not only experts from partners involved in the scoring exercise itself (BASt, CRF, CERTH/HIT and KfV), but also experts from other IN-SAFETY partners (TUDarm, TUDelft, USTUTT and other partners), in order to stimulate the discussion.

Table 16.7 Relative priorities of scenarios in terms of criteria societal point of view

Scenario / Criterion (Weight)	Network effic. (0.171)	Overall safety (0.509)	Soc-pol. accept. (0.082)	Public expend. (0.068)	Envir. effects (0.170)	Overall (Soc.) (1.000)
Scen. 1: In-car VMS dynamic speed limit	0.240	0.033	0.255	0.051	0.513	0.170
Scen. 2: School bus ahead warning	0.187	0.066	0.327	0.154	0.154	0.120
Scen. 3: In-car curve speed warning	0.187	0.220	0.120	0.205	0.077	0.181
Scen. 4: In-car lane departure warning (motorways)	0.075	0.200	0.212	0.179	0.103	0.162
Scen. 5: Overtaking assistant (blind spot vehicle detection) (>1 lane per direction)	0.236	0.180	0.052	0.205	0.103	0.168
Scen. 6: Overtaking assistant oncoming vehicle detection (1 lane per direction)	0.075	0.301	0.033	0.205	0.103	0.200

Overall inconsistency ratio: 0.04
Source: Own computation using ExpertChoice™

Table 16.8 Relative priorities of scenarios in terms of criteria from manufacturer representatives

Scenario / Criterion (Weight)	Investm. risk (0.276)	Liability risk (0.595)	Technic. feasib. (0.128)	Overall (manufact.) (1.000)
Scen. 1: In-car VMS dynamic speed limit	0.245	0.332	0.284	0.302
Scen. 2: School bus ahead warning	0.120	0.377	0.140	0.276
Scen. 3: In-car curve speed warning	0.112	0.053	0.077	0.072
Scen. 4: In-car lane departure warning (motorways)	0.169	0.160	0.249	0.174
Scen. 5: Overtaking assistant (blind spot vehicle detection) (more than 1 lane per direction)	0.241	0.057	0.219	0.129
Scen. 6: Overtaking assistant oncoming vehicle detection (1 lane per direction)	0.112	0.021	0.030	0.047

Overall inconsistency ratio: 0.01
Source: Own computation using ExpertChoice™

obtained through "consensus" and not by using the geometric mean technique as is often done in standard AHP applications (Saaty 1995). These consensus scores are presented in Tables 16.6–16.8 for the respective stakeholders. These results can also be visualised graphically as shown in the corresponding figures (Figs. 16.3–16.5).

Table 16.6 shows the relative priorities of the six scenarios studied in terms of the criteria considered relevant by the users. The scenarios are listed in the first column and the criteria (together with their respective weights) are listed in the first row. The relative priorities are mentioned in the remaining cells of that table. The last column contains the overall or global relative priorities of the scenarios

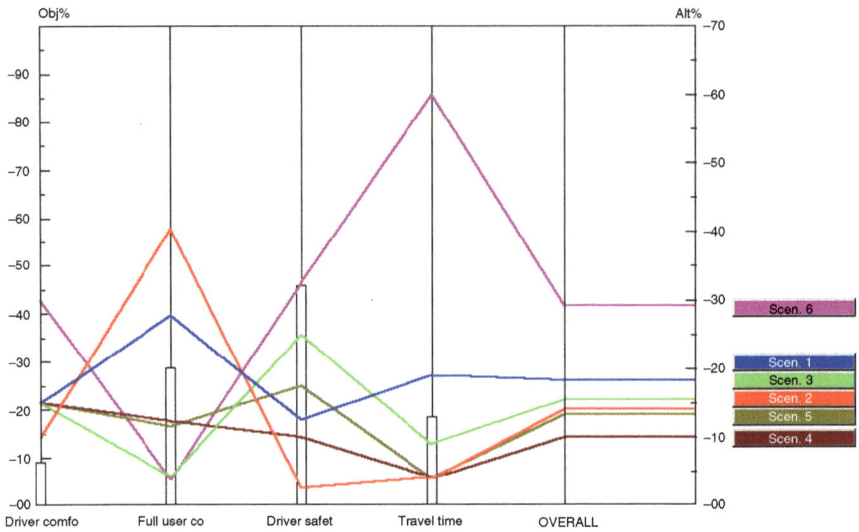

Fig. 16.3 Priorities of scenarios in terms of criteria for "users"
Source: Own computation using ExpertChoice™

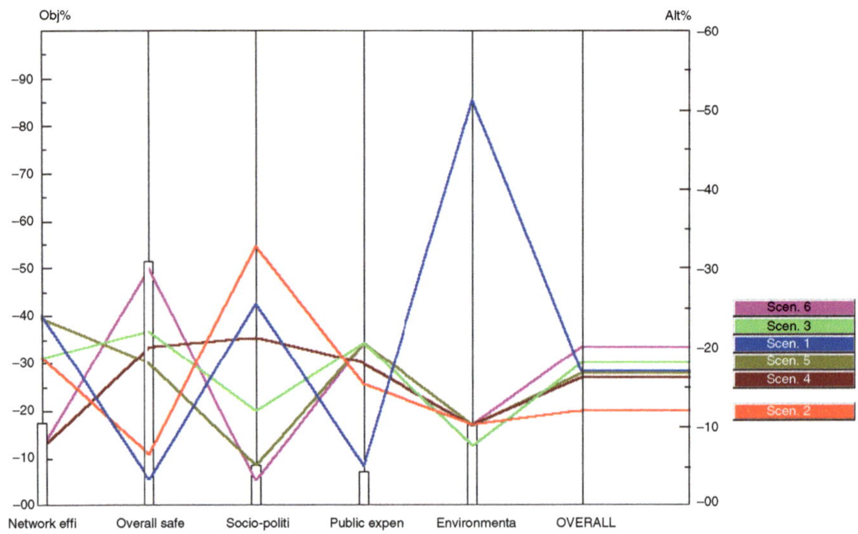

Fig. 16.4 Priorities of scenarios in terms of criteria for "society/public policy"
Source: Own computation using ExpertChoice™

from the point of view of the stakeholder "users". The way in which these were obtained will be explained in the next section.

Figure 16.3 shows the same information, but in a graphical way. The criteria are mentioned on the horizontal axis. The height of the vertical bars represents the

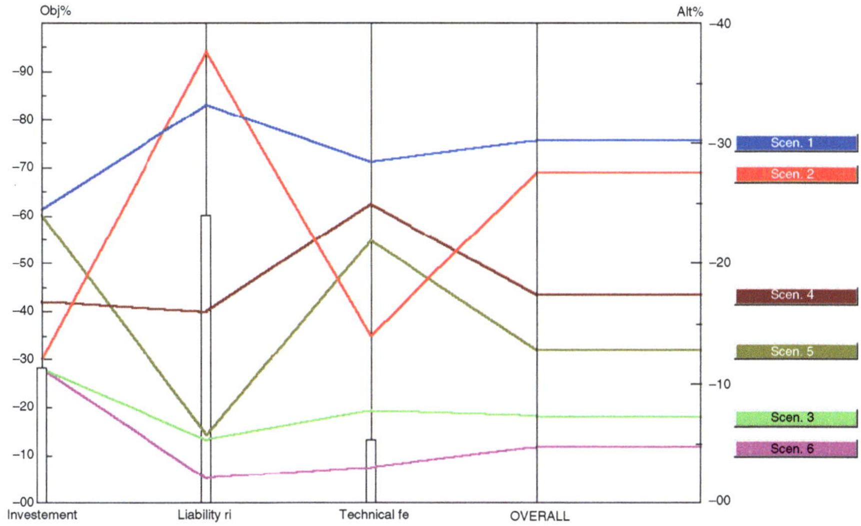

Fig. 16.5 Priorities of scenarios in terms of criteria for "manufacturers"
Source: Own computation using ExpertChoice™

criterion weights. On the right vertical axis are shown the global relative priorities from the users' point of view. The intersection of the lines from left to right with the vertical lines starting at the criterion name represents the relative priority of the scenario for that specific criterion. For instance, the high peak of the line related to scenario 6 when it intersects with the vertical line starting at the criterion "travel time" means that scenario no. 6 has a very high (i.e., a very good) score for the criterion *reducing* "travel time".[13] The overall priorities for each stakeholder are shown at the extreme right vertical axis, but these will be discussed in the next section.

Tables 16.7 and 16.8, as well as Figs. 16.4 and 16.5 are structured in the same way as Table 16.6 and Fig. 16.3 and contain the same information, but in terms of the criteria for "society/public policy" and "users".

It should again be pointed out that the results of this study can only be considered as indicative and any generalisation should be done with caution, since the opinions expressed cannot claim to represent the views of all users, manufacturers and other groups, but only the ones that participated in the study.

[13]Please note that all the criteria are conceived as benefit criteria. A higher score of a scenario on e.g., the criterion travel time (or full user cost) means that this scenario is associated with lower travel times (or lower full user cost) compared to the other scenarios.

16.4 Overall Evaluation of the Scenarios: Deriving Overall Priorities per Stakeholder

In order to derive the overall relative priorities for each stakeholder, the scores of the scenarios on the individual criteria (shown in Tables 16.6–16.8) have to be combined with the respective criterion weights. To this end, the scores on each individual criterion are added after being multiplied by the weight of each individual criterion. These weights are shown on the first row of the aforementioned tables. The results of combining scores with weights give the overall or global relative priorities, which are shown in the last column of Tables 16.6–16.8, for each individual stakeholder.

In Figs. 16.6–16.8 these global relative priorities are represented graphically for the respective stakeholders, namely "society", "users" and "manufacturers". In these figures, the scenarios are ranked in decreasing order of priority.

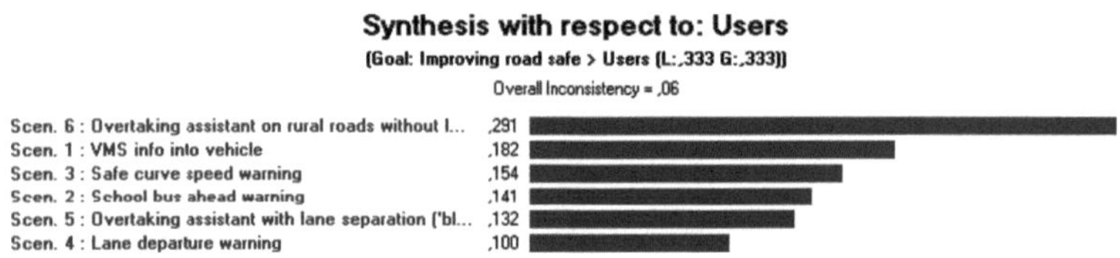

Fig. 16.6 Global priorities of scenarios from participating users' point of view
Source: Own computation using ExpertChoice™

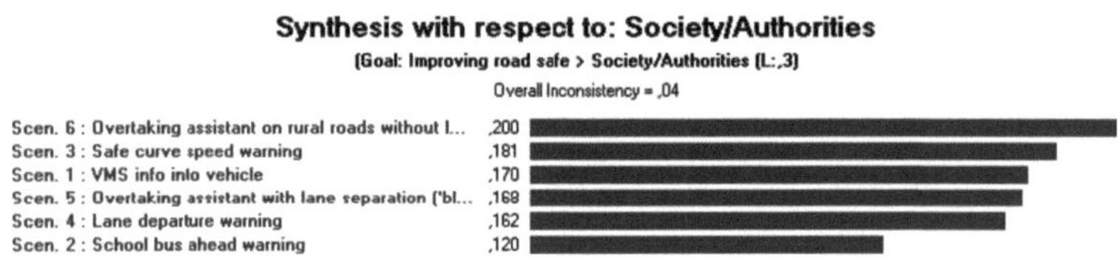

Fig. 16.7 Global priorities of scenarios from society's point of view (based upon participating experts opinion)
Source: Own computation using ExpertChoice™

Fig. 16.8 Global priorities of scenarios from participating manufacturers' point of view
Source: Own computation using ExpertChoice™

From a users' point of view (see Figs. 16.6 and 16.3), the most desirable scenario is definitely scenario no. 6 (i.e., the overtaking assistant on rural roads or oncoming vehicle detection), which obtains an overall relative priority of 0.291. This is mainly due to its high score on driver safety (which is the most important criterion), travel time and driver comfort (as can be inferred from Fig. 16.3). The second most desirable scenario is the scenario no. 1 (i.e., VMS info into the vehicle), which obtains an overall priority of 0.182, because of its relatively good score on the criteria "full user cost", "travel time" and "driver comfort". The least desirable scenario from the users' point of view is the lane departure warning scenario (scenario no. 4), with an overall priority of 0.100. This is due to the fact that this scenario, as compared to the other scenarios, scores not so well at the criteria "travel time" and "driver safety". The intermediate scenarios, i.e., the scenario no. 3 (in-car curve speed warning), no. 2 (school bus ahead warning) and no. 5 (blind spot detection) obtain priorities close to each other, i.e., between 0.132 and 0.154.

From the society's point of view, the final overall relative priorities of the six scenarios studied do not differ from each other substantially, as can be concluded from Figs. 16.7 and 16.4. However, there are two small exceptions. The scenario that is ranked first from society's point of view is the scenario no. 6, namely the overtaking assistant on rural roads (i.e., oncoming vehicle detection), which obtains an overall relative priority of 0.200. This is due to a very good score on the criterion "overall safety", which received a very high weight (as can be inferred from Fig. 16.4). The scenario with the lowest priority from society's point of view is the scenario related to "school bus ahead warning". This scenario obtains an overall priority of 0.120, mainly due to its very low score on the criterion "overall safety". The four remaining scenarios obtain overall relative priorities close to each other, i.e., ranging between 0.162 and 0.181.

From the participating manufacturers' point of view (see Fig. 16.8 or 16.5) the overall relative priorities are quite dispersed. However, two top scenarios can be distinguished, namely scenario no. 1 (VMS info into vehicle) and no. 2 (school bus ahead warning), as well as two bottom scenarios, namely scenario no. 3 (safe curve speed warning) and no. 6 (overtaking assistant with oncoming vehicles) and two intermediate scenarios, namely scenario no. 4 (lane departure warning) and no. 5 (blind spot detection). The two top scenarios obtain overall relative priorities of 0.302 respectively 0.276, mainly due to their good scores on the criteria "investment risk" and "liability risk" (as can be inferred from Fig. 16.5). The two bottom scenarios obtain low scores on all three criteria.

16.5 Discussion of Results Using a Multi-Actor Approach

The relative priorities derived in the former sections and discussed below were obtained through the application of a MAMCA and express the degree to which the various scenarios are expected to contribute to the objectives of the various stakeholders, namely: users, society and manufacturers.

Table 16.9 Priorities of users as compared to manufacturers' and societal priorities

Scenario	Society	User	Manufacturer
Scen. 6 (Overtaking assistant without lane separation: oncoming vehicle detection)	1	1	6
Scen. 3 (Safe curve speed warning)	2	3	5
Scen. 1 (variable message signs info into vehicle)	3	2	1
Scen. 5 (Overtaking assistant with lane separation: blind spot vehicle detection)	4	5	4
Scen. 4 (Lane departure warning)	5	6	3
Scen. 2 (School bus ahead warning)	6	4	2

Source: Own computation

The point of view of society is the most relevant one from a public policy point of view. The two other stakeholders' priorities are also important, but in another context, namely to test to which extent the priorities derived in terms of the societal point of view are compatible with the users' and manufacturers' priorities. In case they are, implementation of the scenarios will be easier and public policy will be facilitated by the actions of the other stakeholders. In case they are not, public policy may be hindered. Extra governmental incentives may, then, be necessary, in order to make the solutions preferred by the public policy makers more compatible with the solutions preferred by the other stakeholders.

In Table 16.9 the scenarios for the stakeholder "society" (extreme left) are, therefore, compared to the priorities of the other two stakeholder groups (right).

The priorities of the society in Table 16.9, when compared with those of the two other stakeholder groups, suggest where government intervention policies may be required to achieve effective implementation of safety systems, namely where a strong discrepancy can be observed in prioritisation among stakeholder groups. In some cases, safety systems may also be introduced autonomously, by market actors. This will occur when the systems have a high market potential, as expressed by their perceived contribution to both user and manufacturer objectives.

The most striking conclusion from Table 16.9 is the high discrepancy among stakeholder priorities as regards the scenario no. 6 (overtaking assistant with oncoming vehicle detection). This scenario is ranked at the top by users and society, but completely at the bottom by manufacturers. Manufacturers consider the risk associated with this scenario as too high, in particular the liability risk, but also the investment risk and the risk of technical non-feasibility. In other words, users and society have a high preference for this scenario, but manufacturers do not. Although this scenario has some market potential, it is not likely to hit the market in the near future. Further research is, therefore, needed to make this application more reliable and to reduce the risks associated with it. Additionally, policy makers should consider what measures could be taken to address the manufacturer's hesitations with respect to the possible liability risks.

Another scenario for which the conclusion from the comparison of stakeholder priorities is similar to that of the overtaking assistant with oncoming vehicle detection is scenario no. 3 (safe curve speed warning). This scenario is ranked second from society's point of view and third from the users' point of view, but only second last from the manufacturers' point of view. Again, manufacturers consider the risks of investments associated with this scenario as too high. This scenario also has quite some market potential, but it is not likely to materialize in the near future either. Here again, further research may be needed to make this application more reliable and to reduce the risks associated with it.

A scenario receiving a good overall priority from the various stakeholders is scenario no. 1 (VMS info into vehicle). This scenario receives a good score from users and from manufacturers (and a relatively good one from the society's point of view as well). This scenario will, therefore, more easily be implemented in the market, solely as the result of market forces.

Scenarios no. 5 (blind spot vehicle detection) and no. 4 (lane departure warning) do not receive too low scores from society's point of view. Although they are ranked fourth and fifth, the difference with the scenarios ranked just ahead is, indeed, rather small. However, these scenarios are not very much preferred neither by users nor by manufacturers.

Scenario no. 2 (school bus ahead warning) is ranked last from society's point of view and is in fourth position from the users' point of view. Although children are the most vulnerable road users in society, accidents with children running out of a school bus only represent a small portion of the total number of accidents. School buses usually already have a high visibility in the road environment. Manufacturers, however, consider this scenario as being of rather low risk in terms of reliability, technical feasibility and investment risk.

It should be noted that the relative priorities derived above were obtained assuming an implementation scheme, called "market trend", whereby installation would be compulsory in all new vehicles. If the starting base would be the implementation scheme called "voluntary", then results may be different, especially from the society's point of view. Scenarios that do not require vehicle-to-vehicle communication may, in that case, obtain a higher priority as compared to scenarios that do need this type of communication.

For a number of criteria, however, hard data regarding the scenarios performance was rather scarce or non-existing. Relying on expert judgment was, therefore, necessary. Indeed, the MCA-AHP is a decision tool which allows various experts to express their opinions regarding the contribution of scenarios to a number of stakeholder objectives (as measured by criteria) and these expert opinions are then synthesized using the pairwise comparison mechanism of the AHP. By confronting the various expert judgments (and making all the pairwise comparisons), subjectivity in the decision-making process is limited or made objective (Forman and Selly 2001). The final synthesis, i.e., the result in terms of final relative priorities, express a consensus of the various experts' opinions (just like in Delphi-poll techniques) and may be used as a basis to identify further research needs, even when hard data are rather scarce. It should indeed be noted that the scenarios studied are highly innovative and that hard data was rather scarce at this stage. The reliability of the priorities derived in this study may indeed by enhanced in the future when more hard data become available and/or when more expert opinions are incorporated in the actual synthesis.

16.6 Conclusions

In this chapter, a MAMCA was performed for the strategic evaluation of a number of innovative systems (called "scenarios") aimed at increasing road safety by creating a more FOR and SER environment.

A limited number of scenarios were selected, based on a preliminary prioritisation, a number of pilot tests and extensive discussions among experts. In the preliminary prioritisation a number of categorical alternatives were derived, by combining typical driver errors with dimensions, along which systems can be developed (such as the vehicle, the infrastructure and cooperative systems, combining both elements).

Three main stakeholders were considered relevant in this MAMCA, namely: users, society and manufacturers. The point of view of the society was considered as the most relevant one from a public policy point of view. The stakeholders "users" and "manufacturers" were included in the analysis, in order to test to which extent the priorities derived in terms of the societal point of view were compatible with those from a users' and a manufacturers' point of view.

The most striking general conclusion from the final prioritisation is that there is a high discrepancy among stakeholder priorities for some scenarios, whereas for other scenarios this discrepancy is rather low. For instance, for scenario no. 6 (overtaking assistant with oncoming vehicle detection) and scenario no. 3 (safe curve speed warning), discrepancy is high. These scenarios are considered to be good in terms of societal objectives, but not in terms of manufacturers' objectives. Manufacturers consider the risk associated with these scenarios as too high. A scenario receiving a good overall priority from the various stakeholders is, however, scenario no. 1 (VMS info into vehicle). This scenario will, therefore, more easily be implemented in the market by market forces, without the need for substantial governmental intervention.

Another striking conclusion is obtained regarding the scenario no. 2 (school bus ahead warning), which is ranked at the bottom from society's point of view. Accidents with children running out of a school bus only represent a small portion of the total number of accidents. Manufacturers, however, consider this scenario as being low risk.

Furthermore, the relative priorities derived through the use of MAMCA express the degree to which the various scenarios are expected to contribute to the stakeholders' objectives. For a number of criteria, however, hard data regarding the scenarios performance was rather scarce or non-existing. Relying on expert judgment was, therefore, necessary. This means that experts had to express their opinions regarding the expected contribution of the scenarios to a number of criteria and these experts' opinions were then synthesized into the final relative priorities.

The final synthesis, i.e., the result in terms of final relative priorities, express a consensus of the various participating experts' opinions (just like in Delphi-poll techniques) and may be used as a basis to identify further research needs. As more hard data become available in the future and more expert opinions can be expressed, the reliability of the results may be enhanced in further studies.

References

K. De Brucker, C. Macharis, A. Verbeke, B. Bekiaris, Integrated multicriteria analysis for advanced driver assistance systems. Final Deliverable of the research project "ADVISORS". FP5 Project GRD11999 10047 (Brussels, Commission of the European Union – Department Transport and Energy (DG TREN), 2002), http://www.advisors.iao.fhg.de > Reports > Deliverables > D6.1. Accessed 31 July 2009

E.H. Forman, M.A. Selly, *Decision by Objectives. How to Convince Others that You are Right* (World Scientific, Hackensack, NJ, 2001)

E.R. Freeman, *Strategic Management. A Stakeholder Approach* (Pitman/Ballinger, Boston, 1984)

C. Macharis, A. Verbeke, K. De Brucker, The strategic evaluation of new technologies through multicriteria analysis: the ADVISORS case, in *Economic Impacts of Intelligent Transportation Systems. Innovations and Case Studies*, ed. by E. Bekiaris, Y.J. Nakanishi (Elsevier, Amsterdam, 2004)

C. Macharis, A. Verbeke, K. De Brucker, E. Gelová, J. Weinberger, J. Vašek (2008) Implementation scenarios and further research priorities regarding forgiving and self-explaining roads. Implementation priorities and policy recommendations. Final deliverable (Del. 5.3) of the IN-SAFETY research project. Brussels, Commission of the European Union – Department Transport and Energy (DG TREN) (Contract No. 506716). http://www.insafety-eu.org/results.html. Accessed 31 July 2009

T.L. Saaty, A scaling method for priorities in hierarchical structures. J Math Pschychol **15**, 234–281 (1977)

T.L. Saaty, Axiomatic foundation of the analytic hierarchy process. Manag Sci **32**(7), 841–855 (1986)

T.L. Saaty, *The Analytic Hierarchy Process* (McGraw-Hill, New York, 1988)

T.L. Saaty, *Decision Making for Leaders. The Analytic Hierarchy Process for Decisions in a Complex World* (RWS, Pittsburgh, 1995)

Chapter 17
Value for Money. Cost–Benefit Analysis

Knut Veisten, Alena Erke, and Rune Elvik

17.1 An Error-Based Approach to Estimating Safety Impacts of ITS-Based Measures

The safety effects of selected ITS-based measures for more forgiving (FOR) and self-explaining (SER) road environments are virtually unknown (Gillen et al. 1999). There are few reports available with estimates of effects, particularly safety effects, from ITS-based systems (Bekiaris and Nakanishi 2004). In such a situation of missing data, one explorative approach is an *error-based* approach, aiming at the identification of target errors and target accidents that a particular measure is meant to correct (Wiethoff et al. 2006).

Most road safety measures aim at avoiding certain types of driver errors – and these are the *target errors* of the measures. Accidents that have been caused by these errors are the *target accidents* of the measures. If all accidents that are caused by a certain error could be identified, and if it could be assumed that the particular measure prevents all errors it is meant to, the exact number of prevented accidents could be identified. However, a precise estimate would require detailed in-depth accident analyses, in which a set of contributing factors would be identified and assessed, which was far beyond the scope of the IN-SAFETY project. Furthermore, it is unlikely that the implementation of a specific measure would really avoid all errors, e.g., because of driver non-compliance (Wiethoff et al. 2006).

Still, an explorative error-based approach using available accident data may be useful, indicating some potential safety effect. In our case, we estimate target accidents as precisely as possible from available accident data, and the resulting numbers of accidents are reduced by estimates of the proportion of accidents that will *not be affected* by the measure. As a simplification, we will assume full compliance, i.e. not entering the assessment of the proportion of errors that will be affected but not avoided by the scenarios due to non-compliance. Thus, the

K. Veisten (✉), A. Erke, and R. Elvik
Institute of Transport Economics (TOI), Oslo, Norway
e-mail: kve@toi.no

E. Bekiaris et al. (eds.), *Infrastructure and Safety in a Collaborative World*, 327
DOI 10.1007/978-3-642-18372-0_17,

estimates from this error-based approach may be considered as maximum potential safety effects of the ITS-based measures (Wiethoff et al. 2006). The error based approach is applied to German accident data.

17.2 A Partial Cost–Benefit Analysis

In our case only safety effects are calculated at the benefit side, so the cost–benefit analysis is only partial. The error of such a partial analysis will of course increase if the omitted effects, on time use, environment, etc., are considerable (Bekiaris and Nakanishi 2004), and this should be taken into account when assessing the resulting estimates.

For the assessment of the measures the reference line will be a "do-nothing"/ "status quo", meaning that none of the ITS-based measures will have any impact over the project horizon. The economic assessment in terms of estimated benefit–cost ratios is given from the anticipated implementation of the measures ("do-something"), as a difference from the reference.[1] The implementation will comprise "immediate (year 0) installation" of ITS infrastructure and a gradual installation of necessary ITS equipment in new cars. The other basic assumptions for the economic analysis are shown in Table 17.1.

The scenarios are evaluated with German data. The numbers of cars in the 15 years from 2008 to 2022 are estimated based on the total numbers of cars in 1999 and 2004. The average annual increase of the total number of cars is 1.24%.

Table 17.1 Basic assumptions for the economic assessment

Parameter	Assumption/description
Time horizon	15 years (2008–2022)
Result year ("year 0")	2007
Discount rate	3%
Implementation	Full scale infrastructure installation from year 1 and equipment installation in all new vehicles entering the market from year 1
Impact delineation	Measures have safety impacts on either motorways or (other) rural roads
Impact measure	Estimated decrease in fatalities/injuries
Costs	Infrastructure and vehicle equipment costs
Benefits	Proposed European valuation (for Germany) of reduced fatalities/injuries

[1] Alternatively, one could consider that the fatalities/injuries in the "do-nothing" reference still would involve a development of the ITS-based vehicle technology. I.e., for those systems that are already in the (car) market or foreseen in the near future, it could possibly be assumed some market penetration rate also in the "do-nothing" reference; and then having a more sharply increasing market penetration rate in the "do-something" project scenarios. COWI (2006) assumed 10% market deployment of in-vehicle lane departure warning in 2025 in the "do-nothing" reference.

Table 17.2 Predicted total numbers of cars, numbers of new cars and penetration rates in 2008–2022 (Germany)

	Total number of cars	Number of new cars	Penetration rate
2008	47,305,759	3,433,948	0.07
2009	47,894,329	3,476,673	0.14
2010	48,490,221	3,519,929	0.22
2011	49,093,528	3,563,723	0.29
2012	49,704,341	3,608,062	0.35
2013	50,322,753	3,652,953	0.42
2014	50,948,860	3,698,402	0.49
2015	51,582,756	3,744,417	0.56
2016	52,224,539	3,791,005	0.62
2017	52,874,308	3,838,172	0.69
2018	53,532,160	3,885,925	0.75
2019	54,198,198	3,934,273	0.81
2020	54,872,522	3,983,223	0.88
2021	55,555,236	4,032,782	0.94
2022	56,246,444	4,082,957	1.00

The annual numbers of new cars are estimated based on the assumption that all cars registered in 2022 will have been registered in the year 2008 or later. This results in an annual renewal rate of 7.3%. The annual numbers of units that are relevant for investments in ITS equipment, the annual number of new cars, are shown in Table 17.2.

The analyses of ITS safety measures will be carried out *as if* they yielded effects on the whole geographical area – on all German motorways or on all (other) German rural roads.

The time horizon reflects approximately the lifetime of cars and, thus, the replacement period of existing cars in year 0. The chosen discount rate is the one currently applied for transport projects in Germany and in several other European countries, while EU proposes a 5% discount rate. Implementation will cover all relevant roads within the chosen area, e.g., all German motorways or all German rural roads. Costs are estimated from various sources, both for the vehicle equipment (information receivers, warning devices, etc.) and infrastructure (road side beacons, VMS), also building on information about other ITS-based systems than the selected four. Monetized benefits of safety impacts are based on recent proposals for the EU countries (Bickel et al. 2006). Official German valuations are lower, and would basically imply lower benefit estimates (Hakkert and Wesemann 2005). In addition to the assessment of single measures, also the simultaneous implementation of all four measures will be assessed. For all assessments a benefit–cost ratio is provided, together with a break-even analysis, whereby (vehicle) cost assumptions are adjusted to yield a benefit–cost ratio equal to unity.

Four of the six IN-SAFETY scenarios are relevant for application with German data, and will be assessed in cost–benefit analyses. These four ITS-based proposed measures, A–D, are presented in Table 17.3.

Table 17.3 The four ITS-based proposed road safety measures (adapted from Kleine and Lotz 2007)

No. of proposed scenario/name	Description	Type of system	Data needed for operation	Data collection for operation	Condition requirement	Reference (baseline situation)	Main contributing factor in target accidents
A. In-car curve speed warning (rural roads)	Self-explaining system Safe curve speed calculated based on curve geometry and weather conditions	Vehicle autonomous	Current speed, curve geometry, environmental data, vehicle characteristics	Digital maps, vehicle sensors	Reliable updated data basis for infrastructure conditions, algorithms for calculation of safe speed	No in-vehicle curve warning	Inappropriate speed in curves on rural roads
B. In-car lane departure warnings (motorways)	Forgiving system Lane departure warnings based on lane markings and road side beacons in road work zones	Warning into vehicle	Lane markings, speed, local conditions (e.g., roadwork)	Vehicle sensors Road side beacons	Lane markings, reliable detection systems	No in-vehicle lane departure warning	Lane departure on motorways
C. Overtaking assistant blind spot vehicle detection (more than 1 lane per direction)	Forgiving system Warning when overtaking while vehicle approaching from behind	Vehicle autonomous	Position and speed of vehicle approaching in blind spot, current speed	Vehicle sensors for detection of vehicles behind	Reliable detection systems	No in-vehicle overtaking assistance	Overseeing vehicle approaching from behind while overtaking
D. Overtaking assistant approaching vehicle detection (1 lane per direction)	Forgiving system Warning when overtaking with oncoming traffic	Vehicle to vehicle communication	Location and speed of own vehicle and oncoming traffic	Vehicle sensors, equipment for vehicle-to-vehicle communication	Reliable detection and communication systems	No in-vehicle overtaking assistance	Overseeing oncoming traffic while overtaking

17.3 Data Retrieval and Gathering

17.3.1 Accident Data

For the error-based approach, registered causes for (injury) accidents are applied. Table 17.4 displays the German accident data base from 2005.

ITS-based measures B (in-car lane departure) and C (overtaking assistant – blind zone) have a potential road safety impact on motorways. Table 17.5 shows registered accident causes on motorways, for injury accidents involving a car, as well as how these are distributed according to injury severity. Road departure, slippery road, and speed are identified as the main causes of accidents on German motorways.

Table 17.6 displays the registered accident causes on German rural roads, for injury accidents involving a car. ITS-based measures A (in-car curve speed warning) and D (overtaking assistant – in front) have a potential road safety impact on rural roads. Accidents in curves are given as a total percentage and distributed as shares between three causes.

Regarding additional causes for injury accidents in curves, ca 42% is related to speed, ca 7% to drunk driving, and ca 1% to game in the curve (the sum exceeds 48% due to rounding). Subtracting causes that are not affected by the specific ITS-based measures, the target accidents and target errors can be singled out.

Table 17.4 Accidents on rural roads and motorways involving a car – Germany, 2005

	Rural roads	Motorways	Sum
Injury accidents	66,464	17,484	83,948
Fatalities	2,288	432	2,720
Serious injuries	22,470	4,383	26,853
Slight injuries	76,226	22,759	98,985

Source: https://www-ec.destatis.de/; "Verkehr im Überblick – Stand 12 Sept 2006 – Fachserie 8 Reihe 1.2 – 2005"; Table: 3.5.4(3), pp. 135–136

Table 17.5 Registered causes for accidents, percentage of all accidents on motorways involving a car – Germany, 2005[a]

	Speed (%)	Slippery road (%)	Drunk driving (%)	Road departure (%)	Head-on collision (%)	Overtaking (%)
Injury accidents	22	35	4.3	43	0.6	4
Fatalities	43	30	7	60	6	1
Serious injuries	31	33	6	60	1.6	3
Slight injuries	17	35	4	35	0.7	4

[a]Calculations are based on https://www-ec.destatis.de/; "Verkehr im Überblick – Stand 12 Sept 2006 – Fachserie 8 Reihe 1.2 – 2005"; Table: 3.5.4(3), p. 136. More than one cause may be attributed to a single accident, such that the sum of all causes is above 100%. Inappropriate (too high) speed as registered cause of accident comprises both optimal and non-optimal conditions, and for speed the figures for different injury severity are estimated based on the "power model" (Elvik et al. 2004)

Table 17.6 Registered causes for accidents, percentage of all accidents on rural roads involving a car, and shares of causes for accidents in curves – Germany, 2005[a]

	Curves (%)	Overtaking (%)
Injury accidents	48	2
Fatalities	43	7
Serious injuries	31	4
Slight injuries	17	3

[a]Calculations are based on https://www-ec.destatis.de/; "Verkehr im Überblick – Stand 12 Sept 2006 – Fachserie 8 Reihe 1.2 – 2005"; Table: 3.5.4(3), p. 135. Accidents in curves are shown as a total and distributed as shares among causes

Table 17.7 Assumed costs (in €) for equipment of vehicles, per unit and summed, for single measures and combination of all measures

Vehicle device	Cost per device	ITS-based measures				
		A	B	C	D	All
Warning device, incl. information processing and interface	150	150	150	150	150	200
Information receiver	100		100			100
GPS, incl. digital map	150	150		150	150	150
Measurement equipment for road and weather conditions	200	200				200
Safe curve speed model and calculation	50	50				50
Detection lane markings	150		150			150
Detection approaching vehicle (from in front/behind)	150			150	150	150
Vehicle to vehicle communication	300				300	300
Sum of units per measure (€)		550	400	450	750	1,300
Total vehicle investment costs (million €), present value (3% discount rate)		24,453	17,784	20,007	33,345	57,799
Total vehicle maintenance costs (million €), present value (3% discount rate)		17,769	0	0	0	17,769

17.3.2 *Cost and Benefit Data*

Cost data are primarily based on COWI (2006) and the US Department of Transportation (DOT 2007). Costs for equipment of vehicles consist of some components that are very similar for most of the ITS-based measures. The calculation of total vehicle costs includes the numbers of units (new cars), unit investment cost, and effective lifetime for infrastructure and vehicle equipment. Costs for maintenance of vehicle equipment are only included for Measure A. There may also be specific project management costs (e.g., planning, project management, regulation) in addition to investment and maintenance of equipments that are omitted from analysis. Table 17.7 displays the vehicle component costs, the sum costs per measure, and the aggregate costs for implementing all four measures.

Table 17.8 Assumed costs for infrastructure, investment/maintenance per unit and present costs

Measure	Unit cost		Lifetime	Present value (3% discount), million €		
	Investment	Maintenance		Investment	Maintenance	Total
A	10 € million input GPS maps	50% of investment cost	15years	10	60	70
B	Lane markings: 2,500 € per km motorway	1,000 € per km motorway	5 years			
	Road side beacons: 4,550 € per beacon	10% of investment cost	5 years			
	Sum			98	128	226
C	No infrastructure	-	-	-	-	0
D	10 € million input GPS maps	50% of investment cost	15 years	10	60	70
All				118	247	365

The unit costs per measure are simply given as sums of the device costs. Measure B (in-vehicle warning when the vehicle is about to depart from the driving lane) is assumed "cheapest" in terms of necessary vehicle equipment, while measure D (in-vehicle warning of vehicle ahead when overtaking on two-lane rural road) is assumed most expensive. Measure A is assumed to have a vehicle component implying maintenance costs (for the particular safe curve speed component) that will be in the same range as the investment costs. Regarding the costs for implementing all four measures together, it is assumed some cost synergy for vehicle equipment – that there is joint production when single measures are put together. Total vehicle investment costs are calculated as a sum of investment costs for new cars, from 2008 to 2022 (see Table 17.2).

Costs assumptions for infrastructure equipment are shown in Table 17.8, including unit costs, maintenance costs and present values of the costs.

Measure B is assumed most expensive in terms of infrastructure investment, while measure C (in-vehicle warning of vehicles in the blind spot when overtaking on motorways) will not need infrastructure investment – it is a vehicle-to-vehicle measure, less dependent on improved GPS maps than the rural road measures A and D.

17.4 Results

17.4.1 Estimated Safety Impacts from the Error-Based Approach

For ITS-based measures B and C, the combination of figures from Tables 17.4 and 17.5 will yield the estimated safety potential. The proportion of injury accidents (fatalities/injuries) that can be prevented by measure B, can be calculated as a sum of the proportions of road departure and head-on collision accidents, scaled down by a proportion *not* involving inappropriate speed (assumedly 50% of these): $(0.43 + 0.006) \times (1 - 0.11) = 0.39$ of injury accidents on motorways. For measure C the proportion that can be prevented is the overtaking accidents: 0.04 of injury accidents on motorways. Multiplying these estimated shares by the figures from the German accident data, yields an estimate of the target accidents, as displayed in Table 17.9.

According to these figures, the ITS-based measure B (in-vehicle lane departure warning) has clearly the largest safety potential, among the measures targeting motorway accidents. However, it should be remarked that the estimated safety impact of measure B may be regarded as "high". COWI (2006) presents lower estimates (ca 25%). The safety potential is given from the case of installation of necessary equipment in all cars in project year 0, thus in the case of gradual installation in new cars the numbers of prevented fatalities and injuries will not be reached before the end of the investment project period (in 2022).

For ITS-based measures A and D, the combination of figures from Tables 17.4 and 17.6 will yield the estimated safety potential. The proportion of injury accidents (fatalities/injuries) that can be prevented by measure A, can be calculated as a product of the proportions of curve accidents and speed as a cause of the accident, scaled down by the proportions *not* involving alcohol and *not* involving game accidents: $0.48 \times 0.42 \times (1 - 0.07) \times (1 - 0.01) = 0.18$ of injury accidents on rural roads. For measure D, the proportion that can be prevented is the overtaking accidents: 0.02 of injury accidents on rural roads. Multiplying these estimated shares by the figures from the German accident data, yields an estimate of the target accidents and safety potential, as displayed in Table 17.10.

Table 17.9 Safety potential of ITS-based measures for accidents on motorways involving a car – error-based approach – Germany, 2005

	Motorway accidents	Measure B		Measure C		Benefit unit value
		Target accidents (%)	Safety potential	Target accidents (%)	Safety potential	
Injury accidents	17,484	39	6,799	4	716	
Fatalities	432	59	240	1	5	1,496,000
Serious injuries	4,383	55	2,321	3	136	209,400
Slight injuries	22,759	31	6,511	4	821	17,100

Table 17.10 Safety potential of ITS-based measures for accidents on rural roads involving a car – error-based approach – Germany, 2005

	Rural road accidents	Measure B		Measure C		Benefit unit value
		Target accidents (%)	Safety potential	Target accidents (%)	Safety potential	
Injury accidents	66,464	18	12,219	2.5	1,676	
Fatalities	2,288	27	623	6	148	1,496,000
Serious injuries	22,470	23	5,178	4	924	209,400
Slight injuries	76,226	17	12,933	3	2,054	17,100

Table 17.11 Economic assessment of ITS-based measures, partial analysis including safety impacts – error-based approach – Germany

	A	B	C	D	All
Infrastructure costs	70	226	0	70	365
Vehicle costs	42,222	17,784	20,007	33,345	75,567
Safety benefits	26,179	11,190	585	5,265	43,219
Benefit–cost ratio	0.62	0.62	0.03	0.16	0.57

[a]Present values, in 2007, of costs and benefits in million Euros

Table 17.12 Vehicle costs of ITS-based measures necessary for achieving break-even (given estimated safety impacts) – Germany

Sum unit costs, vehicle devices	A (€)	B (€)	C (€)	D (€)	All (€)
Main analysis	550	400	450	750	1,300
Break-even	341	247	13	117	565

According to these figures, the ITS-based measure A (in-vehicle curve warning) has the largest safety potential, among the measures targeting rural road accidents. Similarly to the case for motorways, the safety potential is given from the case of installation of necessary equipment in all cars in project year 0, such that in the case of gradual installation in new cars the numbers of prevented fatalities and injuries will not be reached before the end of the project period (in 2022).

17.4.2 Estimated Benefit–Cost Ratio

The estimated costs (infrastructure plus vehicle equipments), benefits (monetised values of reduced fatalities, serious and slight injuries) and benefit–cost ratios are shown in Table 17.11.

It becomes clear from the table that the vehicle equipment costs are driving the results, together with the safety benefits. Infrastructure costs will in all cases have only marginal impact on the estimated benefit–cost ratios. None of the four proposed ITS-based measures reaches the required efficiency level of one (break-even), with our applied assumptions and taking into account only safety impacts. However, measures A and B are close to the efficiency requirement. Table 17.12 displays the unit costs for vehicle devices (and infrastructure) that would be necessary to obtain break-even, compared to the costs applied in the main analysis.

The reductions in vehicle equipment costs indicated for measures A and B, as well as for the combined implementation of all measures, necessary to reach break-even, may be considered as being within a future probable cost interval.

17.5 The Price of Safety

In this chapter an explorative cost–benefit analysis of four ITS-based road safety implementation scenarios has been presented. The results indicate that in-car speed warning and in-car lane departure warning rank highest in an economic assessment which includes only safety impact estimates. None of the measures obtain benefit–cost ratios above unity (break-even) under the main assumptions. However, the cost–benefit analysis was only partial, omitting possible effects on, e.g., time-use and environmental effects. For those measures that also have positive effects on environment (reduced emissions) and/or time use, the benefit–cost ratios may indeed show efficiency.

COWI (2006) and Baum et al. (2006) presented CBA for something similar to our measures B, C and D, i.e., a combined lane-departure warning and lane-change assistant (targeting road departures, side collisions, and head-on collisions). Their cost estimates were 600 € in 2010 and 400 € in 2020, half for the lane-departure warning and half for the lane-change assistant. They applied their analyses to the whole of EU-25, and estimated target accidents/casualties at the European level using accident data from only a few countries (particularly Germany, but also Spain and Denmark). In addition, Baum et al. (2006) assumed that drivers with lane-departure warning and lane-change assistant drove twice as much as other drivers. They estimated a BC ratio at approximately 2, while COWI (2006) presented a best-estimate BC ratio of 1.7. The main reasons for higher BC ratios in these two studies are already indicated; they both applied:

- Larger number of target accidents/casualties, because they aggregated motorways and other rural roads (while we have treated them separately), i.e., higher estimated safety effects.
- Lower vehicle equipment costs.

The CBA analyses of COWI (2006) and Baum et al. (2006) were also partial, including only safety effects. Our results are more pessimistic than those from the two former studies. Clearly, more research is needed for establishing the economic efficiency of ITS-based measures, and there is a particular need for more complete cost–benefit analyses, that will also include effects on time use and other potential impacts, in addition to more precise quantification of safety impacts.

References

H. Baum, T. Geißler, S. Grawenhoff, W.H. Schulz, Cost-Benefit-Analysis of Intelligent Vehicle Safety Systems. - Some Empirical Case Studies, in: Zeitschrift für Verkehrswissenschaft, 77. Jg. (2006), S. 226–254

E. Bekiaris, Y.J. Nakanishi, *Economic Impacts of Intelligent Transportation Systems: Innovations and Case Studies. Research in Transportation Economics, Vol. 8* (Elsevier, Amsterdam, 2004)

P. Bickel, R. Friedrich, A. Burgess, P. Fagiani, A. Hunt, G. De Jong, J. Laird, C. Lieb, G. Lindberg, P. Mackie, S. Navrud, T. Odgaard, A. Ricci, J. Shires, L. Tavasszy, Proposal for harmonised guidelines, Deliverable 5, Developing Harmonised European Approaches for Transport Costing and Project Assessment (HEATCO). Project funded by the European Commission under the Transport RTD Programme of the 6th Framework Programme, 2006

COWI, Cost–benefit assessment and prioritisation of vehicle safety technologies. Final Report, Economic assistance activities, Framework Contract TREN/A1/56-2004, Consultancy within Engineering, Environmental Science and Economics (COWI A/S), Kongens, Lyngby, 2006

DOT (U.S. Department of Transportation, Washington, DC, 2007), http://www.itscosts.its.dot.gov/

R. Elvik, P. Christensen, A. Amundsen, Speed and road accidents: an evaluation of the Power Model. TØI Report 740/2004. Institute of Transport Economics (TØI), Oslo, 2004

D. Gillen, J. Li, J. Dahlgren, E. Chang, Assessing the benefits and costs of ITS projects: volume 1 methodology, Research Report, Institute of Transportation Studies, University of California, Berkeley, 1999

S. Hakkert, P. Wesemann (eds.), *The Use of Efficiency Assessment Tools: Solutions to Barriers, SWOV Report R-2005-02* (Institute for Road Safety Research (SWOV), Leidschendam, 2005)

J. Kleine, C. Lotz, Selected scenarios for WP5. Internal Report WP5, C.N. 506716, IN-SAFETY Project, April 2007

M. Wiethoff, V.A.W.J. Marchau, D. de Waard, L. Walta, K.A. Brookhuis, C. Macharis, C. Lotz, G. Wenzel, E. Ferrari, M. Lu, S. Damiani, Implementation scenarios and concepts towards forgiving roads. Deliverable D1.1, C.N. 506716, IN-SAFETY project, October 2006.

Chapter 18
Anybody Listening?

Marion Wiethoff, Cathy Macharis, and Evangelia Gaitanidou

18.1 A Message for Whom?

Throughout this book, the results of the IN-SAFETY project have been put forward. In this chapter, we will formulate the main policy recommendations that can be extracted from these research results. These policy recommendations are primarily addressed towards the following stakeholder groups:

- Legislation bodies on the EU and national level
- EU and national research funding bodies
- Public and private infrastructure owners and road operators
- Standardisation bodies
- Insurance companies

The private sector is not directly addressed with policy recommendations; nevertheless the IN-SAFETY results are of interest to it as well.

First of all, the motivation of each stakeholder group to act according to political recommendations to achieve more traffic safety is analysed.

18.1.1 Motivation of EU and National Legislation Bodies

Legislation bodies on the EU-level as well as on the national level act on certain political objectives – usually defined in political programs. According to the White Paper "European transport policy for 2010: time to decide" by the EC (2001), a

M. Wiethoff (✉)
Delft University of Technology, Delft, The Netherlands
e-mail: m.wiethoff@tudelft.nl

C. Macharis
Department MOSI-Transport and Logistics, Vrije Universiteit Brussel (VUB), Brussels, Belgium

E. Gaitanidou
Centre for Research and Technology Hellas/Hellenic Institute of Transport (CERTH/HIT), Thessaloniki, Greece

E. Bekiaris et al. (eds.), *Infrastructure and Safety in a Collaborative World*,
DOI 10.1007/978-3-642-18372-0_18,

main political task is to achieve a sustainable transport system. The White Paper provides a statement of requirements on safer roads: "The European Union must, over the next 10 years, pursue the ambitious goal of reducing the number of deaths on the road by half; this by way of integrated action taking account of human and technical factors and designed to make the trans-European road network a safer network."As European accident statistics show, there is still a high number of traffic fatalities per year and country. It is evident that each member state of the European Union considers traffic safety as a high priority political issue.

18.1.2 Motivation of Research Funding Bodies

EU and national research funding bodies investigate on further research needs and co-ordinate research actions to make sure that results are statistically significant and transparent as a basis for standardisation and legislation actions. A main objective of research funding bodies is to get to know more about solved and unsolved questions and missing links between the results of different research projects. They are also interested in ensuring support to their future research programs. The main objective of research funding bodies is to gain reliable results, where good decisions can be based upon. Their investments ought to be decided wisely, taking into consideration the latest research results and needs. European research funding bodies need a good overview about all European research activities. National research funding bodies have a better overview of what research results are needed in order to achieve national policy objectives or what innovation is needed in order to bring national interests forward. They also check the transferability of results on national frameworks.

18.1.3 Motivation of Infrastructure Owners and Road Operators

Systems that provide efficient solutions for traffic management and increase safety are interesting for public and private road operators. Fatalities cause considerable economic damage. Public road operators have to manage the impact of an accident. Their focus lies on accident prevention and managing the accidents impact (traffic jam, resulting accidents). Many systems are proposed by research projects to avoid accidents occurrence and/or reduce accidents severity. For a decision on those investigations, proofed knowledge about costs and efficiency of safety systems is needed. Road operators and infrastructure owners face international cross-border traffic as a new and increasing challenge. Road operators and infrastructure owner also have to consider ITS-systems for cross-border traffic. An important objective is technical compatibility but also comprehensibility for the user.

18.1.4 Motivation of Standardisation Bodies

There are several official standardisation bodies at international level (e.g., ISO; International Organization for Standardization, IEC; International Electrotechnical Commission and ITU; International Telecommunication Union). At European level there are: CEN; European Committee for Standardization, CENELEC; European Committee for Electrotechnical Standardization and ETSI; European Telecommunication Standards Institute. Furthermore, there are the corresponding standardisation bodies also at national level.

Standardisation is important for cost efficient production, for open European wide markets and for a Europe-wide implementation of systems. Standardisation is important, not only from the economic point of view but also concerning the safety aspect.

18.1.5 Motivation of Insurance Companies

Insurance companies are addressed because they are interested in the reduction of follow-up costs of accidents. They also have a special interest in risk assessments (e.g., to calculate subscription fees or incentives). For insurance companies, the amount of risk is less important than its predictability. For marketing reasons, insurance companies tend to more frequently allow discounts of various types, which refer to certain attributes of vehicle or owner. It might be considered to allow discounts for safety features of the vehicle, as it is already done, e.g., for vehicles with electronic stability control (ESC). From the road safety point of view, it is quite favourable that such discounts exist in order to motivate purchasers to spend their money also on additional safety features. To enable insurance companies to calculate discounts on safety equipment on a real micro-economic basis (instead of marketing aspects), the reduced or additional risk of safety equipment has to be predictable.

In order to provide them with policy recommendations, we will first explain the approach and framework we used, to set up these recommendations.

18.2 Approach to Policy Recommendations

According to Marchau et al. (2007) and Walker (2000), policymaking on transport requires an integrated view with respect to the various alternative options, their possible consequences for transport system performance, and societal conditions for implementation. The basis for such a view has been provided by Walker (2000). According to this view, policymaking, in essence, concerns making choices regarding a system in order to change the system outcomes in a desired way (see Fig. 18.1).

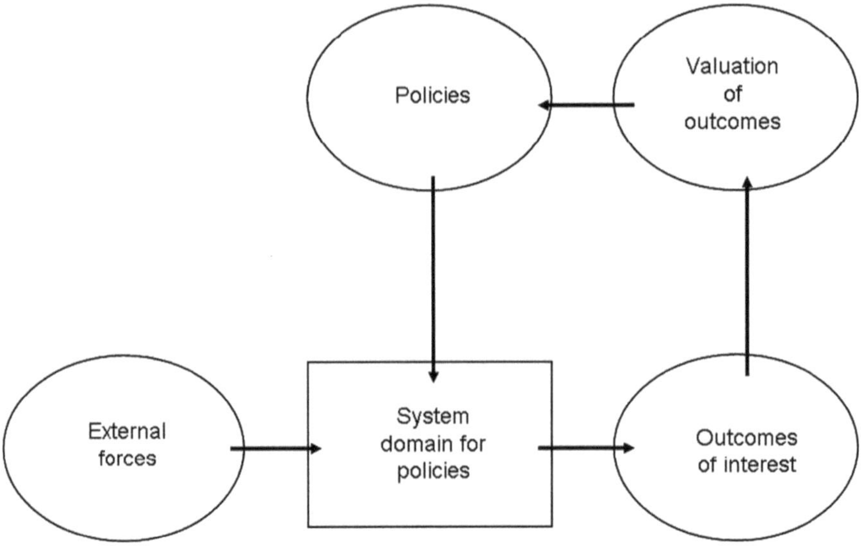

Fig. 18.1 An integrated view of policymaking (Walker 2000)

At the heart of this concept lays the system comprising the policy domain, in our case the transport system. A transport system can be defined by distinguishing its physical components (e.g., loads, vehicles, and infrastructure) and those components' mutual interactions. The results of these interactions (the system outputs) are called outcomes of interest and refer to the characteristics of the system that are considered as relevant criteria for the evaluation of policies. The valuation of outcomes refers to the (relative) importance given to the outcomes by crucial stakeholders, including policymakers. Two types of forces act on the system: external forces and policies. Both types of forces are developments outside the system that can affect the structure of the system (and, hence, the outcomes of interest to policymakers and other stakeholders). External forces refer to forces that are not controllable by the decision-maker but may influence the system significantly, i.e. exogenous influences. A policy is a set of actions taken to control the system, to help solve problems within it or caused by it, or to help obtain benefits from it.

Applying the framework shown in Fig. 18.1 to long-term transport policy making reveals several locations where uncertainties arise. Firstly, the external forces are uncertain, since it is difficult to identify which external developments will be relevant to long-term future transport system performance (e.g., changes in demography, economy, technology) and, perhaps more important, the size and direction of these changes. Second, even if there were certainty about the external developments (that is, we knew how the transport system's external world would develop), there might still be uncertainty about how the system would respond to those external developments, since the key-relationships determining transport system performance are uncertain because (some of) the interactions within the transport system are insufficiently known. Finally, the valuation of the various outcomes is uncertain. Stakeholders tend to have different opinions about the importance of future transport problems. This results in different, often conflicting, opinions regarding the various transport policies (Macharis 2007). As such, the willingness of stakeholders to accept (or reject) outcomes of transport policies is uncertain. In addition, over time, new stakeholders might emerge and/or current stakeholders might leave, and/or the opinions of the current stakeholders might change. Marchau et al. (2007) therefore propose an Adaptive Policy making view. The inevitable policy changes, resulting from changes in the external forces or the transport system, are becoming part of a larger, recognized process and are not forced to be made repeatedly on an ad-hoc basis (Fig. 18.2).

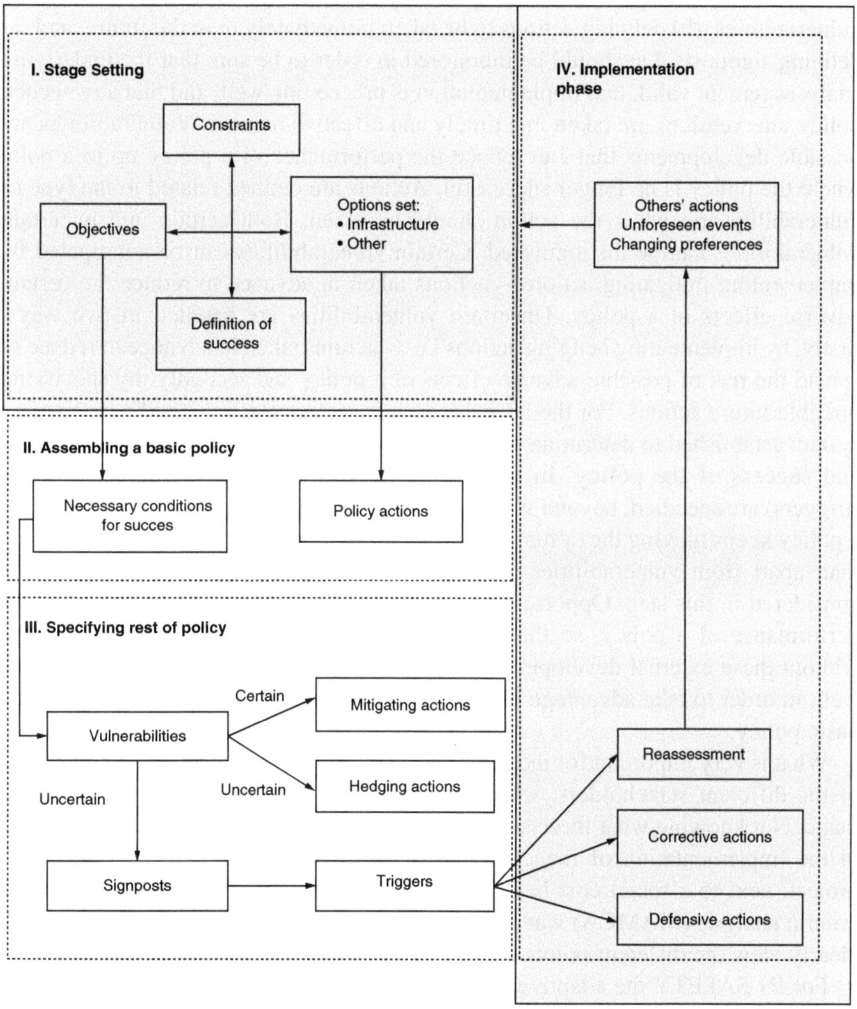

Fig. 18.2 The adaptive policymaking process (Marchau et al. 2007)

This specification should lead to a definition of success, i.e., the specification of desirable outcomes. In the next step, a basic policy is assembled, consisting of the selected policy options and additional policy actions, together with an implementation plan. It involves (a) the specification of a promising policy and (b) the identification of the conditions needed for the basic policy to succeed. These conditions should support policymakers by providing an advance warning in case of failure of policy actions.

In the third step of the adaptive policymaking process, the rest of the policy is specified. These are the pieces that make the policy adaptive. This step is based on identifying in advance the vulnerabilities of the basic policy (the conditions or events that could make the policy fail), and specifying actions to be taken in anticipation or in response to them. This step involves (a) the identification of the vulnerabilities, (b) defining actions to be taken immediately or in the future, and (c) defining signposts that should be monitored in order to be sure that the underlying analyses remain valid, that implementation is proceeding well, and that any needed policy interventions are taken in a timely and effective manner. Vulnerabilities are possible developments that can reduce the performance of a policy up to a point where the policy is no longer successful. Actions are defined related to the type of vulnerability and when the action should be taken. Both certain and uncertain vulnerabilities can be distinguished. Certain vulnerabilities can be anticipated by implementing mitigating actions – actions taken in advance to reduce the certain adverse effects of a policy. Uncertain vulnerabilities are handled in two ways: firstly, by implementing hedging actions i.e., – actions taken in advance to reduce or spread the risk of possible adverse effects of a policy and secondly, by specifying possible future actions. For the latter cases, signposts are defined and a monitoring system established to determine when actions are needed to guarantee the progress and success of the policy. In particular, critical values of signpost variables (triggers) are specified, beyond which actions should be implemented to ensure that a policy keeps moving the system in the right direction and at a proper speed. Note that, apart from vulnerabilities to the basic policy, opportunities might also be considered in this step. Opportunities are external developments that improve the performance of a policy, so that it is more successful than it would have been without these external developments. These opportunities should be monitored as well, in order to take advantage of the developments and, for instance, expand the basic policy.

What is very important for the addressed policy makers is that they keep account of the different stakeholders, which are involved in the choices that have to be made. Not knowing what these stakeholders' objectives are, will create difficulties in the implementation of the chosen policies. That is why in the IN-SAFETY project, next to a social cost benefit analysis (see Chap. 17), a multi actor, multi criteria analysis (MAMCA) was performed (see Chap. 16). This analysis allows to clearly view the different points of view of the measures that can be taken.

For IN-SAFETY the adaptive policy making view has been adopted, focussing upon managing the uncertainties. Therefore, the Policy Recommendations in all sections described in this chapter are presented according to the following schedule:

First, the objective of the policy recommendation is listed, and the basic policy action. Then, a few of the most relevant vulnerabilities are given. For each vulnerability, mitigating or hedging actions are suggested, as well as the possible signposts, triggers or actions.

18.3 Recommendations from Application Guidelines and Further Research Issues

18.3.1 Approach

Many existing guidelines, targeting the self-explanatory and forgiving nature of road environment, were collected with the help of questionnaires. The questionnaires were filled out by experts from various countries. They were asked to briefly describe national guidelines and research needs on how to give roads a more forgiving and self-explaining quality, and to define gaps in knowledge and potential regulation. Also included is knowledge from a detailed literature analysis during the whole IN-SAFETY project.

On the basis of collected responses a concluding matrix of guidelines was created. Furthermore a list of needs for the future research was created.

Geographical focus has been detected for specific guidelines on the:

- International (mostly European)
- National
- Local levels
- A very important organisation scheme was a classification in
- Infrastructure related guidelines
- Guidelines on vehicle autonomous system
- Guidelines about co-operative system (vehicle-infrastructure)

18.3.2 Recommendations

For most ITS applications there is more than one basic solution. As an example, warning a driver approaching a curve may be done by infrastructure-based equipment such as a road-side device that detects an approaching vehicle and activates a VMS in case the vehicle is assumed being too fast. On the other hand, a digital map could provide information on the radius of an oncoming curve, by considering the usual behaviour of the driver, calculating the recommended speed and warning the driver in case of exceeding the personal limit or even the physical limits of his vehicle. In a cooperative solution, an in-vehicle device could receive a speed recommendation from a road-side beacon, considering the usual behaviour of the driver and then providing warning when required.

Apart from existing systems, the technologies available today could and should be used for developing new systems, either enhancing previous ones, or dealing with new functions and preventing different kinds of dangers. Therefore migration strategies ought to be developed how to upgrade existing systems. Within IN-SAFETY, apart from the scenarios that were developed based on existing systems, two additional ones were described and rated, introducing the suggestion of new systems (Macharis et al. 2008). These were the following:

- Overtaking Assistant on roads with lane separation ("Blind spot")
- Overtaking Assistant on rural roads without lane separation
- For both these proposed systems the following could be recommended:
- Further research is needed around the potential of employing innovative technologies dealing with Vehicle to Infrastructure (V2I), Infrastructure to Vehicle (I2V) as well as Vehicle to Vehicle (V2V) communication (Brookhuis et al. 2001).
- Integrated HMI prioritising warnings with different risk origin is needed. Potential for integration of haptic HMI's needs to be further investigated.
- Personalisation of HMI warning strategies needs to be investigated according to drivers individual profile. In this way, specific driver groups (i.e., elderly drivers, novice drivers, etc.) may be addressed. Self-adaptive and self-learning systems, which would adapt different driving patterns should be investigated.
- Intuitive HMIs should be developed addressing all phases of overtaking.
- More research is needed for special infrastructure segments (i.e., curves) and special visibility conditions (which may hinder the full and/or sufficient operation of vision detection systems) (Table 18.1).

Table 18.1 Recommendations from application guidelines and further research issues

<table>
<tr><td colspan="3">Objective: It is very important not only to push and use ITS systems as a very important instrument to improve road safety but also to evaluate them. Little is known about precise number of target accidents, synergies between several systems, costs (public authority, user, society) and a quantitative evaluation of negative and positive impact of systems</td></tr>
<tr><td colspan="3">Pre-condition: The evaluation of ITS systems can not be done theoretical but needs the cooperation between researchers and infrastructure owners and road operators as well as the assistance of automotive industry and insurance companies. They all possess important information about costs, technical feasibility, road safety impacts and much more. It is to be clarified whether all parties are willing to open their databases under real life condition</td></tr>
<tr><td colspan="3">Policy action: The most important research need according to IN-SAFETY is the need to gain more knowledge to evaluate ITS systems. This can be done by evaluating existing ITS systems and evaluation processes after implementing new systems</td></tr>
<tr><th>Vulnerabilities</th><th>Mitigating/hedging actions</th><th>Possible signposts/triggers/ actions</th></tr>
<tr><td>Gained knowledge from project evaluations stays unpublished and therefore is useless for other researchers</td><td>A database about evaluation results could bring the knowledge to a wider user group. The data ought to be in simplified, standardised format</td><td>Research about structure, possible content and how to use the database is necessary. It ought to be discussed with all affected parties</td></tr>
</table>

18.4 Recommendation on Pictograms and Verbal Messages, Horizontal and Vertical Signing

In 2003 the TERN (Trans-European Road Network) covered 15 countries with 11 languages spoken plus 3 additional states which are not EU members. These countries and languages, together with ten "new member states" with nine official languages, were considered with the aim to derive at feasible suggestions of the cross-language and language independent display of information on VMS (Variable Message Signs) and static message boards on motorways.

18.4.1 Approach

The following requirements on VMS have been identified and studied (Simlinger et al. 2008):

- Physiological requirements with regard to conspicuity and discriminability
- Cognitive requirements with regard to understanding
- Technical requirements with regard to the size and quality of the presentation of the information

The elaborated symbols/pictograms, together with Vienna Convention traffic signs, suitable for application on VMS, static signs and in-car navigation displays meet all documented requirements. So does the complementing Latin and Greek "TERN" alphabet versions, which have already been used for text elements in the renderings of the newly designed symbols/pictograms and the modified Vienna Convention traffic signs required on motorways.

Apart from verbal messages elements like place names, specific words and abbreviations have been identified as "Europeanisms", suitable for communication across language barriers.

18.4.2 Recommendations

Simlinger et al. (2008) gives a summary of recommendations on follow-up activities (e.g., Recommendation to the European Commission to amend Annex III of Council Directive 91/439/EEC: Review the viability of 0.5 visual acuity).

Table 18.2 summarizes recommendations from Simlinger et al. (2008) concerning pictograms and verbal messages, horizontal and vertical signing.

18.5 Recommendations for Application of Traffic Simulation and Risk Modelling

Safety and Risk analysis and assessments are helpful to make decisions on safe road and vehicle systems. Simulation models are important for the analysis of existing situation of traffic system, and for the answer of the question "what would happen

Table 18.2 Recommendation on pictograms and verbal messages, horizontal and vertical signing

Objective: International understandable (language independent) traffic signs/information throughout Europe are useful to make "understanding" easier for international traveller and therefore are supposed to increase traffic safety due to less misunderstandings

Pre-condition: The need of revision of the Vienna Convention on Traffic Signs ought to be clarified. Investigations are necessary to specify the negative impact of today's situation in relation to what can be improved with harmonisation. Other positive and negative effects of harmonisation are to be taken into account beside visibility, comprehensibility before a decision on harmonisation can be taken

Policy action: The results of Simlinger et al. (2008) provide a basis for further discussions on Europe-wide harmonization of verbal message elements, traffic signs and VMS elements. It ought to be checked and decided on substitution of signs/symbols/pictograms or adding new signs/pictograms/symbols/Europeanisms to Vienna Convention. Information systems such as in-car traffic signing and information ought to be considered as alternative to infrastructure signing elements

Vulnerabilities	**Mitigating/hedging actions**	**Possible signposts/triggers/ actions**
The IN-SAFETY results of D3.2 are not jet verified in real test installation	Initiations of pilot test and long term research is needed	
Not all Europeanisms proposed in D2.3 are likely to be harmonised, because often the English version of a word, e.g. "exit" is used instead of national wording e.g. "Ausfahrt". Harmonisation of wording could lead to confusion and misunderstanding. Especially the meaning of abbreviations causes confusion (e.g. min = minimum = minute)	Europeanisms should be used wisely. An alternative are bilingual information (analysed in Simlinger et al. (2008)). If the dimension of traffic signs and VMS allows it than bilingual information should be used (e.g. "Police/Polizei")	Although there are many guidelines and recommendations how to design warning messages, alarm sounds and so on, it is impossible to design alarm messages suitable for all drivers and for every situation. Therefore users should be able to personalize their applications, e.g. adjusting alert levels. Individual in-car traffic signing and information ought to be considered in addition to infrastructure signing elements. Further activities in this field are necessary
In D2.3 high requirements were formulated on VMS (several colours, certain minimum dimension, lot of graphical details, animation, freely programmable). Not all today's existing VMS are conforming to those requirements	VMS are often used at accident blackspots. The required and detectable information may vary from point to point. The information shown on VMS can be either recommendation or mandatory (e.g. dynamic speed limit is mandatory)	Taking into consideration the necessary traffic information content and cognition requirements new VMS might became necessary. Individual in-car traffic signing and information ought to be considered in addition to infrastructure signing elements

if...?" Application of traffic simulation and risk modelling are used within IN-SAFETY to analysis reasons for accident blackspots or identifying gaps and imprecise regulations in standards and to evaluate and verify different alternatives of safety measures.

Simulation can help to measure the impact of the implementation of ITS-systems, their effects (e.g., important for CBA and accident risk analysis) and to compare alternative measures for a certain problem. Since traffic safety depends on numerous factors (e.g., human behaviour, infrastructure, legal factors etc.) they all have to be integrated into the analysis.

18.5.1 Approach

Several traffic simulation models were analysed in IN-SAFETY. They can be divided in two groups: microscopic and macroscopic simulation models. The models contain state-of-the-art approaches for simulation of traffic at various stages: from the macroscopic view on networks and the traffic streams on the links down to microscopic approach with the focus on the individual driver and the vehicle. A description on the models, their parameter and methods can be found in Anund et al. (2008a).

The existing models were analysed and additional safety relevant parameters (such as time-to-collision), adaptive objective function, new safety indicators (such as the shape of the headway distribution) were integrated. Sample applications within IN-SAFETY show the potential of the models for safety analyses.

Furthermore existing risk analysis methodology has been further developed in Bald et al. (2008). The so called Darmstadt Risk Analysis Method (DRAM) describes the cause-and-effect chain of critical situations taking into account the uncertainties of the system (especially human behaviour). DRAM is able to analyse complex systems with uncertainty and non-linear relations. The analysis may be done qualitative, quantitative and in a mixed form. A tool called Darmstadt Risk Analysis Tool (DRAT) is provided. DRAT is principally not limited to a certain number of dimensions and elements and so restricted only by available computer memory and calculation time, allowing the model to evolve as needed. Additionally two scenarios are analysed within IN-SAFETY: "approaching a sharp bend" and "lane changing manoeuvres".

18.5.2 Recommendations

Simulation and risk analysis models can help to solve questions without implementation of a system in reality. This can help to save funds and time, as well as to evaluate possible alternative measures. It ought to be kept in mind that for a certain problem an appropriate model is needed (sometimes adaptations of existing simulation and risk analysis models are necessary) as well as a reliable data input.

It is necessary to analyse the long-term effect of new infrastructure, regulations and accessories with all-embracing risk analysis methods which are able to integrate the effects of human behaviour and habits. A new endangerment regarding ADAS systems may arise after the first safety successes have become apparent: if such systems are useful and effective but not reliable, new risks may arise if the user trusts the system but the system fails and the user has no chance to remark the failure in time.

It seems useful to build an overall covering model of the road system as most of the behavioural aspects are cross-linked throughout the system. The modelling process may be started at different points, letting the different parts gradually grow together. The model may temporarily branch if reliable knowledge is not yet available within certain sections. But always, the goal should be to integrate all road related knowledge into one model (and its adjacent database of knowledge). Such a model could be used to enable and simplify the process of problem analysis, discussion of variants and assessment of political recommendations (Table 18.3).

Table 18.3 Recommendations for application of traffic simulation and risk modelling

Objective: IN-SAFETY shows that the use of simulation models and risk analysis tools can help to model the ITS system impact both on traffic conditions and on road safety. The analysis of traffic safety problems with the help of risk analysis models can help to systematically find improvements and knowledge how to avoid safety problems

Pre-condition: An urgent need for all safety analysis based on traffic simulation is research on the relation between actual accident risks and the derived safety indicators. Today, researchers assume that a change in the indicators correspond to a change in accident risks. Reliable parameters, data input for model calibration as well as a detailed description of scenarios and alternatives, which are to be analysed, are important for reliable results of simulation and risk analysis models

Policy action: The use of simulation and risk analysis models as described in Anund et al. (2008a, b) and Bald et al. (2008) in addition to conventional methods to calculate efficiency of certain measure/system is recommended. IN-SAFETY shows a wide variety of use cases of simulation and risk analysis models and their advantages

Vulnerabilities	Mitigating/hedging actions	Possible signposts/ triggers/actions
Simulation models, both micro and macro, as well as risk analysis tools can only produce reliable and realistic results if they are calibrated using realistic and representative data	A questionnaire survey shows the different weights the main factors affecting route choice from the drivers point of view: travel-time, distance and safety level. Other possibilities to gain data input is using results from pilot studies (e.g. Swedish pilot is used as input for so called "RuTSim model"). The development of a worldwide database of knowledge also helps to collect necessary data from several projects. Precondition here is to promote cooperation between projects, establish common procedures, to interact between the researchers and this database	

18.6 Lessons Learnt from Pilot Tests

The Pilots have primarily examined the effectiveness and usability of the selected implementation scenarios and concepts of forgiving and self-explanatory road environments. The results have also been used in order to improve simulation models, risks analysis tools and training schemes of road safety assessment. These tests used road infrastructure elements and test vehicles equipped with ADAS and IVIS functions, as well as IN-SAFETY services and applications as defined and developed within the project.

18.6.1 Approach

The IN-SAFETY Pilots sites were the following:

- Italy (Turin) – field tests (CRF)
- Sweden (Linköping) – simulator tests (VTI)
- Germany (Stuttgart) – field tests (IAT)
- Greece (Athens) – field tests (CERTH/HIT)

The aim of the IN-SAFETY Pilots was to determine the users' acceptance against the introduced IN-SAFETY cooperative solutions, as well as to perform a detailed assessment of their foreseen impacts. The evaluation of the effects of some of the selected scenarios of forgiving and self-explaining roads has been done in four different pilots realized in four selected regions, but also in an advanced moving base driving simulator in Sweden.

The results of pilots were analysed and structured in correspondence to the topics: technical verification, impact assessment, user acceptance, socio-economic, guidelines. Anund et al. (2008b) provides a summary over all pilots and results.

18.6.2 Recommendations

The overall IN-SAFETY pilot results showed that all applications were seen as more useful than the baseline, defined as normal driving without the IN-SAFETY systems. In some cases with rather low usefulness scores there was often a reduced technical performance of the system involved. The pilots have illustrated that there is an impact in several of the IN-SAFETY scenarios. Indicatively we mention some exemplary results.

The Swedish Pilot with a School bus ahead warning on-board system shows a decreasing speed while approaching the bus in comparison to a reference scenario without the system. The average passing speed was about 60 km/h which is far too high to avoid severe accidents. Even in this simulation environment drivers do not react properly on such a warning. Research needs to address the question of long term reaction and how to motivate drivers to act safely. Therefore large field

Table 18.4 Lessons learnt from pilot tests

Objective: Statistically significant results and long term assessments about safety effects, driver behaviour, market penetration, business models and technical feasibility/reliability is very important for further decisions on implementation of ADAS

Pre-condition: The definition of set of systems/technologies to be tested and the test design (parameters, alternatives, testing method, representative sample of participants) is important to prepare field operational test

Policy action: It is recommended to do necessary field operational test and long term assessments

Vulnerabilities	Mitigating/hedging actions	Possible signposts/triggers/actions
Field operational test take a long time (several years). It might happen that new important questions arise and that others are less important than it was predicted before the test started	Both vulnerabilities should be addressed in the field operational test design from the very beginning. The development of a set of scenarios what might happen during test period, could help to estimate the described risk	Pre-test studies and a compilation of methodologies, knowledge and experiences from previous projects can help to avoid problems occurring during the field operational test but also help to identify missing solutions/methods
Different causes might lead to the case that not all questions can be answered in field operational tests		

operational test would also help to collect missing practical experience and data especially how drivers react in long term.

A state-of-the-art LDW system has been investigated in a German field experiment with a subjective assessment. Empirical results from 17 test subjects who drove more than 5,000 km show that lane departure warnings were generally well accepted. Driving with the LDW system tends to reduce the number of lane departures and also educates drivers to use the indicators more often and earlier when changing lanes. However, the differences were not big enough to be statistically significant (Table 18.4).

18.7 Recommendations for Application of the Operators Manual

Already several attempts have been made within related European projects to develop training specialised for operators on the application of innovative ITS systems for their staff (operators) (see Chap. 9). Current trainings usually focuses on handling skills for the management/information system used. Typically, today's training is on the job, using a stepwise approach from just watching experienced operators to working self dependently in times of difficult traffic conditions. None of these training schemes included a reasonable share of general knowledge about traffic management, in-vehicle information systems (IVIS) or advanced driver assistance systems (ADAS).

The training handbook is dedicated to all TMI (Traffic Management Information) and TMC (Traffic Management Centre) operators (e.g., highway and tunnel operators, traffic surveillance centres, traffic information by mass media as radio and internet, urban traffic management and surveillance, etc.), to their staff and to the management as well. Main categories of users are road operators (Urban/Rural/Highway/Ring road), area operators (TMI/TMC-Urban/Rural/Integrated), specific infrastructure operators (tunnel/bridge/other) and generic.

18.7.1 Approach

The training is primarily dedicated to technicians responsible for the development and incorporation of ITS systems on the high level road network and operators which control the systems. The training includes information on installation, use and maintenance of state-of-the-art technology.

Computer-Based Training (CBT) uses the computer for training and instruction. CBT programs are called "courseware" and provide interactive training sessions for all disciplines. CBT was originally introduced on Laserdiscs, then CD-ROMs and, later, online. CBT courseware is typically developed with authoring languages that are designed to create interactive question/answer sessions. Web Based Training (WBT) is disseminated over the internet and provides added value through up-to-datedness and networking. IN-SAFETY's "operators' training manual" is available as web based training.

18.7.2 Recommendations

Similar procedures for traffic management should be applied and rules should be implemented according to common standards on TERN. Harmonised training for TMI/TMC operators throughout Europe lays a basis for approaching this goal without having to harmonise all the official procedures. It may be assumed that decisions taken by different operators that are based on equal information and education are likely to be similar and therefore familiar and understandable for driver from the home country as well as from any other origin.

The IN-SAFETY consortium proposes an "Operators' Training Manual" as a first step towards convergence of operators' training, which has (according to the goals of the IN-SAFETY project) a particular focus on ADAS and IVIS. This manual may serve as a basis for developing a curriculum for operators, which on the long run should be mandatory for all staff providing public information and traffic management.

The "operators' training manual" describes a variety of reasons to use ITS systems. It can be seen as a decision guidance on ITS systems and it is supposed to support improving the service quality by giving background information both for the regular business (strengthening the basis of decision making) and for improvement of existing or development of new services (Table 18.5).

Table 18.5 Recommendations for application of the operators manual

Objective: Training on optimal use of Intelligent Transport Systems and Services (ITS) can save lives, time and money as well as reduce threats to our environment and create new business opportunities		
Pre-condition: ITS already has a presence in everyone's day-to-day mobile activities, for example active support systems such as vision enhancement, lane-keeping assistance and collision warning systems but also collective ITS systems such as coordinated traffic control, ramp metering, variable message signs, and traffic and incident detection systems		
There is much qualitative knowledge about the benefits and positive impact of ITS systems. It is to be seen as a pre-condition to quantitative proof positive and of course also negative (long term) effects of ITS on drivers behaviour, environment, traffic efficiency and road safety. Qualitative results are also important to fall decisions on a rational basis like Cost-Benefit-Analysis		
Policy action: The broader, appropriate use of ITS systems is recommended		
Vulnerabilities	**Mitigating/ hedging actions**	**Possible signposts/ triggers/actions**
This handbook shall support improving the service quality by giving background information. It does not give any recommendation for training on the use of existing traffic management hard- and software systems	Training based on this manual does not replace any part of existing training procedures; it is meant to accomplish the education of new staff and may be implemented as a part of retraining for existing staff	
ITS systems are developing very rapidly. The handbook might not include all ITS systems	The quality of a handbook depends on continuous updates	

18.8 Recommendations from MCA-AHP and CBA Assessment of Selected Systems and Functions

A main objective for societal CBA (socio-economic analysis) is to identify those projects/measures/scenarios that will increase aggregate economic welfare as measured in monetary terms. Societal interest is in that case an aggregation of individual interests. The money measure is given from projects net benefits (total benefits−total costs) or the benefit cost ratio (total benefits divided by total costs).

The objective of the MCA-AHP (multi-criteria analysis (MCA)-analytic hierarchy process (AHP)) approach is to obtain a prioritisation for a number of scenarios contributing to the creation of a more forgiving road (FOR) and self-explaining road (SER) environment. In order to assess not only the policy or societal priorities regarding these scenarios, but to assess also their implementation potential, an analysis needed to be performed for each relevant stakeholder, namely society, users and manufacturers. Within IN-SAFETY we focused on the societal point of view, since this represents the general interest, and should be taken as a starting base for policy purposes. The two other points view, namely those of the users (demand side) and manufacturers (supply side), were considered as important from an implementation perspective.

The aim of the MCA-AHP approach is broader than that of a strict cost-effectiveness analysis (CEA) or cost-benefit analysis (CBA), since in the MCA approach the contribution of the scenarios is assessed not only in terms of safety effects, but also in terms of a much larger number of policy objectives, including inter alia, driver comfort, travel time duration, network efficiency, environmental effects, liability risk, etc.

18.8.1 Approach

Ideally, a CBA should include all possible benefits and costs expected to result from the scenario implementation. However, in many cases all effects are not easily quantified and/or not easily monetised. In IN-SAFETY CBA only safety effects (on expected injuries/fatalities) were included, while other potential effects (on time use, environment, etc) were omitted from the calculations. Thus, the CBA is partial. There are also large uncertainties related to the estimated safety effects (that are based on an "error-based approach" which possibly yields maximum potentials) and the estimated costs (that may change a lot if the market expands). However, even if the analyses in Erke et al. (2008) are tagged to one country, they should rather be regarded as example studies. The analyses are not intended for detailed policy analysis for the selected country, even if the national injury/fatality data applied will influence on the resulting estimates.

The approach followed in order to obtain the prioritisation of scenarios is that a multi-actor MCA (MAMCA) methodology was applied. The MAMCA is a specific methodology within the entire MCA methodology. It is a methodology which makes it possible to obtain a prioritisation in terms of what each stakeholder considered relevant. In this application of MAMCA, three specific stakeholders were identified, namely society, users and manufacturers.

18.8.2 Recommendations

See Table 18.6.

18.9 Conclusions

Within the IN-SAFETY project it became clear that final conclusions are not always possible. In the course of IN-SAFETY it was revealed several times, that on the basis of existing accident statistics, it is hardly possible to estimate the impact of ITS. Current accident data bases also hardly allow for investigation of accident causation. It is even difficult just to determine target accidents for ITS. In addition, exposure data is also missing.

In-depth analysis and field operational testing are needed to answer such questions, however, deriving from the enormous cost for such efforts, the samples are normally rather small and extrapolation to the whole of the European fleet and driver population is rather imprecise. As a consequence, the improvement of

Table 18.6 Recommendations from MCA-AHP and CBA assessment of selected systems and functions

Objective: In preparation of the Cost-Benefit-Analysis today's databases on accidents were analysed and it became clear that knowledge about target accidents and benefits of a system is not yet satisfying. To achieve reliable CBA-results as a basis for policy recommendations and actions better data basis on effects and cost of ITS-systems are needed

Pre-condition: More research on innovative ITS-systems, especially co-operative systems is to be seen as a pre-condition. Without detailed knowledge about system architectures, technology solutions and business models it is impossible to assess costs

Policy action: In European countries accident statistics show different structures, different interpretations of collected data as well as different amount of data. Therefore European and national legislation bodies ought to encourage national road authorities to develop needed databases on a common European level (better standard). A result might even be a guideline for the structure of national accident statistics

Research projects dealing with rather technical details of cooperative systems are needed. On the other hand, since the cooperative systems depend on infrastructure and vehicle systems, public authorities have to play a leading role in a partnership with private sector. Those questions are to be discussed on round tables with all affected parties. Round tables need initialisation, preferably by national road authorities!

Vulnerabilities	Mitigating/hedging actions	Possible signposts/triggers/ actions
Necessary data are not collectable (due to high cost of detailed accident analysis)	Expert group ought to decide on obligatory core information with respect of performable actions (done by police) at accident side	
CBA in IN-SAFETY shows high dependency of cost of scenarios, especially cooperative systems reach low CB ratios	When developing cooperative systems other use cases ought to be integrated. In case that road-side beacons, etc. can be used for more than one use case CB ratios became better. Those use cases ought to be considered in the project (e.g. standardisation of interfaces, communication protocols)	
IN-SAFETY described possible business models which are not to be seen as finalised. They are more a basis for discussions with all affected parties	The discussion on business models ought to be done in parallel to technical developments. Without common decision on business models co-operative systems won't work. Taking international transport into consideration the discussion might be initialised on EU level	The stakeholders controversial rankings of scenarios was shown in Macharis et al. (2008) It is therefore necessary to clarify different standpoints of Stakeholders and develop several business models alternatives, discuss dis-/advantages

official national traffic accident records should be supported and international databases should be extended accordingly.

Databases, which are operated by vehicle manufacturers, could also be used to provide data for this purpose.

Towards it we propose that:

- Insurance companies might introduce ITS solutions which themselves collect data for estimating the particular risk of one vehicle (e.g., mileage) in order to use this data for calculating the premiums. Benefits might be given by the insurance companies to motivate drivers to accept such new solutions, which at the same time promote other very useful ITS (as UNIQA does with its "SafeLine" program promoting e-Call and car finder)

- Insurance companies should collect and provide data for analysis by road safety experts (with a focus on property-damage-only accidents)
- Specific ITS features of the vehicle should be included in police reports on injury accidents
- Alternatively, in-depth accident analysis should be carried out
- Such research should be funded by the relevant bodies

The overall problem of lacking data can be addressed with more co-operations between stakeholders. With their help not only a wider data base on positive and negative impacts of systems can be established but also organisational and operational issues of innovative ITS systems can be discussed. It cannot be expected that all questions concerning cost and effectiveness, technical details or organisational issues can be solved within a research project. European legislation bodies ought to raise awareness of open questions and bring parties like national road authorities, industrial partners, automobile clubs etc. together ("round table principle").

References

A. Anund, T. Benz, E. Gaitanidou, I. Spyropoulou, S. Toffolo, Improved micro and macro simulation models, IN-SAFETY project, Deliverable 3.1, 2008a

A. Anund, A. Tapani, A. Kircher, C. Marberger, E. Bekiaris, A. Vatakis, P. Spanidis, K. Kalogirou, C. Liberto, S. Damiani, M. Raimondi, M. Wiethoff, T. Benz, Pilot results consolidation, IN-SAFETY project, Deliverable 4.2, 2008b

J.S. Bald, K. Stumpf, T. Wallrabenstein, L.T. Huyen, Road risk analysis tools, IN-SAFETY project, Deliverable 3.2, 2008

K.A. Brookhuis, D. De Waard, W.H. Janssen, Behavioural impacts of advanced driver assistants. EJTR **1**, 245–253 (2001)

A. Erke, K. Veisten, R. Elvik, C. Macharis, A. Verbeke, K. De Brucker, M. Wiethoff, E. Bekiaris, A. Anund, M. Winkelbauer, Cost-benefit analysis and cost-effectiveness analysis, IN-SAFETY project, Deliverable 5.2, 2008

European Commission, White Paper: European Transport Policy for 2010: Time to Decide. Brussels, 2001

C. Macharis, Multi-criteria analysis as a tool to include stakeholders in project evaluation: the MAMCA method, in *Transport Project Evaluation. Extending the Social Cost–Benefit Approach*, ed. by E. Haezendonck (Edward Elgar, Cheltenham, 2007), pp. 115–131

C. Macharis, A. Verbeke, K. De Brucker, E. Gelová, J. Weinberger, J. Vašek, Implementation scenarios and further research priorities regarding forgiving and self-explaining roads, IN-SAFETY project, Deliverable 5.3, 2008

V.A.W.J. Marchau, W.E Walker, G.P. van Wee, Innovative long-term transport policymaking: from predict and act to monitor and adapt. Proceedings European Transport Conference (pp. 1–17). Association for European Transport, 2007

P. Simlinger, S. Egger, C. Galinski, Proposal on unified pictograms, keywords, bilingual verbal messages and typefaces for VMS in the TERN, IN-SAFETY project, Deliverable 2.3, 2008

W.E. Walker, Policy analysis: a systematic approach to supporting policymaking in the public sector. J. Multi-Criteria Decis. Anal. **9**, 11–27 (2000)

Chapter 19
Our Future

Evangelos Bekiaris and Evangelia Gaitanidou

19.1 Quo Vadis?

According to (ETSC 2008), 97% of all transport fatalities in EU are caused by road transport. This number becomes more scaring when considered that road transport holds a share of 88% of the total passenger transport, while its fatalities are 100-times more than all the other modes together.

Before the publication of the White Paper on Transport in 2001, and more specifically in 1995, 45,000 deaths and 1.5 million casualties were reported for the EU15 (ETSC 1997). Since then a lot of progress has been made. In today's EU, consisting of 27 Member States, the numbers have significantly decreased. In 2007 (ETSC 2008) 43,000 fatalities of road transport accidents occurred (28,791 for EU15).

As we have reached the deadline-year of 2010, it is noticed that the number have of course been reduced, however the target point is still far from being reached. This situation is clearly illustrated in Figs. 19.1 and 19.2, where the estimated trends in road fatalities as well as the progress of their reduction are shown in numbers.

Hence, the need for setting new targets and imposing further measures beyond the 2010 is becoming more than evident. A proposal of ETSC indicates a proposal of a shared target in each Member State of reduction of deaths by 40% and the injuries with lasting effects by 20% (ETSC 2008). It is although recognized that the problem should probably be treated from a more detailed perspective, in order to deal with its perplexing complexity. Thus, a series of transport related areas have been identified, also through previous research initiatives – including IN-SAFETY, in which more focused research, together with targeted intervention is urgently needed. Some of these areas are indicatively mentioned below. These priorities have been discussed and notified within FERSI (Forum of European Road Safety Research Institutes) Organization.

E. Bekiaris (✉) and E. Gaitanidou
Centre for Research and Technology Hellas/Hellenic Institute of Transport (CERTH/HIT), Thessaloniki, Greece
e-mail: abek@certh.gr

E. Bekiaris et al. (eds.), *Infrastructure and Safety in a Collaborative World*,
DOI 10.1007/978-3-642-18372-0_19,

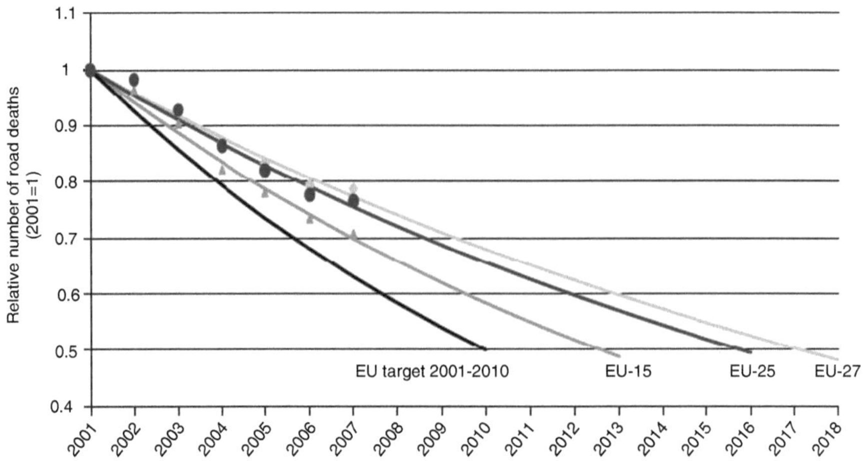

Fig. 19.1 Estimated trends in road deaths in EU27, based on developments 2001–2008 (ETSC 2008)

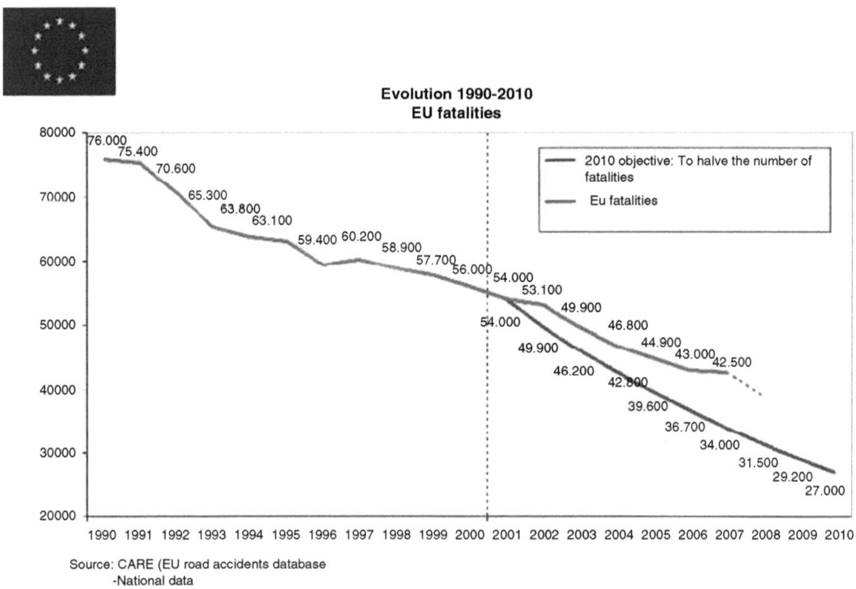

Fig. 19.2 Foreseen versus actual reduction of EU road accidents between 1990–2010 (CARE 2008)

- The prevention of injuries caused by road traffic accidents, as the monetary valuation of injury prevention is currently significantly lower that this of deaths. On the other hand, the cost to society caused by the consequences of injuries may be well higher than the one of deaths, taking into account e.g. causation of permanent impairment.

- Prevention of deaths of children in traffic accidents. According to the European Road Safety Observatory (ERSO, http://ec.europa.eu/transport/road_safety/specialist/index_en.htm), more than two fifths of child deaths are encountered in car accidents, while as pedestrians, this number reaches 25%. Already some countries, as Great Britain, have set specific targets relating to the reduction of the number of children killed or seriously injured in traffic accident.
- Power Two Wheelers (PTW) constitute another group which should be in focus. In 2006 almost 16% of the number of road fatalities had PTW as victims. This number is really large if one considers that they account only for 2% of the kilometers annually driven on the roads (ETSC 2008).
- Similarly to PTW, the risk of death in a traffic accident for pedestrian is about 9 times higher than for car occupant, while for cyclists about seven times higher, rendering these two road user groups in need for particular care (ETSC 2008).
- Urban safety is a definite priority, as 60% of the European population live in cities (Eurostat 2008 and about 2/3 of accident and 1/3 of deaths occur in urban environments. The Green Paper of 2007 "Towards a New Culture of Urban Mobility" (COM (2007) 551) is another indication of urban safety being a priority.
- Training novice drivers and retraining schemes of private and professional vehicle drivers is also a major need. Appropriate enforcement policies and measures are also a necessity, towards reducing the large number of fatalities in young ages due to traffic accidents.
- Elderly road users, either as drivers but mostly as pedestrians, hold a higher risk of about 16% for being killed in a road accident in comparison to a younger person, mostly because of their increased exposure and vulnerability to injury due to age decline (AGILE 2001).

From all the above, and not only, it is evident that research in the field for road safety is – and should remain – an open issue for the EU. According to ERTRAC (2004) accident prevention and mitigation could contribute to 55–65% and 35–45% respectively to the overall gain in fatalities and injury reduction. Moreover, the new priority areas of research, such as environmental issues, offer a new field for action to transport safety research.

19.2 Sustainable Safety and Environmental Protection: Two Sides of the Same Coin

As nowadays the problems of society continuously become more complex and the need for a better quality of life, in an era which is menaced by economic and environmental crisis, has become a priority, a holistic perspective also for transport safety is the upcoming trend. The term of "Sustainable Safety" is often being used, meaning that in order to provide increasingly safer road traffic system, not only for the present but also for its future users, an approach is needed, which encompasses combined, multidisciplinary and integrated actions, leading to long lasting safety improvements. For example, local enforcement by more police on the road is not an economic viable action sustainable on the long-term (when the number of controls is decreased the accidents will return), whereas electronic and automated means of enforcement can be considered as sustainable measure. In this context lies also the connection of road safety and environmental protection, as well as energy efficiency.

Up to 2008, in the policy documents of the EC, the reference to "zero vision" meant to achieve zero deaths on the road. As of 2008, the same term is used, but it now implies the goal of zero pollution because of circulating vehicles. This sharp change in policy priorities is also reflected in change of balance in research budgets and agendas. Are however traffic safety and environmental protection really different and competing goals? The misunderstanding emanates from a recent tendency to consider Energy Efficiency and Environmental Protection research fields as separate and independent territory from Road Safety and to treat them as such in the context of research agendas and priority implementation plans. However this perception is not correct. Historically, there have been many traffic accidents (such as the PRESTIGE maritime accident but also the big fires at Gotthard and Mont Blanc tunnels) which constituted a major environmental threat, maybe more considerable than this of the collective use of oil for transportation. Moreover, accidents are to blame for serious traffic congestion and disruption of the transportation network, which consecutively results in higher fuel consumption and greenhouse gas emissions. On the other hand, the introduction of new types of fuels and vehicles may cause significant impact on traffic safety. For example, electric or hybrid vehicles that run silently up to a certain speed (when running on the electric motor, roughly up to 70 km/h for hybrids) may endanger pedestrian that didn't "hear them coming" or provide false concept of speed to novice drivers (i.e. IMMACULATE project results). However, at the same time, eco-driving, due to the use of low speeds and the conservative nature of driving behaviour that it implies, is also improving the proactive safety of the driver. Thus, the connection between safety and environmental awareness becomes evident, and these two constitute a combination that holds a high potential for improving a twofold target of accident and CO_2 reduction immensely. At the bottom line, it seems that road safety and environmental protection are the two sides of the same coin and their combined treatment can only lead to benefits for both areas, but also for the economy and the society as a whole.

19.3 Road Safety in a Changing World

Road safety is the amalgam of diverse inter-related issues with emphasis on human safety and trouble-free mobility. The future needs to be both holistically and in detail sketched by setting clear and pragmatic research priorities. Strategic planning will provide the medium for establishing essential short-term steps which accommodate for current research aims and initiatives.

Decreasing rates of deaths due to road related accidents are a positive outcome of current policies; however, people still die and get seriously injured on road. Although, halving of casualties is the EU's target for 2010, novel and more refined objectives are required for the next 20 years. In a shorter term perspective, specific objectives and research priorities should be defined (i.e. in a 5 year horizon), in accordance with existing strategic plans and using as a starting point recent research results.

Interestingly, Transport is the second or third priority (in terms of both problem and potential solution) of all Grand Challenges that Europe faces, namely Climate Change, Energy, Water and Food, Public Health, Ageing Societies and Globalisation (ERA Vision 2020, http://ec.europa.eu/research/era/pdf/2020-vision-for-era_en.pdf). And, as a matter of fact, Traffic safety is at the heart of most of these challenges, as (only indicatively):

- Road accidents (and especially these of dangerous goods or within special infrastructure to major environmental losses and energy waste (from the subsequently created traffic jams). This is further analysed in Sect. 2 of Chap. 13.
- Deaths and injuries from traffic accidents are one of the major menaces to Public Health and a key cost outlier.
- Elderly of today and tomorrow are not the ones of yesterday; currently both men and women drive cars and wish to do so for as long as possible, to keep independent mobility and their social life. The corresponding impact to traffic safety is as yet under-researched.
- Globalisation changes rapidly and radically the traffic participants synthesis, leading to more and more people on the road that do not understand the text at VMSes and do not know the local traffic arrangements and drivers habits.

Thus, traffic safety should remain high in the research and political agenda.

From all the above it can be concluded that road safety is still an open issue for EU in total and each Member State separately. Substantial steps forward have been made in the past decade towards reducing the number of injuries and fatalities caused by road traffic accidents. The numbers today are significantly lower than the ones at the beginning of the Century; however, unfortunately the target that has been set for the first decade of the 2000 has not been reached. Many research initiatives have been implemented focusing as much to the improvement of the infrastructure as also to the use of innovative technologies and the introduction and adoption of new policies. However, for many user groups and road safety areas the relevant research is still at a germinal stage.

The work that has started has to continue, incorporating also new perspectives that have come along in the meantime. The economic crisis and the environmental alertness introduce a new axis for road safety research, where the combined efforts towards common goals set by the different fields, within the concept of Sustainable Safety, is a promising dimension for a holistic approach of the major problems in modern European societies. The lessons learned and the new horizons that emerge form the basis for the continuation of research and the broadening of target fields, towards the transportation safety of the future, that would encompass, not only the reduction of fatalities in terms of people killed or seriously injured, but also the broader consequences of accidents to the environment and the quality of life of the European citizens.

Acknowledgments This chapter has included valuable input from FERSI members. FERSI is the Forum of European Road Safety Research Institutes. It was established in 1991 with the objective of encouraging collaboration between European road safety research institutes. Such collaboration was, and continues to be, necessary to ensure that the problems of road safety in European countries are researched by the best available expertise, and that the results of the research are implemented in the most appropriate and effective way, both at national or at European level, (www.fersi.org).

References

AGILE (QLRT-2001-00118) Technical Annex, 2001

COM, 551 Final, Green Paper, Towards a new culture for urban mobility, Commission of the European Communities, Brussels, 25 September 2007

ERA Vision 2020 http://ec.europa.eu/research/era/pdf/2020-vision-for-era_en.pdf

ERSO http://ec.europa.eu/transport/road_safety/specialist/index_en.htm

ERTRAC (2004) Strategic Research Agenda, December 2004

ETSC, *A strategic road safety plan for the European Union* (European Transport Safety Council, Brussels, 1997)

ETSC, Road safety as a right and responsibility for all, in *A Blueprint for the EU's 4th Road Safety Action Programme 2010–2020*, European Transport Safety Council, Brussels 2008

CARE, Road Safety evolution in the EU, European Road Accident Database, 2008

Eurostat, *Eurostat Regional Yearbook 2008*, ISBN 978-92-79-08212-2 (Eurostat statistical books, European Communities, Luxembourg, 2008)

European Communities, *White Paper, European Transport policy for 2010: Time to Decide* (Commission of the European Communities, Brussels, 2001)

List of Editors

Dr. E. Bekiaris, Dr. Mech. Engineer of the National Technical University of Athens, is research director in Hellenic Institute of Transport, Head of the Department "Driver and Vehicle." He has acted as project coordinator of 16 research projects, and technical coordinator in another 8. In total he has participated in 77 Research projects so far. He is visiting professor at the University of Newcastle upon Tyne in the UK, is giving lectures in master courses in the Aristotle University of Thessaloniki in Greece and has been visiting professor in University of Trento in Italy. His field of expertise ranges from Road Safety to specialized telematics applications for private vehicles, public transportation, even ships. He has also profound experience in technology for the integration of people with special needs, with emphasis on accessible transportation systems. Dr. Bekiaris has been invited as expert evaluator of proposals in four different programmes of the European Commission. He has also been invited to speak in front of the European Parliament on the subject of mobility and transportation of disabled people. He has been member of the organizing committee of seven international conferences and he has chaired relevant sessions. Editor and co-author in various books. He has published 41 articles in international scientific journals and 177 conference presentations. He is the editor-in-chief of the European Transport Research Review – An Open Access Journal of the European Conference of Transport Research Institutes (ECTRI). He is the chairman of the Forum of European Road Safety Research Institutes (FERSI).

E. Bekiaris et al. (eds.), *Infrastructure and Safety in a Collaborative World*,
DOI 10.1007/978-3-642-18372-0,

Dr. Marion Wiethoff is associate professor at Delft University of Technology, Department of Transport and Logistics and fellow researcher of TRAIL. She is experimental psychologist and her background is psycho-physiological research. Her expertise is in the field of safety assessment of intelligent systems in road traffic, and increasing mobility for people with special needs through ICT and Intelligent transport. She has graduated and finalised her PhD at Groningen University. The subject of her PhD thesis was the application of psycho-physiological methods in measurement of mental effort, as also applicable to research in driver performance. She has carried out the more applied part of field research at Sheffield Social and Applied Psychology Unit in the UK, for a period of 3 years. She has contributed to various European projects and national projects as a researcher (e.g. MUSiC, ASK-IT, IN-SAFETY). She was coordinator for ADVISORS project (DGTREN FP5 GRD1 2000 1004). Furthermore, she has supervised various PhD students on design methodologies for ICT systems, safety research on specific advanced driver assistance systems and MSc students on transport technologies and policy making and industrial design projects. She has been teaching and coordinated university courses on transport models, research methodology, policy making in transport, and also various courses on social skills development. She has published various articles in international journals on mental effort and psychophysiology, advanced driver assistant systems and road safety and improving mobility for people with special needs. She co-edited books on automation of car driving and new developments in technology at the work place.

Evangelia Gaitanidou has a Diploma in Civil Engineering from the Aristotle's University of Thessaloniki, Greece, where she also obtained her MSc diploma on "Planning, Organization and Management of Transportation Systems" and is currently a PhD candidate in the area of Road Safety. She works in the Hellenic Institute of Transport since 2004, as an associated researcher, responsible for the Vehicle Safety sector since 2009. She has so far participated in about 15 EU funded projects in FP5 (IMMACULATE), FP6 (IN-SAFETY, ASK-IT, AIDE, PReVENT, SUPREME, HUMABIO, RIPCORD-ISEREST, PEPPER, DRUID etc) and FP7 (TeleFOT, ACCESS2ALL, 2DECIDE, BESTPOINT) in most of which holding a significant role (assistant Coordinator/Technical Manager, Quality Manager, WP/Task leader). She has 15 publications in National and International peer reviewed conferences and is co-author of a chapter in a book. She is also acting as managing editor and reviewer in the European Transport Research Review (ETRR) Journal. Her main fields of interest lie in the areas of: Road Safety, Telematics Applications for Transport, Transport simulation modeling, Transportation of E&D, Mobility for All, Advanced Driver Assistance Systems (ADAS), etc.

List of Preface Authors

Prof. George A. Giannopoulos is the director of the Hellenic Institute of Transport of the National Center for Research and Technological Development (CERTH) since its founding in 2000 and member of its board of directors since then and vice president since November 2008. Since August 2010 he is president of the CERTH and Director of its Central Administration. Until July 2010 he was professor at the Aristotle University of Thessaloniki where he was director of the Laboratory of Transport Engineering for 20 years, and director of the postgraduate course on Transport Systems for 7 years. He studied civil engineering at the National Technical University of Athens (Diploma of Civil Engineering) and Transportation Planning and Engineering at Imperial College University of London (PhD, DIC, and MSc degrees). Since 1989 he is continuously representing Greece in various committees, working groups, and other bodies of the EU (indicatively: the DRIVE and the Advanced Transport Telematics programme in the 1990s, the ISTAG, SST, ERTRAC, and EIRAC, more recently). Since 2008 he is chair of the Transport Advisory Group of DG RTD and MOVE. He has been deputy of the Greek Minister of Transport at the European Conference of Ministers of Transport (today International Transport Forum – ITF of the OECD), and head of Greece's delegation for the negotiations between Greece and the European Economic Community for

Greece's entry as a member. In 1997 he founded, and was until 2004 chairman, of the South East European Transport Research Forum (SETREF) an international non-governmental organization devoted to promoting transport research in South East Europe, with more than 30 organizations members in 12 countries of the area. He is founding member and first president (for 2003–2007) of the European Conference of Transport Research Institutes (ECTRI) an organization with 27 members – major European Transport Research Organizations established in almost all European countries.
He has been awarded several international distinctions among which:

- Honorary PhD of the University of Kingston, London, UK.
- Certificate of Appreciation from the US Transportation Research Board (US Academy of Sciences).
- Personality of the year (2006) award from the Network of Black Sea Universities.

He is the author of more than 150 publications in scientific magazines, and of 10 books, 2 of which in English. Areas of interest/expertise: transport planning, transport policy, freight transport/logistics, public transport systems, and ITS (intelligent transport systems) applications.

Sylvain Haon is the executive director of the Polis Network. Polis gathers European cities and regions to support innovation for improving urban and regional mobility. Sylvain is currently leading the working group of ERTRAC (European Road Transport Advisory Council) on urban mobility. Before joining Polis, he was head of SNCF Directorate for European Affairs office in Brussels.

Jean-Pierre Medevielle has been deputy general director of INRETS (The French National Institute for Transport and Safety Research) for the last 15 years, after having been the head of INRETS Lyon Centre for 7 years. Before, he had been a city and transportation planner and a research program manager (transportation, urban services, ICTs, life sciences and marine sciences). He is a member of various committees of the US Transportation Research Board, and has been the French Governmental Focus Point for Enhanced Safety Vehicle Conferences and International Harmonized Research Agenda (for Road Vehicle Safety) for 4 years. He has been a member of the FERSI Board and he was the general rapporteur for the ad hoc group preparing the surface transport research including the automotive and rail domains of the Fifth Framework Research and Development Program (FRDP), and he was very involved in the setting of transport issues within the Sixth and Seventh FRDPs. He is a member of the Bureau of the OECD-ITF Joint Transport Research Centre. For the last 12 years, he also has been a member of the International Program Committee of the Intelligent Transport Systems World Congresses. He has been the Secretary General of ECTRI (European Conference of Transport Research Institutes) for January 2003 to January 2007 and has been appointed as the chief executive officer of Europe Research Transport from January 2003 to April 2007. He is a member of the Steering Committee of the European Technology Platform ERTRAC and the eSafety Forum and a member of the International Cooperation and vice chairman of the RTD eSafety Working Groups and eSafety Intelligent Car Initiative. Being the coordinator of the HUMANIST Network of Excellence from 2004, he has been elected as the chairman of the HUMANIST Association. He is also a member of the Council of EURNEX Network of Excellence.

List of Authors

Tom Alkim is an advisor for the Rijkswaterstaat Centre for Transport and Navigation which is part of the Dutch Ministry of Transport, Public Works and Water Management, The Netherlands.

Dr. Anna Anund is a researcher in traffic safety in the Swedish National Road and Transport Research Centre (VTI), Sweden.

Prof. Dr.-Ing. J. Stefan Bald holds the chair "Road and Pavement Engineering" at Technische Universität Darmstadt, Germany, since 1999.

Dr.-Ing. Thomas Benz works in PTV's Research and Innovation Division, Germany.

Prof. Karel Brookhuis is part-time professor at the Department of Experimental Psychology, and part-time professor at the Department Transport Policy and Logistics of Delft University of Technology, Germany.

Eleni Chalkia is a Research Associate in the Centre for Research and Technology Hellas/Hellenic Institute of Transport (CERTH/HIT), Greece

Dr. Manfred Dangelmaier is director of the Business Unit "Engineering Systems" and Head of the Competence Center Virtual Environments including the Virtual Reality Lab and Vehicle Interaction Lab at Fraunhofer IAO in Stuttgart, Germany.

Prof. Dr. Klaas De Brucker is associate professor at the Hogeschool-Universiteit Brussel (HUB), Belgium.

Johan De Mol is a senior researcher in the Institute Sustainable Mobility of Ghent University, The Netherlands.

Dr. Dick De Waard is a research fellow/university lecturer at the Department of Psychology at the University of Groningen in the Netherlands.

Stefan Egger is an information designer, researcher at the International Institute for Information Design (IIID) in Vienna, Austria.

Dr. Rune Elvik is chief research officer at the Institute of Transport Economics, Oslo, Norway.

Dr. Alena Erke is a psychologist, researcher in Norwegean Institute of Transport Economics, Norway.

Dr. Christian Galinski is the director of the International Information Centre for Terminology (Infoterm), Austria, since 1986 and (till 2008) Secretary of ISO/TC 37.

Maria Gemou is a research associate in the Hellenic Institute of Transport, Greece, and PhD delegate in Transport Telematics.

Dr.-Ing. Le Thu-Huyen is a lecturer in University of Transportation and Communication, Vietnam, since 2003.

Kostas Kalogirou is research associate in Hellenic Institute of Transport, Greece.

Cathy Macharis is a professor at the Vrije Universiteit Brussel.

Dr. ir. Vincent Marchau is an associate professor in Delft University of Technology, The Netherlands.

Stella Nikolaou is a software developer in the Hellenic Institute of Transport, Greece. She is also acting as the General Secretary of FERSI (Forum of European Road Safety Institutes) as from May 2008.

Dr. Mary Panou is a researcher at the Hellenic Institute of Transport, Greece.

Dr. Nick Reed works in the Human Factors and Simulation Group at TRL, UK, since January 2004.

Dr. Karin Siebenhandl is head of Research Center Know-Comm at Danube University Krems, Austria.

Michael Smuc is a research associate in the Danube-University Krems (Research Center KnowComm), Austria, since 2007.

Pavlos Spanidis is a research associate in the Hellenic Institute of Transport, Greece.

Dr. Ioanna Spyropoulou is a lecturer at the National Technical University of Athens, Greece.

Dr. Alan Stevens is the chief research scientist and research director, Transportation at TRL, UK.

Dipl.-Ing. Katja Stumpf works as an research assistant for "Road and Pavement Engineering" at Technische Universität Darmstadt, Germany, since 2005.

Dr. Andreas Tapani is a researcher in traffic and transport analysis and adjunct senior lecturer in the Swedish National Road and Transport Research Centre (VTI), Sweden.

Silvana Toffolo is project manager at IVECO, External Relations & Communication CSST, Italy.

Jan-Willem van der Pas is a researcher/PhD candidate at Delft University of Technology, Department of Transport Policy and Logistics, The Netherlands.

Dr. Knut Veisten is senior research economist at the Institute of Transport Economics (TOI), Norway. He holds a PhD in environmental economics from the Norwegian University of Life Sciences.

Sven Vlassenroot is a PhD-researcher at Institute of Sustainable Mobility (IDM) of Ghent University, Belgium, and Section Transport Policy and Logistics' Organisation (TLO) of Delft University of Technology.

Dipl.-Ing. Tim Wallrabenstein works as research assistant in the Department of Civil Engineering and Geodesy, Road and Pavement Engineering at Technische Universität Darmstadt, Germany, since 2006.

Dr. Guenter Wenzel is a senior researcher at the Institut für Arbeitswissenschaft und Technologiemanagement IAT, University Stuttgart, Germany.

Florian Windhager is member of the scientific staff at the Danube University Krems (Research Center KnowComm), Austria.

Martin Winkelbauer is holding a position as a senior researcher in the Austrian Road Safety Board, Austria.

Prof. Frank Witlox is professor of economic geography at the Geography Department of Ghent University, Belgium, and associate director of the Globalization and World Cities Group and Network (GaWC) at Loughborough University. He is also a visiting professor at the Institute of Transport and Maritime Management Antwerp (UA-ITMMA).

Prof. George Yannis is associate professor at the School of Civil Engineering at the National Technical University of Athens, Greece.

Index

MIX
Papier aus verantwortungsvollen Quellen
Paper from responsible sources
FSC® C105338

If you have any concerns about our products,
you can contact us on
ProductSafety@springernature.com

In case Publisher is established outside the EU,
the EU authorized representative is:
Springer Nature Customer Service Center GmbH
Europaplatz 3, 69115 Heidelberg, Germany

Printed by Libri Plureos GmbH
in Hamburg, Germany